The Rapidly Transforming Chinese High-Technology Industry and Market

CHANDOS
ASIAN STUDIES SERIES:
CONTEMPORARY ISSUES AND TRENDS

Series Editor: Professor Chris Rowley,
Cass Business School, City University, UK
(email: c.rowley@city.ac.uk)

Chandos Publishing is pleased to publish this major Series of books entitled *Asian Studies: Contemporary Issues and Trends*. The Series Editor is Professor Chris Rowley, Cass Business School, City University, UK.

Asia has clearly undergone some major transformations in recent years and books in the Series examine this transformation from a number of perspectives: economic, management, social, political and cultural. We seek authors from a broad range of areas and disciplinary interests: covering, for example, business/management, political science, social science, history, sociology, gender studies, ethnography, economics and international relations, etc.

Importantly, the Series examines both current developments and possible future trends. The Series is aimed at an international market of academics and professionals working in the area. The books have been specially commissioned from leading authors. The objective is to provide the reader with an authoritative view of current thinking.

New authors: we would be delighted to hear from you if you have an idea for a book. We are interested in both shorter, practically orientated publications (45,000+ words) and longer, theoretical monographs (75,000–100,000 words). Our books can be single, joint or multi-author volumes. If you have an idea for a book, please contact the publishers or Professor Chris Rowley, the Series Editor.

Dr Glyn Jones
Chandos Publishing (Oxford) Ltd
Email: gjones@chandospublishing.com
www.chandospublishing.com

Professor Chris Rowley
Cass Business School, City University
Email: c.rowley@city.ac.uk
www.cass.city.ac.uk/faculty/c.rowley

Chandos Publishing: is a privately owned and wholly independent publisher based in Oxford, UK. The aim of Chandos Publishing is to publish books of the highest possible standard: books that are both intellectually stimulating and innovative.

We are delighted and proud to count our authors from such well known international organisations as the Asian Institute of Technology, Tsinghua University, Kookmin University, Kobe University, Kyoto Sangyo University, London School of Economics, University of Oxford, Michigan State University, Getty Research Library, University of Texas at Austin, University of South Australia, University of Newcastle, Australia, University of Melbourne, ILO, Max-Planck Institute, Duke University and the leading law firm Clifford Chance.

A key feature of Chandos Publishing's activities is the service it offers its authors and customers. Chandos Publishing recognises that its authors are at the core of its publishing ethos, and authors are treated in a friendly, efficient and timely manner. Chandos Publishing's books are marketed on an international basis, via its range of overseas agents and representatives.

Professor Chris Rowley: Dr Rowley, BA, MA (Warwick), DPhil (Nuffield College, Oxford) is Subject Group leader and the inaugural Professor of Human Resource Management at Cass Business School, City University, London, UK. He is the founding Director of the new, multi-disciplinary and internationally networked *Centre for Research on Asian Management*, Editor of the leading journal *Asia Pacific Business Review* (www.tandf.co.uk/journals/titles/13602381.asp). He is well known and highly regarded in the area, with visiting appointments at leading Asian universities and top journal Editorial Boards in the US and UK. He has given a range of talks and lectures to universities and companies internationally with research and consultancy experience with unions, business and government and his previous employment includes varied work in both the public and private sectors. Professor Rowley researches in a range of areas, including international and comparative human resource management and Asia Pacific management and business. He has been awarded grants from the British Academy, an ESRC AIM International Study Fellowship and gained a 5-year RCUK Fellowship in Asian Business and Management. He acts as a reviewer for many funding bodies, as well as for numerous journals and publishers. Professor Rowley publishes very widely, including in leading US and UK journals, with over 100 articles, 80 book chapters and other contributions and 20 edited and sole authored books.

Bulk orders: some organisations buy a number of copies of our books. If you are interested in doing this, we would be pleased to discuss a discount. Please contact Hannah Grace-Williams on email info@chandospublishing.com or telephone number +44 (0) 1993 848726.

Textbook adoptions: inspection copies are available to lecturers considering adopting a Chandos Publishing book as a textbook. Please email Hannah Grace-Williams on email info@chandospublishing.com or telephone number +44 (0) 1993 848726.

The Rapidly Transforming Chinese High-Technology Industry and Market

Institutions, ingredients, mechanisms and modus operandi

NIR KSHETRI

Chandos Publishing
Oxford · England

Chandos Publishing (Oxford) Limited
TBAC Business Centre
Avenue 4
Station Lane
Witney
Oxford OX28 4BN
UK
Tel: +44 (0) 1993 848726 Fax: +44 (0) 1865 884448
Email: info@chandospublishing.com
www.chandospublishing.com

First published in Great Britain in 2008

ISBN:
978 1 84334 464 3 (hardback)
1 84334 464 5 (hardback)

© N. Kshetri, 2008

British Library Cataloguing-in-Publication Data.
A catalogue record for this book is available from the British Library.

All rights reserved. No part of this publication may be reproduced, stored in or introduced into a retrieval system, or transmitted, in any form, or by any means (electronic, mechanical, photocopying, recording or otherwise) without the prior written permission of the Publishers. This publication may not be lent, resold, hired out or otherwise disposed of by way of trade in any form of binding or cover other than that in which it is published without the prior consent of the Publishers. Any person who does any unauthorised act in relation to this publication may be liable to criminal prosecution and civil claims for damages.

The Publishers make no representation, express or implied, with regard to the accuracy of the information contained in this publication and cannot accept any legal responsibility or liability for any errors or omissions.

The material contained in this publication constitutes general guidelines only and does not represent to be advice on any particular matter. No reader or purchaser should act on the basis of material contained in this publication without first taking professional advice appropriate to their particular circumstances.

Typeset by Domex e-Data Pvt.Ltd.
Printed in the UK and USA.

Contents

List of figures and tables	ix
About the author	xi
Preface	xiii

1 Intellectual property rights issues in China's high-technology industry 1

    Introduction 1

    The theoretical framework 3

    China's response to IPR issues 9

    Discussion 18

2 The Chinese e-business industry 21

    Introduction 21

    The institution-technology diffusion nexus: a theoretical perspective 22

    Institutional processes in the Chinese e-business industry 24

    Discussion 33

    Notes 36

3 Cyber-control in China 37

    Introduction 37

    The theoretical framework: institutional field 39

    Cyber-control in China as an institutional field 41

    Discussion 48

    Conclusions and implications 51

    Notes 51

4 The Chinese government's influence on the third-generation cellular standard: motivators, enabling factors and mechanisms 53

    Introduction 53

	The emergence of TD-SCDMA	55
	The Chinese government's motivations for promoting TD-SCDMA	57
	Enabling factors	62
	Mechanisms	63
	Conclusion	66
	Notes	69
5	**Drivers of foreign multinationals' research and development activities in China**	**71**
	Introduction	71
	Foreign firms' R&D in China: a brief survey	73
	Host country characteristics affecting foreign MNCs' R&D activities	75
	Institutional factors	82
	Spillover effects	85
	Conclusion	88
6	**Forces shaping the development of the Chinese technology workforce**	**89**
	Introduction	89
	The Chinese technology workforce: a brief survey	90
	Four mega-effects in the Chinese technology workforce landscape	96
	Discussion and implications	105
	Where is the future leading?	106
	Notes	106
7	**The Chinese software industry: structural shifts**	**107**
	Introduction	107
	Forces of structural changes in an industry	110
	Structural shifts in the Chinese software industry	111
	Overcoming competitive disadvantage	118
	Conclusion: where is the future leading?	119
	Note	121

8	**Diffusion of open source software: institutional and economic feedback**	**123**
	Introduction	124
	OSS as a hybrid between the private investment model and the collective action model	125
	The transformation of institutions, enterprises and markets	128
	Drivers of OSS diffusion in China: institutions, industry and market	130
	Conclusion	141
	Notes	142
9	**Drivers of broadband diffusion in China**	**143**
	Introduction	143
	Broadband diffusion in China: a brief survey	143
	Discussion and conclusion	154
10	**The Chinese internet protocol television market**	**155**
	Introduction	155
	Development of a technology industry: a theoretical framework	157
	Factors driving the development of the Chinese IPTV industry	158
	Discussion and implications	168
11	**China's nanotechnology prowess**	**171**
	Introduction	171
	The Chinese nanotechnology industry and its drivers: a brief survey	172
	Kaleidoscopic comparative advantage and the Chinese nanotechnology industry	176
	Concluding remarks	179
12	**Chinese technology enterprises in developing countries: sources of strategic fit and institutional legitimacy**	**181**
	Introduction	181
	Chinese high-technology products in the developing world: a brief survey	183
	The theoretical foundation	189
	Chinese high-technology products in the developing world	191

	Conclusion	199
	Notes	199
13	**Internationalisation of firms in the Chinese cellular industry**	**201**
	Introduction	201
	Internationalisation pattern of the Chinese cellular industry: a brief survey	202
	Inward–outward connection in the internationalisation process: relevant theories and past research	207
	Internationalisation of the Chinese cellular industry	209
	Discussion and conclusion	215
	Notes	216
14	**Chinese high-technology firms' outward-oriented mergers and acquisitions: a case study**	**217**
	Introduction	217
	The Chinese context from the standpoint of M&A activities	218
	A note on TCL	219
	TCL's outward M&A activities: motivators	223
	Cases related to TCL's M&A activities in Europe	225
	Discussion and conclusion	230
	Notes	233
15	**Concluding remarks**	**235**
	The rapidly transforming Chinese high-technology industry and market	235
	The growth of the Chinese technology industry: current debate	236
	Global economic and political implications of China's technology prowess	237
	How should the world respond?	241
Bibliography		**243**
Index		**311**

List of figures and tables

Figures

1.1	Institutional influences and diffusion of intellectual property rights in China	10
5.1	Factors influencing multinational companies' research and development activities in developing countries	75
8.1	Institutions, industry and market related factors influencing OSS diffusion pattern in China	130
9.1	A comparison of broadband and related technologies in China and India	146
9.2	A proposed framework to explain broadband diffusion in developing countries	147
10.1	A framework for understanding the development of the Chinese IPTV industry	158
12.1	Mechanisms associated with the performance of Chinese technology companies in the developing world	192
13.1	Internationalisation of the Chinese cellular industry: actors, factors and moderators	210

Tables

2.1	Institutional factors affecting the diffusion of e-business in China	25
4.1	Emergence of TD-SCDMA: a timeline	56–7
5.1	A comparison of China with other developing and developed economies in terms of R&D related indicators, 2004	82

6.1	China and India in terms of major indicators related to the high-technology workforce	91
6.2	Some qualitative indicators related to the Chinese high-tech workforce	94
6.3	Four mega-effects in the development of China's high-tech workforce	97
6.4	Cumulative number of Chinese leaving overseas for study	99
7.1	A comparison of China and India in terms of major indicators related to the software industry	109–10
7.2	Market sizes of major software destinations of China and India	120
8.1	PIM, CAM and characteristics of Linux projects in China	127
8.2	Red Flag's collaborations in Linux-related projects: selected examples	140
9.1	A comparison of indicators related to broadband in China and India	150
10.1	IPTV in China: A timeline and future perspective	156
11.1	The Chinese nanotechnology industry: some notable achievements	174
11.2	China's advantages in the nanotechnology industry	177
11.3	Some barriers facing the Chinese nanotechnology industry	178
12.1	The operations of selected major Chinese high-technology players	184–5
12.2	Sources of value of Chinese technology products in other developing countries	186
13.1	International activities of selected players in the Chinese cellular industry	204–5

About the author

Nir Kshetri is an assistant professor at Bryan School of Business and Economics, The University of North Carolina-Greensboro. Nir holds a PhD in Business Administration from the University of Rhode Island; an MBA from Banaras Hindu University (India); and an MSc (Mathematics) and an MA (Economics) from Tribhuvan University (Nepal). His undergraduate degrees are in Civil Engineering and Mathematics/Physics from Tribhuvan University. Nir's previous positions include faculty member at the Management School, Kathmandu University (Nepal), visiting lecturer at the Management School, Lancaster University (UK) and visiting professor at the European Business School in Paris. During 1997–99, Nir was a consultant and a trainer for the UN Food and Agricultural Organization, the GTZ (German Technical Cooperation) and the Agricultural Development Bank of Nepal.

Nir's works have also been published in journals such as *Foreign Policy, European Journal of Marketing, Journal of International Marketing, Journal of International Management, Communications of the ACM, IEEE Security and Privacy, IEEE Software, Electronic Markets, Small Business Economics, Electronic Commerce Research and Applications, IT Professional, Journal of Developmental Entrepreneurship, First Monday, Pacific Telecommunications Review, Journal of Asia Pacific Business* and *International Journal of Cases on Electronic Commerce*. He has also contributed chapters to several books including *In the wave of M&A: Europe and Japan* (Kobe University, RIEB Center, 2007), *M-commerce in North America, Europe and Asia-Pacific: Country Perspectives* (Idea Group Publishing, 2006), *Encyclopedia of Information Science and Technology* (Idea Group Publishing, 2005), *Indian Telecom Industry – Trends and Cases* (The ICFAI University Press, 2005), *The Internet Encyclopedia* (John Wiley & Sons, 2004); *Wireless Communications and Mobile Commerce* (Idea Group Publishing, 2003); *The Digital Challenges: Information Technology in the Development Context* (Ashgate Publishing, 2003); *Architectural Issues of Web-enabled Electronic Business*

(Idea Group Publishing, 2003), and *Internet Marketing* (2nd edition, Schaeffer-Poeschel, 2001). Nir has presented over 60 research papers at various national and international conferences in Canada, China, Greece, India, Italy, Japan, South Korea, Sweden, Thailand, the Philippines and the USA. He has also given invited talks at Cornell University, Duke University, Kobe University, University of Maryland (College Park) and Temple University. In 2008, the Kauffman Foundation awarded Nir a grant to study entrepreneurial firms in OECD economies.

Nir was awarded Pacific Telecommunication Council's 2008 Meheroo Jussawalla Research Paper Prize for his work on the Chinese IPTV market. Nir was the runner-up in the 2004 dissertation competition of the American Marketing Association's Technology and Innovations Special Interest Group and the winner of the 2001 Association of Consumer Research/Sheth Foundation dissertation award. He also won the Pacific Telecommunication Council's essay competition in 2001, having come second place in the same competition in 2000. In May, 2006, the Information Resources Management Association (IRMA) presented Nir with the Organization Service Award for the Best Track Chair in the IRMA 2006 International Conference.

Nir's works have been featured in *Foreign Policy*'s Global Newsstand section (a publication of the Carnegie Endowment for International Peace) and in *Providence Journal*. In autumn 2004, he was pictured on the front page of *Global Perspective*, a publication of the Fox School's Temple CIBER and Institute of Global Management Studies. Nir has been quoted in magazines and newspapers such as *Telecommunications*, *Greensboro News and Record* and *High Point Enterprise*.

The author may be contacted at:

Bryan School of Business and Economics
The University of North Carolina at Greensboro
Greensboro, NC 27402-6165
USA
Tel: +1 336 334 4530
Fax: +1 336 334 4141
E-mail: *nbkshetr@uncg.edu*

Preface

This book is about a number of unique, idiosyncratic and unusual features in the Chinese high-technology landscape, which has undergone rapid and extraordinary shifts in the past three decades. The growing size, sophistication and impact of the Chinese technology industries are now being felt at the global level. This book examines how China's transition to market economy has been a complex interaction of Chinese institutions, industries, markets and other ingredients, and how this has shaped the country's technology trajectory.

Global technology analysts disagree as to the nature of the achievement and global significance of the Chinese high-technology industry. Some consider today's Chinese high-tech firms as comparable to those of Japan in the 1950s and 1960s and South Korea in the 1970s (Greenfeld, 2003; Zainulbhai, 2005). Others maintain that it may be more useful to regard China as a normal emerging economy like Brazil or India (Nolan, 2001; Gilboy, 2004; Mooney, 2005). This book contributes to this debate by analysing the Chinese high-technology industry and market as well as associated institutions, ingredients, mechanisms and modus operandi from various angles, perspectives and focal points.

Building on the institutions and other ingredients related to the Chinese technology industry, this book seeks to understand the status and locus of the Chinese high-technology industry. Trajectories and drivers of key information and communications technology industries such as broadband, WiFi, internet and software are examined. The book also investigates Chinese high-technology companies' modus operandi both abroad and at home. A close look is taken at a new breed of high-technology companies that have challenged established multinationals from the industrialised world.

In sum, by providing a comprehensive overview of the ingredients, institutions, mechanisms and modus operandi related to the Chinese high-technology industry and market, this book aids in better understanding and analysing China's technological transformation and

Chinese firms' competitiveness. The book also provides academic, managerial and policy implications associated with the rapid transformation of the Chinese high-technology industry.

This book is primarily targeted at academic specialists, practitioners, professionals and policy makers interested in the evolution of ICT industries in developing countries. Undergraduate and graduate students are also targeted.

As for the ideas presented in this book, I am indebted to a number of people for comments, suggestions, support, encouragement and feedback. My major debt is to my PhD adviser Nikhilesh Dholakia, who has provided me with constant intellectual stimulation, support and encouragement. I have also benefited greatly from interacting with my colleagues Ralf Bebenroth, Nicholas Williamson, David Bourgoin, D. Li and Maggie Cheung. Glyn Jones, Managing Director, Chandos Publishing and Chris Rowley, Asian Studies Series Editor have been constructive, supportive, helpful and encouraging in guiding and managing this project. My warm thanks to copy editor Neill Johnstone, whose thorough and careful reading, meticulousness and close attention to details have helped make the book more clear, consistent and precise. I also received help and support from my graduate assistants Andreea Schiopu and Crystal Pierce. Last, but not the least, my life's companion and best friend Maya supported in countless ways while I worked on this project, as she always does. I'd like to dedicate this book to her with a lot of love and affection.

Intellectual property rights issues in China's high-technology industry

The growth of high-technology industry is tightly linked to the nature of the intellectual property rights (IPR) regime in an economy. In this regard, the Chinese IPR landscape is characterised by seemingly contradictory trends. A number of indicators point to the fact that China has made a significant stride towards creating formal institutions to address IPR issues. An important question, however, is whether we can really take such indicators as 'proof positive' that Beijing possesses the ability and willingness to protect intellectual property any more than we can draw the opposite conclusion based on the fact that over 90 per cent of the software used in the country is pirated. Competing interests of local, national, international and global institutional actors determine the locus and trajectory of IPR diffusion in developing economies such as China. This chapter examines how institutions with different and opposing perspectives are shaping the diffusion of IPR in China. We draw upon two research streams to conceptualise the changing IPR landscape in China. The first stream of research – which we describe as strategic decoupling – focuses on how Chinese decision-makers deal with legitimacy-related pressures when they have to appease the conflicting and competing demands of diverse institutional actors. The second stream of research – selective adaptation – is devoted to understanding how the Chinese balance local needs with IPR related pressures.

Introduction

The growth of high-technology industry is tightly linked to the nature of the IPR regime in which it finds itself. In this regard, the Chinese IPR landscape is characterised by seemingly contradictory trends. Following the 1978 economic and political reforms, China enacted thousands of new laws to protect intellectual property (Pei, 1998; Meredith, 2003), and abolished or amended many laws in this area to comply with World Trade Organization (WTO) obligations (Hughes, 2005). Since the late 1990s, there have been a series of anti-piracy raids in the country (*Economist*, 1999; Weidenbaum, 2006a). In 2005, more than 13,000 intellectual property related cases were filed in Chinese courts (AFX News, 2006a) and in many cases, the country's courts ruled against violators of IPR (Weidenbaum, 2006a). Most impressive of all, in July 2007, Chinese authorities teamed up with the US FBI, which led to the arrest of 25 people and the seizing of more than US$500 million worth of Chinese-made counterfeit software being distributed worldwide (Barboza and Lohr, 2007).

Chinese technology firms have also significantly increased their spending on foreign intellectual property imports (Lague, 2006a, 2006b). For example, Lenovo, the Chinese PC manufacturer, increased the proportion of computers it sold with preinstalled Windows software from 10 per cent in 2005 to 70 per cent in the first half of 2007 (Greene, Einhorn and Hamm, 2007). In July 2007, Microsoft chief Bill Gates noted that proportion of new PCs shipped with legitimate software in China exceeded 40 per cent (Kirkpatrick, 2007).

Faced with evidence like these, optimists have noted China's significant progress on the IPR front (Carney, 1999). An important question, however, is whether we can really take the above trends as 'proof positive' that Beijing possesses the ability and willingness to protect IPR any more than we can draw the opposite conclusion based on the fact that over 80 per cent of the software used in the country is pirated.

Competing interests of local, national, international and global institutional actors determine the locus and trajectory of IPR diffusion in developing economies. Given such complexities, the question remains as to how best to conceptualise the diffusion pattern of IPR in China. In this chapter, we argue that we can reconcile apparently contradictory observations in the Chinese IPR landscape by examining the phenomenon from the standpoint of institutional theory.

Prior research has indicated that the degree to which ideas such as IPR are diffused and translated into local practices is a function of local institutions including the nature of power struggles and rivalry, leadership support and implementation capacities (Scott et al., 2000). Although previous studies have extended our understanding of the influence of culture and mediating role of the government in the formation of institutions related to business systems in China (Redding, 2002), what is not clear is whether and how the findings of these studies can be applied in the context of complex phenomena such as the diffusion of IPR. Likewise, while some scholars have examined the role of legal institutions in socioeconomic and political changes in Asia (Jayasuriya, 1999; Pistor and Wellons, 1999), the complex institutional dynamics related to IPR are left largely unexamined.

Clearly, there is insufficient research into how institutions with different perspectives are shaping China's response to IPR. To more fully understand China's responses to IPR, this chapter integrates and applies findings in the literature on institutional theory. The framework proposed in this chapter identifies clear contexts and attendant mechanisms related to IPR diffusion in China.

In the remainder of the chapter, we first briefly review the theoretical foundation. Then, we translate the theories within the context and limits of China and attempt to explain the nature of the institutions that influence China's response to IPR with some propositions. The final section provides some conclusions.

The theoretical framework

To better understand the institutional dynamics associated with China's response to IPR, institutions that lie at the centre of the IPR field as well as competing institutions residing within populations need to be analysed (Hoffman, 1999). In the context of IPR diffusion in China, these include international institutions such as the WTO and the World Intellectual Property Organization copyright treaties, regulatory authorities in China such as the State Council and Ministry of Information Industry, and other constituencies within China such as competitive private firms, government-owned enterprises, reformist leaders, leftist opposition and ordinary citizens.

Given the complex institutional dynamics, what might be the likely nature of IPR diffusion in China? Let us begin with the organisational field formed around the issue of IPR (Hoffman, 1999). Like other 'issue-based' fields, IPR can be viewed as 'arenas of power relations' (Brint and Karabel, 1991: 355) in which various players and constituencies with competing interests and disparate purposes negotiate over issue interpretation and engage in institutional war (White, 1992; Hoffman, 1999: 351–2).

We define institutions as 'rules of the game in a society' (North 1990: 3), or 'a set of socially prescribed patterns of correlated behavior'

(Bush, 1987: 1076). Institutionalists use the concept of diffusion to refer to the spread of institutional principles and practices in a society (Strang and Meyer, 1993; David and Foray, 1994). We draw upon two research streams to examine the diffusion of IPR in China. The first stream of research – which we refer to as strategic decoupling – focuses on how decision-makers deal with legitimacy-related pressures when they have to appease and serve conflicting demands. The second stream of research – selective adaptation – is devoted to understanding the question of how Asian societies engage in selective adaptation to balance local needs with the pressures of complying with practice rules imposed from outside.

Strategic decoupling

Although most studies on strategic decoupling are conducted at the firm level, the notion that the functioning of a state can be conceptualised as that of a corporation is gaining popularity (Kotler, Jatusripitak and Maesincee, 1997). It can thus be argued that findings on strategic decoupling can be extended to a nation as the unit of analysis.

Isomorphism, that is, engagement in actions consistent with the responses of other actors in the environment (George et al., 2006), is arguably positively related to legitimacy (Deephouse, 1996). Organisations that are able to acquire legitimacy from external institutional actors, on the other hand, are likely to gain resources as well as maintain control over the environment (George et al., 2006). Put differently, an organisation can increase its chance of survival and/or growth by engaging in actions that are approved by powerful institutional actors, used by organisations that are perceived to be 'successful' (Newman, 2000), or have the backing and approval of professions in their industry or field (Meyer and Scott, 1983; Baum and Oliver, 1991; Sitkin and Sutcliffe, 1991; Ruef and Scott, 1998; H. Aldrich, 1999). Note that isomorphic actions are not necessarily the most efficient actions (George et al., 2006), although non-isomorphic responses that deviate away from 'established structures, practices, and utterances of other actors in the environment' (George et al., 2006) are likely to face resistance.

In many cases, however, actors with conflicting demands are to be appeased and served. In China, for instance, competitive firms favour IPR protection (Gilboy, 2004). Most state-owned enterprises, however, are less innovative and thus constitute strong anti-IPR lobbies (Stevenson-Yang and DeWoskin, 2005). Similarly, China's reformist leaders recognise the importance of institutions consistent with free-market principles, the rule of law and a transparent legal system (Dorn, 1998; Lin, Cai and Li, 1996: 226). These leaders think that IPR compliance and WTO membership can help them to support such institutions and encourage innovations, pursue free market capitalism and eliminate inefficient state enterprises (Lardy, 2002). They think that given the strength of leftist opposition, global pressures are needed to achieve such goals (Lardy, 2002). Chinese leftist leaders, on the other hand, perceive complex institutions built to promote efficient behaviour (Kennedy, 2005), such as improved legal institutions, as potential challenges to the legitimacy of the Chinese Communist Party regime (Potter, 2004: 478). Indeed, many Chinese government officials and policy-makers also consider that China's integration with the global market is associated with significant socioeconomic costs (Lung, 1999; Heer, 2000). China's IPR related responses thus entail complex processes related to control in domestic affairs and in the international arena.

For Chinese decision-makers, a proper decoupling of responses is needed to maintain control over the domestic and international agenda. Decision-makers simultaneously utilise different combinations of actions in parallel that reflect their mixed reading of the environment (George et al., 2006). Chinese leaders, for instance, emphasise different dimensions of their rule-making, sanctioning and monitoring activities (Scott, 1995, 2001) in dealing with different institutional actors. Faced with challenges to comply with human rights and international trade agreements, for example, the Chinese government often replies that it has adopted Western rule of law principles and is further strengthening them (State Council Information Office, 2000; WTO, 2001; Potter, 2004). On the other hand, China's compliance with WTO requirements is interpreted locally as consistent with underlying policy imperatives rather than because of international pressure (Kong, 2001).

The various institutional actors differ in terms of the power they have to affect an organisation's outcome. Different theoretical contributions and various empirical studies have led to the accepted view that the exact nature of decoupling is a function of the relative powers of competing organisational and institutional interests (Pfeffer, 1981a, 1981b; March and Olsen, 1989; Oliver, 1991; Westphal and Zajac, 1994, 1998, 2001; Zajac and Westphal, 1995). These studies also provide support for the notion that substantial responses cannot be made to appease two sets of actors that diametrically oppose one another. More to the point, the substantive response relates to the threat or opportunity associated with the actor that is perceived to be more powerful and the symbolic response relates to the threat or opportunity associated with the actor perceived to possess less power (George et al., 2006).

To better understand the nature of symbolic and substantive response, let us consider the following question: what might symbolic actions look like? In some cases, symbolic actions constitute measures that just satisfy the minimum expectation of the less powerful actor. For instance, Elsbach and Sutton (1992) found that decision-makers in activist organisations responded to control-related threats and managed to hold powerful actors in the controlling environment at bay by modifying their structures and practices just enough to meet the normative expectations of the key actor. Just like these activist organisations, Chinese decision-makers have managed control-related threats by taking sufficient measures to satisfy the WTO and other international constituencies. For instance, China carried out a series of anti-piracy raids (*Economist*, 1999) and abolished or amended many intellectual property laws to comply with WTO obligations (Hughes, 2005). Some observers have, however, noted the lack of substance in China's intellectual property measures. Some argue that the 1997 raids on pirate CD factories were carried out by the Chinese government because the USA put China on the 'Super 301' list, which increased the risk of trade sanctions (*Economist*, 1999). Likewise, following the WTO accession, China refused to offer further concessions on market access, arguing that it had already complied with WTO commitments (Dreyer, 2002). A visible example is the Government Procurement Law enacted in January 2003, which requires government departments to procure domestic goods and services where possible (Ebusinessforum, 2004). As government procurement is excluded from the scope of multilateral trade rules governing the WTO, so China is not obliged to open government procurement to foreign companies. Faced with such examples, some commentators also argue that China is unwilling to enforce laws enacted to comply with WTO requirements (Meredith, 2003).

In other cases, formal structures that are not implemented in practice represent symbolic actions. To respond to various institutional actors and maintain the coherence of internal functioning, some organisations decouple formal structures from their activities and practices (George et al., 2006: 357). For instance, in China, about 3,200 new laws related to intellectual property protection and private ownership were passed from 1978 to 1994 (Pei, 1998), including the first copyright law enacted in 1990 (Meredith, 2003). Likewise, 2,600 legal statutes and regulations were abolished or amended by mid-2002 to comply with WTO demands (Hughes, 2005); this significantly changed China's formal structure regarding the protection of intellectual property. Nonetheless, the country's activities and practices regarding intellectual property deviate significantly from the formal structures, as evidenced by piracy rates that can exceed 90 per cent. Some observers suspected that the CD presses that were confiscated in the 1997 raids were used in government-owned factories to manufacture CDs containing pirated software (*Economist*, 1999).

Selective adaptation

A theory capable of explaining and predicting China's response to IPR issues probably could not rely exclusively on Chinese decision-makers' cognitive understanding and assessment of the environment and their decoupling of responses. A second stream of research, which is devoted to understanding how societies in the Asia-Pacific region engage in selective adaptation to balance local needs with the pressures of complying with rules imposed from outside, supplements and enriches institutional theory on strategic decoupling. Selective adaptation is typically framed as a process by which exchanges of non-local rules across cultural boundaries are mediated by and interpreted in terms of local practices, conditions, imperatives and norms (Potter, 2001).

The literature suggests evidence of selective adaptation in China's borrowing of Western laws, practices and ideas (Lubman, 1999; Peerenboom, 2002; Potter, 2004). Selective adaptation is made possible by ways in which governments and elites express their own normative preferences in the course of their interpretation and application of practice rules (Balbus, 1977) such as those pertaining to IPR. This line of research also shows that the nature of selective adaptation is a function of perception, complementarity and legitimacy (Potter, 2004).

Perception

The processes and results of selective adaptation are a function of how Chinese leaders view the 'purpose, content and effect' of foreign and local institutional arrangements related to IPR (Potter, 2004). From China's standpoint, how might the country be affected were it to comply with international pressures related to IPR? Regarding the purpose of these international pressures, many Chinese government officials think that IPR is a part of a US attempt to link China to the global economy that benefits US companies such as Microsoft at the cost of Chinese consumers (Newsweek, 2001). Furthermore, China's policy-makers know that the country's size and strategic importance allow it to limit the imposition of foreign regulatory norms.

In terms of local institutional arrangement, strategic technological policies are devised by national elites, government planners and university scientists (Feigenbaum, 1999). Younger specialists, including those with degrees and/or entrepreneurial experiences in the West, play little role in national technological decisions (Feigenbaum, 1999: 97). Chinese decision-makers thus know that their decisions will be subject to very little or no resistance.

One should not, however, assume that the Communist Party has complete control over local communities and businesses. There is some evidence of the state's declining power over the Chinese economy and society (Garten, 2003; Crane, 2004); similarly, decentralisation of formal decision-making mechanisms has been reported (Abramson, 2006). Local support is thus becoming more critical in the implementation of intellectual property laws. It would also be incorrect to assume that members of local communities and national elites in China share the same beliefs and attitudes toward IPR issues. Experts also argue that economic reform will increase the influence of businesses (Kennedy, 2005).

Complementarity

The basic idea behind complementarity is simple: it describes a situation in which seemingly contradictory phenomena can be combined so that they reinforce each other effectively while preserving the essential characteristics of each component (Bohr, 1963; Rhodes, 1986; Potter, 2004). Put differently, complementarity affects the potential for foreign rules and local norms to be mutually supporting (Potter, 2004: 479). A community's norms and practices can be adjusted to satisfy outside expectations while protecting local needs. For instance, the Chinese government may engage in substantive actions to protect IPR such as enacting and enforcing laws and buying foreign intellectual property (satisfying outside expectations) if local companies possess capabilities to develop intellectual property or to benefit from foreign intellectual property (protecting local needs).

Legitimacy

Legitimacy is related to the extent to which the environment constrains an organisation with respect to its choices (George et al., 2006). Legitimacy concerns the extent to which members of local communities support the purposes and consequences of selective adaptation (Weber, 1978; Wilson, 1992; Rose, 2000; Scharpf, 2000) of IPR. The effectiveness of selectively adapted legal forms and practices depends to an important degree on local acceptance (Potter, 2004).

Organisations that depart from established ways of doing things may have their legitimacy questioned by important external actors (Elsbach and Sutton, 1992; Browning and Folger, 1994). For instance, ordinary citizens perceive IPR enforcement tools as supports to foreign software companies. Taiwan offers a case in point. The Taiwanese government's attempt to force students using pirated versions of Microsoft Windows to pay up was perceived as supporting a foreign company rather than its own citizens (Kshetri, 2004).

The government's legislative measures alone are thus insufficient to solve IPR problems (Shen, 2005). More to the point, local authorities do not necessarily share the same interests as the central government (Yang, 2002). Intellectual property enforcement is also hampered by attitudes of the people involved in the enforcement of the mechanisms (Shen, 2005). While provincial and local authorities have taken symbolic actions to appease the central government by amending and abolishing local laws and regulations not complying with the WTO, it is far from clear whether local authorities are taking substantive actions by cooperating with central authorities on enforcement measures (Yang, 2002).

Taken together, these two research streams – strategic decoupling and selective adaptation – suggest that decision-makers' cognitive assessment of different institutional actors, their normative assessment and the nature of local legitimacy, influence the locus and trajectory of IPR adoption. The research on strategic decoupling frames China's response to IPR as a key legitimacy-seeking process for decision-makers, entailing managing resources or controlling related threats. Selective adaptation evaluates each IPR related measure in terms of how key decision-makers perceive it, how the response complements local norms and whether it acquires local legitimacy.

China's response to IPR issues

In this section we translate the above considerations within the context and limits of China. As a visual aid, Figure 1.1 schematically represents the hypothesised relationship among various constructs. Each arrow in

Figure 1.1 Institutional influences and diffusion of intellectual property rights in China

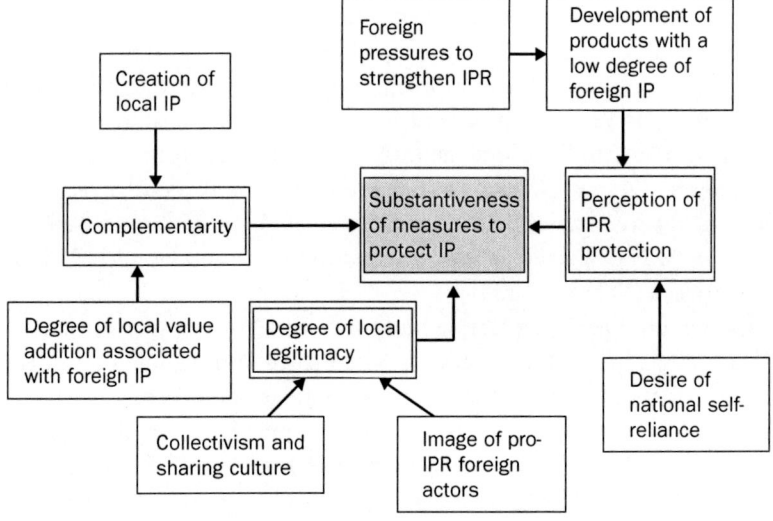

the model represents a proposition as an effect related to institutional influences on China's IPR landscape.

Achieving complementarity with IPR protection

When determining whether an economy would respond favourably to IPR, it is necessary to examine the structure of its trade of intellectual property related products. Experts argue that the TRIPS agreements work against developing countries as most are net intellectual property importers (May, 2006). Not surprisingly, most of China's actions regarding IPR compliance in the past have lacked substance, seeming to be somewhat being symbolic in nature and designed to appease Western multinationals and international agencies. Indeed, as noted above, some analysts suspected that CD presses confiscated in the 1999 anti-piracy raids were put back into service in government-owned factories to manufacture CDs containing pirated software (*Economist*, 1999).

China's progress on the intellectual property front is likely to strengthen local support for IPR and thus lead to more substantive IPR protection measures. In recent years, there have been concomitant changes in China's creation of local intellectual property and more substantive actions for intellectual property protection. The number of patent applications received by the Chinese government increased from 80,000 in 1995 to over 120,000 in 1998 (Patterson, 2001). During July 2006 to December 2007, China's State Intellectual Property Office (SIPO) accepted over one million patent applications (*China Daily*, 2007a). As of December 2007, Chinese applicants accounted for 61 per cent of the most recent million patent applications received by the SIPO, compared with 48 per cent of the first million applications (*China Daily*, 2007a).

Similarly, the number of patents applied by Chinese inventors at US patent offices increased six-fold in the 1990s (*Economist*, 2005a). According to the World Intellectual Property Organization, in 2004, China accounted for 130,000 patent applications and ranked fifth globally (Einhorn, 2006a; Weidenbaum, 2006a). The Chinese government is adding about 200 new research centres every year (Barshefsky and Gresser, 2005). Frans van Empel, chief technology officer for Philips in Shanghai predicts that 'China will go from being on the receiving side of technology to the generating side' (Einhorn, 2003a). China has also initiated aggressive approaches to set its own technical standards and to enhance value from its intellectual property (Suttmeier, 2004).

Although commentators such as Kenneth DeWoskin, a senior consultant with PricewaterhouseCoopers, argue that 'China's interests are in reducing the cost of IP, not in protecting it' (Roberts, 2006), a rapid increase in domestic intellectual property creation has led to more substantive measures to protect IPR. Under China's recently enacted new piracy laws, buyers of pirated goods can be fined from five to ten times the value of the goods, while manufacturers face jail time and equipment confiscation (Kanellos, 2002a). The government has provided a significant empowerment to regulatory agencies involved in IPR issues such as the State Administration of Industry and Commerce, the State Administration of Press and Publications, the Intellectual Property Rights Office and the State Pharmaceutical Administration (Yang, 2002). Similarly, China announced its plans to open special centres in 50 cities by 2006 to handle IPR infringement complaints as well as to provide consulting services (MacLeod, 2006). More recently, Pfizer successfully went against a major Chinese ministry-level government agency to defend its Viagra patent (Boswell and Baker, 2006).

With a rapid increase in the creation of intellectual property by local firms, these firms are actively participating in substantive measures that could help strengthen the country's intellectual property regime. According to the Chinese Supreme Court, in 2005, over 16,000 civil cases and 3,500 criminal cases related to IPR violations were handled by Chinese courts and more than 2,900 people were jailed (Culpan, 2006).

The number of cases involving IPR protection including patents, trade secrets and counterfeit goods increased by 21 per cent in 2005 (AFX News, 2006a). What is more, 95 per cent of China's IPR related cases in 2005 were brought by Chinese companies (Culpan, 2006).

The Chinese nanotechnology industry provides a visible example to illustrate how local intellectual property creation leads to substantive actions to protect IPR. Chinese scientists are capable of producing carbon nanotubes 60 times faster than their US counterparts (Stokes, 2005). The Nanometer Technology Center established in Beijing is actively involved in protecting IPR (Singer, Salamanca-Buentello and Daar, 2005). In sum, the abovementioned new measures relating to intellectual property seem to be much more than merely symbolic.

With respect to intellectual property, from a developing country's perspective, what matters in local firms' success is not simply how much intellectual property is generated but whether and to what extent the firms possess the skills to develop products based on domestic or foreign intellectual property. Many Chinese technology firms are characterised by a high degree of dependence on foreign intellectual property. For example, Qualcomm's CEO has said that much of his company's research and development has enabled Chinese companies to build cellphones without having to spend their own money on research and development (Mehta, 2006).

When discussing what Chinese technology players have achieved in respect of complementarity with foreign intellectual property, the sport of judo offers one of the best analogies. In the handset industry, for example, Chinese manufacturers have won market share from larger and stronger foreign competitors by using these competitors' own strength against them. While Europe and US-based multinationals such as Alcatel, Motorola and Nokia targeted large Chinese cities with their handsets developed from European GSM standards, Chinese players capitalised on their own knowledge of the country's distribution networks to sell phones based on the same technology for smaller Chinese towns and rural locations – the blind spot of the multinationals. These two apparently contradictory phenomena – Chinese technology firms' dependence on foreign intellectual property and their business successes – are operating together and are mutually reinforcing, thus achieving complementarity (Bohr, 1963; Rhodes, 1986; Seliktar, 1986).

Chinese firms' ability to achieve complementarity with foreign intellectual property has led to substantive action to protect intellectual property. In early 2006, a new law was announced in China, requiring local computer manufacturers to pre-install their products with licensed operating systems (Lague, 2006b). Had local firms not possessed the capability to develop products based on foreign intellectual property, the new law would have faced strong resistance. Thanks to the new laws, major Chinese computer manufacturers such as Lenovo, the TCL Group, Tsinghua Tongfang and the Founder Technology Group signed licensing deals with Microsoft (*Business Week*, 2006; Lague, 2006a, 2006b). Lenovo alone signed a US$1.2 billion Windows deal with Microsoft in April 2006. Thanks to Microsoft's dominant market share and network effect (Li, Lin and Xia, 2004), the deal is expected to help Chinese companies such as Lenovo provide consumers with what they want, thereby enhancing their brand image. Lenovo's president and chief executive recently put the issue this way: 'We will support whatever the customer demands, but there's much bigger demand for this genuine Microsoft experience' (Bishop, 2006). An IDC analyst noted: 'It helps to make the brand look more like a worthy brand. It helps to position them in a better light' (Biggs, 2006).

At the same time, Chinese technology players are also taking a number of IPR protection measures in international markets. For instance, in 2006, Netac, the Shenzhen-based manufacturer of flash drives, sued PNY Technologies in a US federal court for patent infringement (von Krogh and Haefliger, 2007). Likewise, as of 2007, ZTE filed trademark applications in 155 countries and regions and registered in 108 countries. Similarly, Baijia, a Chinese noodle maker, fought a trademark infringement case in Munich, and has registered its trademark in 20 countries, with applications pending in 48 countries and regions (Ollier, 2007).

Chinese government officials and national elites' perception of IPR

A number of institutional theorists have conceptualised legitimacy as being related to the potential loss or gain of resources (Tolbert and Zucker, 1983; Covaleski and Dirsmith, 1988; Galaskiewicz, 1991; Haunschild and Miner, 1997; Arthur, 2003). Put differently, legitimacy-related pressures could affect an organisation's access to resources (George et al., 2006).

In the case of China, gaining legitimacy from international institutional actors requires a stricter enforcement of IPR laws. Given the size of the domestic market and the dominance of Microsoft Windows, China's compliance with IPR protection is likely to result in a significant outflow of intellectual property royalty (the country's loss of resources).

To take one example, China is the world's largest maker of DVD players. Adopting its own technology, China could save US$2 billion a year in royalties (Calbreath, 2004).

It is thus pertinent to ask what measures are needed to gain international legitimacy by minimising the country's loss of resources? China has found some alternatives to minimise its dependence on foreign intellectual property. For instance, Chinese policy-makers are encouraging a software industry based on open source software (OSS) in order to cultivate domestic software vendors and minimise the country's dependence on foreign software (Wilson and Segal, 2005; Lague, 2006a; *Nanfang Daily*, 2006). Making piracy illegal results in increased spending of foreign currency on software imports. OSS, on the other hand, can be copied and used freely with no intellectual property problems. Governments in developing countries have opted for the promotion of OSS rather than forcing illegal users to pay for commercial software.

To ensure that substantive actions taken to appease the WTO and multinationals do not lead to a significant loss of domestic resource, China started OSS development in 1998. The Ministry of Information Industry (MII) established an open source alliance to bolster its support for the Linux system (Shen, 2005). Red Flag Linux was established in August 1999. The Beijing Software Industry Production Center was established to organise Linux development. Building on Red Flag and Cosix Linux, and coordinating the efforts of a hundred software engineers across 18 organisations, Yangfan Linux was launched by the centre in early 2002. In 1999, Linux development was the only software project on the government's list of top technology priorities. Similarly, in 2004, Linux internet server software and Linux mobile phone software were among the 19 projects given financial support from the fund established by the State Council in 1986 to encourage research and development in IT (*SinoCast China*, 2004a). In March 2006, the Menglan Group announced its plan to sell a laptop for 1,500 yuan (US$187) to tap demand in the nation's rural areas. The computer, called Longmeng, uses Linux software and does not support Microsoft's Windows (Ong, 2006).

Government agencies have been among the early adopters of OSS. A report by the International Data Corporation (IDC) showed that government procurement was a major factor behind the rapid growth of the OSS industry in China (Guangzhou Daily, 2006). Yangfan Linux was installed in 2,800 government computers in Beijing during the first six months of its launch. OSS has also been state-owned firms' field of competence (Yu and Xin, 2003).

Threats to resources drive decision-makers to find alternatives that may be socially unacceptable and/or may pose other risks (March and Simon, 1958; Ocasio, 1995). To take one example, in China, as noted above, many government agencies switched to OSS, which is less proven than Microsoft products. Likewise, the Chinese government also moved beyond the bound of social acceptability by mandating an ill-fated nationwide switch from Windows to Red Flag Linux (Raymond, 2004). These two examples show the Chinese government's risk-taking behaviours in terms of intellectual property issues in an attempt to deal with resource-related threats.

As organisations depart from established traditions, they create a framework for new legitimate forms through non-isomorphic change (Covaleski and Dirsmith, 1988; Garud, Jain and Kumaraswamy, 2002; Cardinal, Sitkin and Long, 2004; George et al., 2006). In China's case, OSS has been new legitimate software. Whenever the Chinese government has been threatened on the intellectual property front, it has taken more substantive actions to develop the OSS industry (Meredith, 2003). An expert argued that 'every dollar Microsoft spends to fight pirates equals US$10 in advertising for Linux in China' (*Newsweek*, 2001). The Shanghai Education Commission (SEC) provides a case in point. Following claims of piracy, Microsoft demanded that the SEC buy licences for Microsoft Office software on computers in the city's public schools. However, the SEC told schools to discontinue the use of Microsoft Office on thousands of PCs and replace it with OSS software from Beijing-based Kingsoft (CNET News, 2003a).

As noted above, to further integrate in the global economy and to attract foreign investment, China needs to acquire legitimacy from international agencies such as the WTO and foreign investors such as Microsoft. Such legitimacy helps the country exert a high degree of control over the international agenda and increases the likelihood of achieving desired outcomes (George et al., 2006).

As noted above, isomorphism, that is, engagement in actions that are consistent with the responses of other actors in the environment (George et al., 2006) is arguably positively related to legitimacy (Deephouse, 1996). One way to achieve legitimacy is thus to copy the behaviours of successful actors such as other WTO members. As a WTO member, China is compelled to adhere to the TRIPS agreement. The TRIPS agreement requires China to provide adequate legal and enforcement tools to prevent piracy. Unlike other new members, China is subject to an annual review of its WTO obligations for the first 10 years of its membership. Many countries are likely to put pressure on China to act on intellectual property infringement. Conversely, the failure to 'incorporate practices widely adopted by other organisations may be perceived as a risk to an organisation's legitimacy' (Goodstein, 1994: 359). Organisations that do not engage in isomorphic practices are also deprived of operating resources and possible threats to their survival (Singh, House and Tucker, 1986; D'Aunno, Sutton and Price, 1991).

These losses and gains in legitimacy-related controls are often presented in terms of managing the conflicting demands of the numerous constituents of the environment (George et al., 2006). More to the point, control in the global arena requires taking more substantive measures to protect intellectual property, which hinders acquiring legitimacy in the domestic sphere. As noted above, many Chinese government leaders and policy-makers think that China's integration with the global market will result in significant socioeconomic costs (Heer, 2000).

If the Chinese government departs significantly from established ways of enforcing piracy laws, its legitimacy will be questioned by its citizens (Elsbach and Sutton, 1992; Browning and Folger, 1994). As noted above, ordinary citizens perceive IPR enforcement tools as benefiting foreign software companies rather than ordinary citizens (Kshetri, 2004).

Development of products that have a low content of foreign intellectual property helps manage the conflicting demands of pro- and anti-IPR constituents. More to the point, the growth of the domestic OSS industry has allowed Beijing to take substantive actions to appease pro-IPR actors without losing resources or control over the domestic agenda.

A recent substantive action to protect intellectual property is a law announced in March 2006 which requires local computer manufacturers to ship their products with pre-installed licensed operating systems (Lague, 2006b). In the past, Chinese PC manufacturers shipped computers without pre-installed operating systems, which encouraged users to install pirated copies of Windows (*Business China*, 2001a; *Business Week*, 2006). New regulations also require all government computers to be loaded with legitimate software (Lague, 2006b).

It should be noted, however, that the government only requires *an* operating system to be preinstalled – not specifically Windows. As noted above, for most government agencies, OSS happens to be the legitimate software. Analysts suggest that the new rules could help fuel the growth of OSS such as Linux (Lague, 2006a; Bishop, 2006). OSS development has thus allowed the Chinese government to appease pro-IPR constituents and at the same time retain credibility with local consumers.

Selective adaptation is a function of the ways in which Chinese government officials and national elites express normative preferences in interpreting and applying foreign ideas such as IPR (Balbus, 1977). Such preferences can be mapped with normative institutions which introduce 'a prescriptive, evaluative, and obligatory dimension into social life' (Scott, 1995: 37). With respect to these actors, it is important to note that the principle of national self-reliance dominated the Chinese economic system during the Mao Tse Tung era (Terrill, 1977). This was in part due to a sense of national pride and a fear of dependence on foreign countries. Notwithstanding a significant shift in attitude toward technology imports and foreign investment in recent years, there is still a high level of advocacy for national self-reliance and domestic development of technology among Chinese policy-makers, researchers, scientists and military leaders due to the lack of significant alteration in China's political structure (Solinger, 1995: 127; Simon, 2001).

Chinese government officials and national elites thus do not like to see the country's IT sector dominated by foreign players (Meredith, 2003). Indeed, foreign technology imports and the amount of royalties paid to foreigners have been a focus of major concerns among Chinese national elites and government officials (Einhorn, 2004a). Chinese scientists and engineers have made several attempts to create Chinese standards in computer operating systems and audio-video compression to third-generation (3G) data standards (CNETAsia, 2003a). They want to achieve self-reliance and reverse the flow of fees by exporting Chinese standards.

Local legitimacy regarding IPR protection measures

For the sake of argument, let us assume that intellectual property laws are strictly enforced in China. Can such an approach gain local legitimacy? Note that legitimacy is a function of ideology as well as local socioeconomic conditions (Potter, 2004). For one thing, incompatibility with Chinese culture poses a problem in the acceptance of an intellectual property regime in China. This can be attributed to China's history and culture (Spierer, 1999; Shen, 2005), especially the Confucian heritage, which emphasises the importance of the family and society over the rights of the individual person (Hofstede and Bond, 1988). Learning by copying and modifying is emphasised in various aspects of individual and business activities in the country (Chae and McHaney, 2006).

There is thus an interesting contrast between the Western and Chinese views on intellectual property. Alford (1995: 25) argues that the disagreement between Confucianism and Western concepts explains the difference in the systems of copyrights and attitudes towards piracy. He argues that the concept of intellectual property has never previously existed in the entire history of China. In the colonial period, Western attempts to impose intellectual property laws on China failed as the Marxist principle further reinforced Confucianism (Alford, 1995). Alongside the cultural tradition, intellectual property enforcement is also likely to be hampered by the attitudes of those involved in law enforcement (Shen, 2005).

The Chinese arguably have a tendency to distrust 'out-groups' (Fukuyama, 1995; Osland and Cavusgil, 1996). Negative images of pro-IPR foreign constituents such as Microsoft are thus likely to hinder local legitimacy of substantive actions to protect IPR. For instance, in 1999 Microsoft China's general manager, Juliet Wu, left the company and wrote a book describing the company as 'arrogant', 'selfish' and an 'enemy of Chinese consumers' (Meredith, 2003). Shortly after Wu's book was published, a number of government departments reportedly blacklisted Microsoft (Meredith, 2003). In early 2002, out of seven government software contracts, six went to Chinese vendors. Faced with examples like this, experts say that if Microsoft intensifies its anti-piracy crackdown in China, the obvious reaction would be people's rapid switch to OSS (Piller, 2006).

Microsoft faces an image problem in China due to unfavourable press coverage of its aggressive pricing and anti-piracy campaigns (*Business China*, 2001a). For example, when Microsoft Windows 98 was released, its list price of US$241 translated to four months' salary for the average Chinese worker. One internet entrepreneur even sued Microsoft for unfair pricing (Smith, 2000). Starting the late 1990s, Microsoft began suing Chinese computer manufacturers that were shipping PCs with pirated Windows, in addition to other Chinese companies using such software (Meredith, 2003). In 1998, a Chinese court ruled in Microsoft's favour in copyright violations cases against two Chinese companies (Shen, 2005). Microsoft's litigious policy increased public perception that the company was attempting to dominate the market, eliminate competition and bully small Chinese firms (*Business China*, 2001b; Meredith, 2003).

Discussion

In the context of a developing economy, IPR is a very complex issue. In this chapter we have attempted to synthesise the theoretical research on institutions to investigate China's response to international pressure to comply with regulations regarding IPR. In this way, we have identified the core patterns in China's adoption of IPR. The literature on strategic decoupling helps explain how Chinese decision-makers weigh loss or gain of resources as well as control in the domestic domain and the international environment. This approach views adoption of IPR laws in a way that resolves potential conflicts between anti-IPR campaigns and pro-IPR lobbies. For instance, to appease these two sets of actors, intellectual property laws are formally adopted but not enforced. Selective adaptation, on the other hand, examines each IPR related response in terms of factors such as normative preferences of Chinese decision-makers, possibility of complementarity with local norms and practices, and the importance of local legitimacy. In terms of normative preference, most government officials and national elites differ from reformist leaders in terms of the lens through which they view the symbiotic relations associated with dependence on foreign intellectual property and ability to generate domestic technological standards. The development of the OSS industry has helped to achieve complementarity between foreign intellectual property laws and local norms. In this regard, ironically, it is the OSS industry that has benefited from foreign pressure to comply with IPR. Taken together, these two explanations shed some light on China's response to IPR issues.

Where is the future leading? This chapter noted a number of measures taken by the Chinese government that were merely symbolic in nature and lacked substance. It would be erroneous, however, to think that such symbolic IPR related responses will persist for a long period. Theorists have provided evidence which indicates that a symbolic change at one point may lead to more substantive changes subsequently (Edelman, 1990: 1436–37; Oakes et al., 1998; Guthrie, 1999: xii; Forbes and Jermier, 2002: 208; Campbell, 2004: 43). There are thus reasons to believe that China's IPR related responses may move towards the substantive end of the symbolic-substantive continuum.

2

The Chinese e-business industry

The Chinese e-business industry has a number of idiosyncratic and unusual features. Institutional factors such as strong nationalism, the state's entrenchment in the economy, political cognitive and political normative factors, regulative uncertainty, professional associations' roles, and the importance of business and social networks are deeply reflected in China's e-business development pattern. We argue that by approaching the Chinese e-business industry from the standpoint of institutional theory, we can capture the complex factors facilitating and hindering China's rapidly growing e-business industry. The findings presented are broadly consistent with existing theories on institution-technology diffusion nexus. Nonetheless, this chapter reveals unique mechanisms associated with Chinese institutions in shaping the country's e-business landscape.

Introduction

The scale of e-business development in China is large enough to be noticed at the global level. China's first online transaction took place in 1996 (Business Wire, 2007). Estimates suggest that there were 135 million internet users in 2006 (AFX International Focus, 2007) and the country's online transactions crossed US$127 billion that year (*Asia Pulse*, 2006a; Rein, 2007). According to a report released by the Chinese government in January 2008, there were 210 million internet users in the country[1] (Jesdanun, 2008).

By mid-2006, about one-quarter of China's internet users were regularly engaged in e-commerce activities (Xinhua News Agency, 2006a). Estimates have suggested that in the first 10 months of 2007, e-shopping accounted for 0.8 per cent of the total national consumption, which was significantly greater than the 0.25 per cent recorded in 2006

(CCID, 2007). A study released by CCID Consulting in December 2007 estimated China's search engine market at 2.93 billion yuan (US$407 million), which was 77 per cent bigger than in 2006 (CCID, 2007).

Like many other economic sectors (Terrill, 2005), the internet and e-business industries in the Chinese economy have many unusual and idiosyncratic features. The effects of institutional factors such as strong nationalism, the state's entrenchment in the economy, political cognitive and political normative factors, regulative uncertainty, professional associations' roles, and the importance of business and social networks are deeply reflected in China's e-business development pattern.

Despite a sizable and growing body of research devoted to the Chinese internet industry, we lack satisfactory explanations for the development pattern of the Chinese e-business landscape. While internet control in China has been examined from various angles, it is not clear whether and how the Chinese government's internet control may influence the e-business industry. The literature also lacks satisfactory explanations of companies' highly government-centric activities related to e-business. Moreover, the effect of social and business networks on organisations' e-business adoption is not examined. These are only a handful of the numerous institutional issues facing the Chinese e-business industry. We argue that approaching from the standpoint of institutional theory, we can capture the complex factors facilitating and hindering the country's rapidly growing e-business industry.

The institution-technology diffusion nexus: a theoretical perspective

In this chapter, we employ the institutional pillars proposed by Scott (1995, 2001) as an analytical tool. This approach allows us to analyse the impacts of a wide range of factors on an organisation's e-business adoption as well as the width and depth of adoption. Scott defines institutions as 'multifaceted systems incorporating symbolic systems – cognitive constructions and normative rules – regulative processes carried out through and shaping social behaviour' (Scott, 1995: 33). He argues that proper understanding of an organisation requires decoding social and cultural contexts that create regulative, normative and cognitive infrastructures, which 'constrain and support the operations of individual organizations' (Scott, 1995: 151). Institutions can be described in terms of three pillars: regulative, normative and cognitive (Scott, 1995, 2001).

Regulative institutions and technology diffusion

Regulative institutions consist of 'explicit regulative processes: rule setting, monitoring, and sanctioning activities' (Scott, 1995: 35). Regulatory bodies (e.g. the Ministry of Information Industry) and existing laws and rules influencing an organisation's e-business behaviour fall under this pillar. These institutions focus on the pragmatic legitimacy concerns in managing the demands of regulators and governments (Kelman, 1987). The state is arguably the most important external institutional actor and powerful driver of institutional isomorphism as a violation of laws can have harsh sanctions (Bresser and Millonig, 2003).

Normative institutions and technology diffusion

Normative components introduce 'a prescriptive, evaluative, and obligatory dimension into social life' (Scott, 1995: 37) and help us understand how 'values and normative frameworks structure choice' (Scott, 1995: 38). To be successful, practices should be consistent with value systems of the national cultures (Schneider, 1999). Normative institutions are concerned with procedural legitimacy and require e-commerce providers and online shoppers to embrace socially accepted norms and behaviours (Selznick, 1984). Elements of normative institutions also include trade associations or professional associations that can use social obligation requirements to induce certain behaviour within the e-business industry.

Cognitive institutions and technology diffusion

Scott (1995: 40) suggests that 'cognitive elements constitute the nature of reality and the frames through which meaning is made'. Although all components of institutions are intertwined with culture (Neale, 1994: 404), cognitive institutions are arguably most closely associated with culture (Jepperson, 1991). Cognitive legitimacy concerns are based on subconsciously accepted rules and customs as well as some taken-for-granted cultural accounts of technology use (Berger and Luckmann, 1967). Although carried by individual members, cognitive programmes are elements of the social environment and are thus social in nature (Berger and Luckman, 1967). Compliance in the case of cognitive legitimacy concerns is due to habits (Grewal and Dharwadkar, 2002).

To put things in context, when internet users and online sellers behave in certain ways, they may not even be aware that they are complying.

The nature of and relationship between the three pillars

It is worth noting that each institutional pillar both reflects as well as determines the nature of the other pillars (Hayek, 1979). It is therefore often difficult to to isolate them in the real world. North (1996: 344) defines institutions as 'formal constraints (rules, laws, constitutions), informal constraints (norms of behaviour, conventions, and self-imposed codes of conduct), and their enforcement characteristics' and observes that informal rules provide legitimacy to formal rules (North, 1994). Likewise, Axelrod (1997) comments on the relationship between regulative and normative institutions, describing that 'social norms and laws are often mutually supporting. This is true because social norms can become formalized into laws and because laws provide external validation of norms.'

Of these three pillars, normative and cognitive components are more likely to explain interorganisational differences as regulative influences are likely to apply uniformly in a given industry (Miller, 1996: 287). For instance, the government's internet control and regulative uncertainty affect all organisations more or less equally.

E-business organisations, however, differ on normative and cognitive institutions. For instance, organisations may have different approaches to ethical decision making in their interactions with other businesses and the government (Whitcomb, Erdener and Li, 1998). For instance, while Yahoo complied with the Internet Society of China's (ISC) codes of conduct on self-censorship, Google and Altavista did not. Likewise, organisations differ in terms of their attitudes towards *guanxi* networks and the need to enhance 'Chineseness' in their e-business models.

Institutional processes in the Chinese e-business industry

Table 2.1 presents various constructs used in this chapter and mechanisms associated with their impacts on organisations' e-business adoption in China.

Table 2.1 Institutional factors affecting the diffusion of e-business in China

Construct	Mechanisms
Government control	Hinders effective uses of the internet, leads to higher costs, and acts as a barrier to consumers' internet adoption
Regulative uncertainty	Difficulties associated with enforcing contracts increase the reluctance of consumers and businesses to engage in e-business
Professional associations	Influence e-business companies to serve government's interest more than consumers'
Social networks	Importance of social dimension in business reduces a company's propensity to adopt e-commerce. Strong *guanxi* compensates for risks associated with legal uncertainty
Chineseness in e-commerce products and technology	High degree of nationalism and a lack of trust regarding 'out-groups' results in Chinese consumers' bias towards products and technologies characterised by Chineseness

Government control

China's state strategies toward information and communications technology (ICT) have been to balance economic modernisation and political control (Kalathil, 2003). The internet, 'the greatest democratiser' (Pitroda, 1993), has posed a severe threat to authoritarian regimes. According to Reporters Without Borders, 'China was one of the first countries to realize it couldn't do without the internet and so it had to be brought under control' (McLaughlin 2005). Arguably Beijing focused its attention on the internet before other developing countries in order to maintain control (Press, Foster and Goodman, 1999; Yang, 2001; Zhang, 2001).

Although the government actively monitors and controls politically objectionable content on the internet (McLaughlin 2005), it has a more relaxed approach to culturally sensitive materials. Popular portals, for instance, feature topics ranging from pollution to homosexuality (Kalathil, 2003). As noted above, although the state policy has been to encourage ICT use in economic modernisation (Press, Foster and Goodman, 1999; Yang, 2001; Zhang, 2001), some government officials have explicitly expressed negative views on even some productive uses of the internet. To take one example, an official of the Zhejiang provincial government commented that internet ads contain 'distorted, misleading and even illegal information' and have 'greatly endangered customers' interests' (Chinese Education and Research Network, 2001).

Although the government's role in shaping internet usage in China is admittedly simplistic, and previous studies have extended our understanding of this phenomenon, what is not clear is how such controls influence the e-business industry. Here is why government control affects organisations' e-business adoption. First, portals and search engines have to undergo self-censorship; concurrently, some are banned. A web portal attracts visitors and satisfies their information needs and in the process presents advertising banners to them (Adar and Huberman, 2000). If a portal does not have content useful to consumers, its ability to attract visitors will be limited and those visiting the portal will spend less time there. Organisations' ability to use the internet in advertising is thus hampered by the government's control.

Second, there are reports that the government has targeted individual websites. According to the Berkeley China Internet Project, the government hides websites containing phrases such as 'freedom', 'democracy', 'China-liberal' and 'falun' (Foushee, 2006). There are also reports that the Chinese government reportedly sends viruses to attack banned sites (Guillén and Suárez, 2005). These actions hinder e-business adoption of target organisations.

Third, government control increases costs related to e-business. For instance, private websites are required to hire censors such as 'cleaning ladies' or 'big mamas' to filter and quickly remove offensive material from bulletin boards and chat rooms (Kalathil, 2003). Chinese internet cafés are required to install software that prevents access to up to 500,000 banned sites (Hermida, 2002). In addition to monetary costs, there are psychological and time costs. For instance, a new rule announced in mid-2005 requires all websites to be registered (Magnier and Menn, 2005).

Fourth, some government measures influence organisations' e-business adoption indirectly by acting as barriers to consumers' internet use. For instance, the government closed 150,000 unlicensed internet cafés in 2002 (Hermida, 2002). In this regard, it is worth noting that in early 2008, about one-third of Chinese internet users were accessing the internet through cybercafés (Jesdanun, 2008). A slow internet adoption rate among Chinese consumers results in lower e-business adoption among organisations.

Regulative uncertainty

An even stronger barrier to the adoption of e-business in China is perhaps the lack of clear policies (*Asia Pulse*, 2006a). Some analysts argue that the government tends to favour Chinese businesses at the cost of foreign companies 'via a byzantine system of laws and regulations' (Ackerman, 2006). Pointing out problems faced by organisations because of a lack of specificity, at least one business group has asked the Chinese government to issue clearer regulations (Kalathil, 2003).

Some argue that, in China 'the law is marginalised and the legal system relegated to a lowly position in a spectrum of meditative mechanisms, while at the same time available for manipulation by powerful sectors within the state and the society at large' (Myers, 1996: 188). This is even more so for laws related to e-business. As of 1999, there was no national legislation addressing the topic of online contracts (M. Aldrich, 1999). Authoritarian regimes, in general, are slow to enact laws to recognise digital and electronic signatures (DES). Encryption software, which is an essential component of DES, allows confidential messages to be sent, thus making it difficult, even impossible, for the government to detect politically and culturally objectionable content transmitted through the internet (Kshetri and Dholakia, 2001).

A study by Gibbs et al. (2003) indicates that adoption of the internet as a transaction medium in China has been hindered by the lack of legal protection for consumers and sellers for faulty products and negligent payments respectively. Another study has shown that 'transactional and institutional trust' resulting from weak rule of law has been a major impediment to e-commerce in China (Efendioglu and Yip, 2004). It is also important to note that China's recently passed law on electronic signatures lacks guidelines related to identification requirements for purchasers using an electronic signature (Srivastava and Thomson, 2007).

Chinese professional associations

Savan (1989: 179) defines 'the professions' as groups that apply 'special knowledge in the service of a client'. A profession is also self-regulated by a code of ethics (Claypool, Fetyko and Pearson, 1990; Cohen and Pant, 1991) and is characterised by its role as a moral community (Camenisch, 1983). These codes require members to maintain higher standards of conduct than what is required by law (Backoff and Martin 1991), help make professional norms visible (Frankel, 1989), and act as a vehicle to assure the public and clients that members are competent, have integrity, and maintain and enforce high standards (Ward et al., 1993). Apart from convincing external parties of the integrity of the profession, codes play an important role in forcing members to question their values (Meyer, 1987).

In China, professional associations in the realm of e-business are more government-centric and less consumer-centric than in the West. Among the many examples of professional associations that engage in e-business activities, one is particularly telling: the government-backed Internet Society of China (ISC). ISC was formed in May, 2001 with more than 130 members and is sponsored by network access carriers, internet service providers (ISPs), facility manufacturers and research institutes.[2] The ISC has asked internet companies to sign a voluntary pledge on 'Self-discipline for China's Internet Industry' that commits them to investigate and block websites that have politically and culturally sensitive contents. In March, 2002, the ISC distributed the pledge, signed by over 120 internet portals (Stout, 2002). The pledge commits signatories not to disseminate information 'that might threaten state security or social stability' (*Economist*, 2002a). On a more prosaic plane, consider the following recent statement from Hu Qiheng, chair of the ISC: 'It may not be popular everywhere to say this, but I think it is important for the government to monitor and police the internet' (Crampton, 2006).

In a conference in Hangzhou, executives of China's search engines and web portals argued that they need to monitor contents and remove those objectionable to the government (McLaughlin 2005). Among foreign affiliates, this aspect is especially evident on Yahoo's Chinese website, which chooses major headlines from government-owned newspapers as most foreign sources of news about China are banned (Yee, 2001). For instance, using simplified Chinese, *The Economist* (2002a) used the Yahoo website to search for 'Falun Gong'; the search retrieved more than 180 news items from the official media, but only one website – that of an anti-Falun Gong group. In 2005, Yahoo reportedly provided private e-mail information to the government, which led to the imprisonment of a Chinese journalist (McLaughlin 2005). Yahoo chief Jerry Yang said that this decision was necessary in order for the company to be able to do business in China (McLaughlin 2005). This and other cases suggest that, as a measure of 'strategic isomorphism' (Deephouse, 1996), many foreign companies are undertaking government-centric activities.

Portals and search engines not following the pledge, such as Google and AltaVista, were blocked in China in 2002 (Singer, 2002). Subsequently, however, Chinese authorities have won agreements from technology companies including Google and Microsoft for filtering and screening out sensitive words (French, 2006). For instance, in China, Microsoft blocks bloggers from posting politically objectionable words and Google shuts down when a user looks for sensitive words (McLaughlin 2005).

The above discussion makes it clear that the ISC's activities differ drastically from comparable professional e-business associations in the West. For instance, the UK Mobile Marketing Association issued its code of conduct in December 2003, specifying what times of day mobile marketers can target consumers (Balmond, 2003). Likewise, in the USA, in early 2001, technology-industry lobbyists and consumer and civil-liberties activists including the American Civil Library Association, Electronic Privacy Information Center and Consumer Federation of America circulated a letter to members of Congress and the president calling for a stronger set of privacy rules (Benson and Simpson, 2001). While these activities are designed to protect consumer privacy, the ISC's actions have promoted the government's interest. For instance, Hu Qiheng, chair of the ISC, has defined internet crime to include 'acts counter to the interests of the Chinese government' (Crampton, 2006).

At least three factors contribute to professional associations engaging in government-centric activities. First, the state's deep entrenchment in the economy means that the government can play a critical role (Pei, 2006). For instance, according to Union Bank of Switzerland, the state accounts for at least 70 per cent of the Chinese economy, compared with less than 7 per cent in India (Pei, 2006). As of 2001, 70 per cent of large and medium-sized 'corporatised' enterprises had Communist Party members on their boards of directors (Pei, 2006). The country's ISPs are controlled by state-run companies (McLaughlin 2005).

Second, to succeed in China, e-business organisations need to make room for civil servants and high-ranking government officials (Einhorn, Webb and Engardio, 2000; Sikorski and Menkhoff, 2000). It is important to note that most internet regulations are guidelines only and do not represent formal laws (Shie, 2004). The regulatory vacuum thus makes it important to have good relationships with government officials. For organisations involved in e-business, it is thus important to consider the government's interest.

Third, in China, the concepts of customer service and privacy, which are the focus of most Western professional associations related to e-business, are not well developed. Terrill (2005) goes even further, arguing that 'because China remains an authoritarian state, we cannot know what the Chinese people want'. Organisations involved in e-business in China thus face little or no pressure related to customer service and privacy.

Social and business networks

As researchers such as Myers (1996) and McIllwain (1999) make clear, there are fundamental differences in how Westerners and Chinese view relational ties and social networks. Perhaps the most notable feature of social networks in China is the concept of social capital in the form of *guanxi*, which is related to generalised reciprocity or 'mutual trust and commitment among interrelated actors that are independent of any specific transaction' (Sandefur and Laumann, 1998: 491). In Confucianism, business relationships are not simply economic but also have social dimensions (Hofstede and Bond, 1988; Putnam, 2000; Romar, 2004). This is different from the West where people form 'transactional bonds to obtain transactional benefits' (Myers, 1996).

Most e-commerce transactions 'attempt to substitute for social information' (Steinfield, 2002: 8) and threaten to undermine the established interpersonal networks (Gibbs, Kraemer and Dedrick, 2003). The importance placed on the social aspect in trading (Driedonks et al., 2005), the precedence of established relationships such as *guanxi* (McKinsey, 2001) and preference for face-to-face communications thus tend to hinder organisations' adoption of e-business (Moodley and Morris, 2004). A study by Lim et al. (2004) found that for countries with lower uncertainty avoidance, collectivism is negatively related to internet shopping. Moreover, in collectivist societies, people rely on direct sources of information. Thus, information flowing through the internet tends to weaken the traditional forms of relationships (Dunfee and Warren, 2001).

At the organisational level, however, the strength of *guanxi* is a key determinant of business success. *Guanxi* arguably provides 'access to resources, which are controlled by powerful elites, who can arbitrarily allocate them' (Myers, 1996: 185). The regulative uncertainty further increases the importance of *guanxi*. In the West, 'the subject of an agreement and the content of an obligation are delimited by a state sponsored value system and a breach is enforced by reference to a state maintained mechanism extrinsic to the parties' (Myers, 1996). In China, by contrast, the terms are informal and subject of agreement is constrained by the parties' intrinsic value system (McIllwain, 1999).

How does strong *guanxi* help in the Chinese cyber-world? We discussed above how internet control affects organisations' adoption of e-business. It is well known among regulators and businesses that enforcement of internet regulation does not always take place (Kalathil, 2003). For example, while other Chinese-based search engines became victims of the government crackdown in 2001, Baidu, thanks to its strong *guanxi*, did not experience any problem (Hachigian, 2001). In the Chinese digital world, strong *guanxi* is thus a critical resource (Peng and Luo, 2000); it is useful in acquiring legitimacy (Suchman, 1995; Ahlstrorn and Bruton, 2001), which minimises legal uncertainties (Lo and Everett, 2001). Organisations involved in e-business have realised the importance of maintaining good relations with civil servants and officials, as well as with members of the local power elites (Sikorski and Menkhoff, 1999; Einhorn, Webb and Engardio, 2000; Kalathil, 2003).

Foreign companies have realised that *guanxi*-building can protect them in 'a gray legal environment' (Lo and Everett, 2001). Local or regional business partners and other prominent tycoons (Sikorski and Menkhoff, 1999; Einhorn, Webb and Engardio, 2000) are important components of foreign companies' *guanxi* in China. It is important to note that Chinese partners are likely to have intimate contacts and connections with critical government officials (Osland and Cavusgil, 1996). Moreover, Chinese, including many government officials, have a tendency to distrust the 'out-group' and are more trusting of Chinese representatives (Fukuyama, 1995; Osland and Cavusgil, 1996). Most successful foreign companies have good *guanxi* (e.g. AOL's stake in Chinadotcom, and CMGI and Intel's collaboration with Pacific Century CyberWorks (Einhorn and Yang, 2000; Computer Weekly, 2001).

Chineseness in e-business products and technology

Past research has found that Chineseness in e-business related products and technologies co-varies positively with Chinese consumers' likelihood of doing business with a company. A *McKinsey Quarterly* (2006) article notes: 'Consumers in China ... have strong national pride, so multinational companies could lose important segments by seeming too foreign'.

Now, consider the digital world. In 2000, 78 per cent of Chinese internet users viewed Chinese language information and 71 per cent viewed domestic information (CNNIC, 2001). A similar study conducted on Indian internet users indicated that only 41 per cent of online Indians prefer Indian language websites (Barnwal, 2006). Another survey found that in 2001, nine of the ten most popular sites for Chinese surfers were China-based (Hormats, 2001). According to iResearch, 80 per cent of Chinese internet users use the local search engine, Baidu, compared with Google's share of 36 per cent and Yahoo's 26 per cent

(Ilett, 2007). Likewise, Alibaba.com, a Chinese business-to-business site, is the biggest e-commerce marketplace in China (Maidment, 2008).

Indeed, Chinese-language content on the internet has been a major factor contributing to a rapid growth of overseas Chinese visiting Chinese language websites (Hormats, 2001). Moreover, Chinese customers associate an organisation with a .cn address with a higher level of serious commitment to doing business in China (Tindal, 2003). Moreover, translation software tends to have a higher error rate for Chinese compared with other languages (Tindal, 2003). Companies outside China are increasingly realising the importance of having locally-built services in order to succeed in China (Secured Lender, 2004).

Many studies have shown that consumers have some degree of bias towards domestic products. Although many factors may affect such dynamics (some of which are described by Balabanis et al., 2001), mechanisms associated with Chinese consumers' bias towards Chinese products in the digital world are mostly left unexamined. An exception is Zhou and Hui's (2003) study which found that 'symbolic benefits' associated with products offered by Chinese companies rather than improved quality were primary motivational forces behind Chinese consumers' shift towards local products.

Nationalism and patriotism appear to be important triggers for Chinese consumers' bias towards e-commerce products characterised by a high degree of Chineseness. Although attachment to one's nation leads to actions 'which are disinterested or self-sacrificing' (Salmon, 1995: 296), the effects of patriotism are highly visible among Chinese. Researchers have made an intriguing argument as to why the Chinese display a very high degree of nationalism. The Communist Party arguably bolsters its legitimacy (Elliott, 2006) through 'intensive inculcation of nationalism via the Chinese press and education system' (Kurlantzick, 2005), thus invoking a deep sense of 'Chineseness' among citizens (Ong, 1997; Barme, 1999; Hansen, 1999). Regarding the formative dynamics of Chineseness, Sautman (2001) has documented how China has adapted a body of complex scholarship to achieve this goal. In a review of the literature, Sautman (2001) concludes: 'nowhere is this more pronounced than in China, where these disciplines [archaeology and paleoanthropology[3]] provide the conceptual warp and woof of China's "racial" nationalism'.

The Chinese Communist Party, however, does not have complete control over nationalism. With the emergence of ICT, popular nationalists in China are increasingly acting independently of the state (Gries, 2005). For instance, after the September 1999 bombing of the Chinese Embassy in Belgrade and following the collision of a US surveillance plane and a Chinese fighter in 2001, many private websites were flooded with extreme nationalistic posts (Kalathil, 2003).

Up to this point, we have concentrated on the roles of nationalism and patriotism in affecting Chinese consumers' bias towards products characterised by a high degree of Chineseness. Local brands also enjoy a higher degree of consumer trust than foreign ones (Schuiling and Kapferer, 2004), and such a tendency is higher in China than in many other countries. Theorists and empiricists have found evidence which indicates that China is a low-trust society characterised by a tendency to distrust 'out-group' people and trust only 'in-group' people (Fukuyama, 1995). In the e-business world, such a tendency produces further bias against foreign products.

Discussion

The foregoing discussion provides a framework for understanding institutional processes affecting e-business in China. The findings are broadly consistent with existing theories on the institution-technology diffusion nexus. Nonetheless, this chapter has revealed unique mechanisms associated with Chinese institutions in shaping the country's e-business landscape.

In particular, the regulative pillar's influence on the other two pillars is more salient in China than in many other countries. According to a report from Reporters Without Borders, internet control in China is done through 'a clever mix of investment, technology and diplomacy' (McLaughlin 2005). The government-backed ISC is an example of such diplomacy. Moreover, the regulative uncertainty has increased the importance of *guanxi* as well as the importance of complying with the codes of conduct issued by the ISC. The bottom line is that regulative uncertainty has a strong influence on the degrees of legitimacy associated with other institutional components.

The evidence provided in this chapter indicates that the practices of organisations are sometimes non-isomorphic with respect to institutions. For instance, Google's and AltaVista's actions before 2002 were non-isomorphic with respect to Chinese normative institutions. Similarly, the practices of Google, Microsoft, Yahoo and many foreign affiliates in China are non-isomorphic with respect to institutions in their home countries. Early works of institutionalists indicated that organisational structure and practices tend to be isomorphic, that is, consistent with regulatory, cognitive and normative institutions (Meyer and Rowan, 1977; Powell and DiMaggio, 1991) and those of other actors (George et al., 2006: 353). Subsequent theoretical and empirical evidence has suggested that organisations do not always engage in predictable and isomorphic actions (Hoffman, 1999; George et al., 2006). They often face pressures for non-isomorphic responses that 'involve departure from established structures, practices, and utterances of other actors in the environment' (George et al., 2006: 353).

Like other organisational fields, e-business activities can be viewed as 'arenas of power relations' (Brint and Karabel, 1991: 355) in which various players engage in institutional war (White, 1992; Hoffman, 1999). Although organisational isomorphism is positively related with legitimacy (Deephouse, 1996), when an organisation is seeking to acquire legitimacy from multiple sources with conflicting demands, its responses are likely to be non-isomorphic with respect to some of the sources. The degree of a response's isomorphism/non-isomorphism is a function of organisational perception of gain or loss of control and/or resources associated with the response (George et al., 2006). To gain access to the US$125 billion Chinese e-commerce market, Western organisations are willing to take actions that are non-isomorphic with respect to the institutions in their home countries.

Various components of institutions, despite their connotation of persistence (Parto, 2005), durability (Hodgson, 2003) and stability (Scott, 2001: 48), are subject to change in evolutionary time (Parto, 2005). Zucker (1988: 26) draws an analogy with physics to describe institutional change mechanisms. He argues that institutions continuously undergo change due to entropy, a tendency toward disorder or disorganisation. An implication of the entropy-like characteristics is that people can modify and reproduce (Scott, 2001) institutions. Some signs of institutional changes likely to affect the Chinese e-business industry have begun to appear.

One important institutional change is the decreasing importance of *guanxi*. Some analysts argue that institutional modernisation is likely to make *guanxi* less important (Guthrie, 1998; Ahlstrom et al., 2005; Herrmann-Pillath, 2006). Previously, regulative uncertainty increased the importance of *guanxi* in China. Because resources have traditionally been controlled by powerful actors who can 'arbitrarily allocate them' (Myers, 1996: 185), some organisations have therefore sought to develop *guanxi*. In recent years, however, China has devoted more resources and has taken other measures to strengthen the country's laws (Weidenbaum, 2006b). In 1979 there were two law schools and fewer than 2,000 lawyers in China; this has now increased to over 300 law schools and an estimated 120,000 certified lawyers (Weidenbaum, 2006b). Consequently, doing business in China is becoming more and more predictable (Carney, 1999). This trend is making *guanxi* less significant.

Diffusion of Western-style business practices among Chinese organisations is also weakening the significance of *guanxi*. For instance, Chinese organisations are rapidly adopting practices such as performance-based pay and results-oriented reporting (Weidenbaum, 2006b). One can therefore reasonably expect a decline in the importance of personal relationships in workplaces. Indeed, in recent years, an ever-increasing number of employees in Chinese companies are changing jobs (Tsang, 1998; Dunfee and Warren, 2001; *Economist*, 2005; *Knight Ridder Tribune Business News*, 2005). Organisations' and individuals' increasing dependence on formal institutions, on the other hand, is likely to further strengthen such institutions. More to the point, regulative institutions are likely to play more important roles in shaping the Chinese e-business landscape.

Changes can also be expected in other institutional components. For instance, China's commitment to free markets, globalisation of its companies and increasing integration in the global economy may erode the perceived legitimacy of internet control. Likewise, the Chinese government may face pressure to address issues related to regulative uncertainty.

The theory presented in this chapter also contains some practical implications. First, a company's success in the Chinese e-business market is a function of its ability to integrate Chineseness in e-business technologies and products. Methods to enhance Chineseness in e-business designed for the Chinese market include Chinese domain names, Chinese language websites, and forming joint ventures with Chinese companies.

Second, the above discussion indicates that the strength of *guanxi* networks is positively related to a company's e-business performance. For foreign companies to succeed in China, it is thus important to form *guanxi* networks. This is especially important given the uncertain Chinese regulatory environment.

Third, as noted above, while actions taken by foreign companies to gain legitimacy in China are isomorphic with respect to some Chinese institutions, they are non-isomorphic with respect to other powerful institutions and constituencies, especially in their home countries. For instance, Amnesty International has accused US-based internet companies such as Google, Microsoft and Yahoo of violating the Universal Declaration of Human Rights in their agreement with the Chinese government to censor internet use in China (US Fed News Service, 2006a). Western technology companies' government-centric activities in China thus may lead to a possible consumer backlash and even legal sanctions in their home country. As noted earlier, the appropriate level of isomorphism/non-isomorphism with respect to a given institution is a function of the resources associated with it and the importance of maintaining control over the institution.

Finally, as Chinese technology companies are rapidly expanding their businesses in other developing countries (McLaughlin 2005), e-business in China is also likely to play an important role in shaping the rapid growth of e-business in the developing world. A clearer understanding of institutional processes in Chinese e-business is thus important to devise a strategy to compete in other developing countries' e-business markets.

Notes

1. The USA was estimated to have 215 million internet users that time.
2. At present, there are more than 150 members, most of which are organisations (see *http://www.isc.org.cn/English/*).
3. Archaeology is the study of ancient societies and cultures; paleoanthropology is the study of the human fossil records.

3

Cyber-control in China

In this chapter, we propose a framework for identifying clear contexts and attendant mechanisms associated with authoritarian regimes' internet control measures. We build on the concept of the 'field' formed around internet control in authoritarian regimes. Viewing cyber-control as an 'institutional field' allows us to examine the evolution of regulative, normative and cognitive institutions. We have advanced a model that explains how an institutional field evolves. The findings are broadly consistent with existing theories on field formation. Nonetheless, this chapter has revealed unique processes and mechanisms associated with cyber-control-led field construction and reconstruction in authoritarian regimes.

Introduction

Although more than three dozen governments in the world control their online environment (Mackinnon and Palfrey, 2006), few have done so more skilfully than China. China's state strategies toward information and communications technology (ICT) have been to balance economic modernisation and political control (Kalathil, 2003). According to Reporters Without Borders, 'China was one of the first countries to realize it couldn't do without the internet and so it had to be brought under control' (McLaughlin, 2005). Arguably, Beijing thus focused its attention on the internet before other developing countries because it spotted a need to maintain control (Yang, 2001; Zhang, 2001). Estimates suggest that tens of thousands of government agents are engaged in various cyber-control activities (Stevenson-Yang, 2006). In 2007, the Chinese government forced the closure of 44,000 websites; some 868 people were arrested on internet pornography charges and about 2,000 people engaged in activities relating to internet pornography were penalised (e-commercetimes, 2008). According to Reporters Without Borders, 50 Chinese 'cyberdissidents' were in prison in January 2008 (Jesdanun, 2008). New regulations, effective from 31 January 2008, require China's online video providers to censor all video clips with 'anti-Beijing content' (Einhorn, 2008). These are labelled by Reporters Without Borders as 'unprecedented censorship measures'.

In recent years, Chinese authorities have also won agreements from foreign technology companies including Yahoo, Google and Microsoft for filtering and screening out sensitive words (McLaughlin, 2005; French, 2006). In addition to social and political effects, government control of the internet has also been a major concern for the growth of e-business in China (Kshetri, 2007).

Issues related to control of the online environment in authoritarian regimes are important to the interests and objectives of a diverse set of players such as Western governments, human rights groups (e.g. Amnesty International), Chinese dissident groups, Falun Gong, anti-Falun Gong groups and businesses involved in the internet value chain. Perhaps more important is that among these players, who tend to have competing interests and disparate purposes (Brint and Karabel, 1991; Greenwood and Hinings, 1996; Hoffman, 1999), the balance of power and the patterns of interaction are rapidly altering over time. New organisations or populations are entering in the dialogues related to cyber-control and some of the existing organisations are exiting (Brint and Karabel, 1991; Hoffman, 1999: 355). This means that the 'content, rhetoric, and dialogue' (Hoffman, 1999: 355) among constituencies that have interests in internet control in China are shifting over time.

A clearer understanding of internet control measures in authoritarian regimes is important from theoretical, managerial and policy standpoints. First, the relation between internet diffusion and the growth of democracy is unclear (Wilson and Segal, 2005). In the USA, for instance, critics are concerned about the internet-led polarisation of the country and argue that the internet's effect on democracy is unclear (Benkler, 2006). The internet diffusion-democracy nexus, especially in authoritarian regimes, seems to be less clear than it might at first appear. One upshot of such a tendency is that there are some well-founded rationales for and against Western companies' involvement in internet-related activities in authoritarian regimes, as well as a number of misinformed and ill-guided viewpoints. To take one example, when the French telecom company, Alcatel, began merger talks with Lucent in

2006, some US lawmakers criticised Alcatel's ties with Iran. Alcatel had upgraded Iran's telecom network and provided the country with its first high-speed DSL internet connections (Bremmer, 2006). One view thus held that Alcatel had helped the Iranian regime. Nonetheless, one might argue the contrary, insomuch as wiring the country has allowed Iranians to communicate with one another and the outside world more easily, thus helping to promote democracy in the country (Bremmer, 2006). Why is the relationship between internet diffusion and democracy in authoritarian regimes so unclear? Much depends on the precise circumstances. In particular, whether the internet forces authoritarian regimes to promote democracy or whether it will be controlled like other mass media (Hachigian, 2001; Stevenson-Yang, 2006) has been a pressing theoretical and policy issue that adjoins the larger social and political concerns of democratic society.

In the remainder of the chapter, we first provide a brief review of the theoretical foundation. We then translate the theories within the context and limits of cyber-control in China. Next, we discuss responses of various institutional actors from the standpoint of internet control in China. The final section provides conclusion and implications.

The theoretical framework: institutional field

At the centre of our argument is the concept of the field formed around internet control. Hoffman (1999: 352) argues that a field is 'formed around the issues that become important to the interests and objectives of specific collectives of organizations'. To put things in context, these organisations include authoritarian governments attempting to control the internet, human rights groups such as Amnesty International, Western governments, professional associations and businesses involved in the internet value chain. Accordingly, the 'content, rhetoric, and dialogue' (Hoffman, 1999: 355) among these constituencies influence the nature of internet control measures.

An important point to bear in mind is the 'evolving' rather than 'static' nature of fields (Hoffman, 1999: 352). Institutional theorists make an intriguing argument as to how a field evolves. It is important to note that an organisational field is a dynamic system characterised by the entry and exit of organisations or populations and/or interaction patterns among them (Brint and Karabel, 1991; Hoffman, 1999: 355). Like other 'issue-based' fields, internet control can also be viewed as 'arenas of power relations' (Brint and Karabel, 1991: 355) in which various players and constituencies with competing interests and disparate purposes negotiate over issue interpretation and engage in institutional war (White, 1992; Hoffman, 1999: 351–2). Institutional evolution thus takes place by an alteration of the interaction patterns and power balances among organisations (Brint and Karabel, 1991; Greenwood and Hinings, 1996; Hoffman, 1999).

As observed above, internet control in authoritarian regimes is not yet fully institutionalised as it is not 'uncritically accepted' and is not considered to be a 'natural and appropriate arrangement' (Tolbert and Zucker, 1996; Greenwood, Suddaby and Hinings, 2002). Tolbert and Zucker (1996) refer to this as semi-institutionalisation. An important question then is how an idea (e.g. cyber-control) moves from preinstitutionalisation towards full institutionalisation. A valuable lead into this question is provided by Oliver (1991: 151), who suggests that organisational actors' strategic responses to institutional processes vary from 'passivity to increasing active resistance: acquiescence, compromise, avoidance, defiance and manipulation'. To put things in context, in China, internet companies' responses to the government's pressure to create a controlled cyber environment vary widely. Moreover, some organisations have changed their strategies to deal with government pressures. Among foreign affiliates, Yahoo followed the strategy of 'acquiescence' from the beginning, obeying rules and norms, and cooperating with the government. Some portals and search engines such as Google and Altavista, on the other hand, 'defied' or actively resisted the institutional processes and were blocked in the country in 2002 (Singer, 2002). Put differently, they exited the Chinese cyber-control field in 2002. Subsequently, however, Chinese authorities have won agreements from Google for filtering and screening out sensitive words (French, 2006). For instance, Google in China shuts down when a user looks for sensitive words (McLaughlin, 2005). Google has thus reentered the Chinese cyber-control field.

According to Hoffman's (1999) model, institutional evolution entails transitions among Scott's (1995, 2001) three institutional pillars – regulative, normative and cognitive. Building a regulative/law pillar system is the first stage of field formation. It is followed by a formation of normative institutions and then cognitive institutions (Hoffman, 1999).

Cyber-control in China as an institutional field

Formation of regulative institutions

A focus on monitoring and sanctioning rather than rule setting activities

With regard to regulative institutions, the Chinese cyber-control field seems to be concentrated on 'monitoring, and sanctioning' rather than 'rule setting' activities (Scott, 1995: 35). For Chinese authorities there are a wide range of reasons for prosecuting and punishing individuals for online 'crimes' (e.g. protecting the value of Chinese assets exposed to global capital markets, preventing unauthorised political organisation, etc.) (Stevenson-Yang, 2006). Chinese cyber police have intensified their prosecution of internet content violations. A number of democracy organisers, human rights activists, Falun Gong members, scholars and other dissidents have been arrested for their alleged involvement in online crimes (Hachigian, 2001).

Indeed, in China, 'the law is marginalised and the legal system relegated to a lowly position in a spectrum of meditative mechanisms, while at the same time available for manipulation by powerful sectors within the state and the society at large' (Myers, 1996: 188). This is even more so for laws related to control on the internet. Most of China's internet regulations are guidelines only and do not represent formal laws (Shie, 2004). Stevenson-Yang notes:

> Published regulations are convenient as a means of expressing policy, but in the end, political policy, not statute, is the true law of the land in China. The case against editors from *Southern Metropolis News* in 2004 provides an example: the three were convicted of bribery and embezzlement, but their real offenses, it was universally believed, were reporting on Severe Acute Respiratory Syndrome, or SARS [on the internet]. (Stevenson-Yang, 2006)

Compared with other developing countries, China was slow to enact laws to recognise digital and electronic signatures (DES) (Aldrich, 1999). Encryption software, which is an essential component of DES, allows confidential messages to be sent, thus making it difficult, even impossible, for the government to detect politically and culturally objectionable content transmitted through the internet (Kshetri and Dholakia, 2001).

Cyber-control in China is thus characterised by a lack of clear rules and policies. In early, 2002, several foreign search engines, most notably Google and AltaVista were blocked for several days without explanation (BBC News, 2002a, 2002b; Weaver, 2002). Pointing out problems faced by organisations because of a lack of specificity, one business group asked the Chinese government to issue clearer regulations (Kalathil, 2003). Stevenson-Yang notes:

> International companies may or may not agree with this set of prohibitions, but they do not generally object to them. Rules, as long as they are knowable and actionable, are easily accepted as a sovereign nation's prerogative. But the heart of China's information control system lies in the prescriptive regime – the one designed to manage the process of creating public speech and, above all, designed to ensure that public speech does not have damaging effects. (Stevenson-Yang, 2006)

The state's entrenchment in the economy

In China, the state's entrenchment in the economy (Pei, 2006) makes it easier to monitor internet control measures. As of 2001, 70 per cent of large and medium-sized 'corporatised' enterprises had Communist Party members on their boards of directors (Pei, 2006). The country's ISPs are controlled by state-run companies (McLaughlin, 2005). China is thus mediating ownership to capitalise on the internet's economic benefits and at the same time combat its risks (Hachigian, 2001). A group at the Ministry of Information Industry (MII) coordinates relevant ministries' efforts in detecting objectionable content and then implements a block. To do this, the group passes the information to the nine state-owned internet access providers that control China's internet traffic (Stevenson-Yang, 2005).

Development of government ICT capability

To increase the effectiveness of monitoring activities, the Chinese government has also increased its ICT capability. As China is harnessing the network for government use (Hachigian, 2001), its e-government performance is rated ahead of industrial countries such as Switzerland, the UK, Singapore and Germany (West, 2002). In its e-government activities, the Chinese government employs a variety of ICT tools, such as the short messaging system (SMS); in some cases, these are replacing more traditional media (Kshetri, Dholakia and Awasthi, 2003).

Tens of thousands of government agents pretend to be dissidents and participate in chat rooms, speaking out against the government, thus many internet users are afraid to engage in online conversations on subjects such as democracy, Japan, religion and other sensitive topics (Mallaby, 2006; Stevenson-Yang, 2006). Indeed, China has the largest cyberpolice force in the world (Mallaby, 2006).

Until the late, 1990s, blocking was mostly in the form of access prevention to certain sites. Later, those sites were made accessible but keyword searches on 'sensitive issues' were made unavailable. According to the Berkeley China Internet Project, the government hides websites containing phrases such as freedom, democracy, China-liberal, and falun (Foushee, 2006).

Decoupling symbolic and substantive actions

The regulative institutions are also characterised by a decoupling of symbolic and substantive actions. One official, who supervises internet affairs for China's State Council has said that China's efforts to regulate web content are aimed primarily at pornography or other content harmful to teenagers and children (Kahn, 2006). In reality, however, access to pornography, despite being technically illegal, is not actually blocked (*Los Angeles Times*, 1997).

Beijing is also decoupling illegitimate internet control measures from formal structures. It is important to note that organisations' responses to institutional pressures are not always legitimate. The study by Elsbach and Sutton (1992: 716), for instance, indicated that activist organisations 'decoupled illegitimate actions from formal organizational structures by performing these actions as anonymous individuals' or as part of a temporary group operating under different names. As authoritarian regimes tend to have fewer checks and balances to prevent misuse of power (Popov, 2006; Stoner-Weiss, 2006), such regimes are likely to be involved in illegitimate internet control measures. Moreover, because of the anonymity inherent in the internet it is easy to decouple illegitimate internet control measures from formal structures. There are even reports that Chinese authorities send viruses to attack banned sites (Guillén and Suárez, 2005).

Formation of normative institutions

Professions and trade associations are important forms of normative institutions. Savan (1989: 179) defines 'the professions' as groups that apply 'special knowledge in the service of a client'. A profession is also self-regulated by a code of ethics (Claypool, Fetyko and Pearson, 1990) and is characterised by its role as a moral community (Camenisch, 1983). These codes require members to maintain higher standards of conduct than required by law (Backoff and Martin 1991), to help make professional norms visible (Frankel, 1989) and to act as a vehicle to assure the public and clients that members are competent, have integrity, and maintain and enforce high standards (Ward et al., 1993). Apart from convincing external parties of the integrity of the profession, codes play an important role in forcing members to question their values (Meyer, 1987).

Chinese professional associations that are part of the normative institutions related to the cyber-control field seem to have idiosyncratic and unusual features. The government-backed Internet Society of China (ISC), which was formed in May 2001 with more than 130 members, is a highly visible example. ISC is sponsored by network access carriers, ISPs, facility manufacturers and research institutes.[1]

The formation of the ISC can be considered as a part of Beijing's manipulation strategy, that is, a 'purposeful and opportunistic attempt to co-opt, influence, or control institutional pressures and evaluations' (Oliver, 1991: 157). To increase their influence, organisations such as the ISC may 'attempt to persuade organizational constituents to join the organization or its board of directors (Oliver, 1991: 157). The ISC, for instance, asked internet companies to sign a voluntary pledge on 'Self-discipline for China's Internet Industry' that commits them to investigate and block websites with politically and culturally-sensitive content. By March 2002, the pledge had been signed by over 120 internet portals (Stout, 2002). The pledge commits signatories not to disseminate information 'that might threaten state security or social stability' (*Economist*, 2002a).

The ISC's activities differ drastically from e-business related professional associations in the West. For instance, the UK Mobile Marketing Association issued its code of conduct in December 2003, specifying what times of day mobile marketers can target consumers (Balmond, 2003). Likewise, in the USA, in early 2001, technology-industry lobbyists and consumer and civil-liberties activists including the American Civil Library Association, Electronic Privacy Information Center and Consumer Federation of America circulated a letter to members of Congress and the president calling for a stronger set of privacy rules (Benson and Simpson, 2001). While these activities are designed to protect consumer privacy, the ISC's actions have promoted the government's interest. For instance, Hu Qiheng, chair of the ISC, has defined internet crime to include 'acts counter to the interests of the Chinese government' (Crampton, 2006).

Prior research has found that authoritarian regimes tend to hinder the growth of 'civic organisations' characterised by equal, voluntary and participatory norms (Putnam, 1993; Un, 2006). Howard (2003: 11–12, 62) has noted that in the former Eastern Bloc, participation in voluntary associations in post-communist societies is significantly lower than in older democracies. In general, in Asian societies such organisations are 'seen in familial, authoritarian terms rather than in terms of voluntary, supposedly "horizontal" associations' (Chandler, 1996). In China, special interest groups and non-government entities are loosely organised (Li, Lin and Xia, 2004) and there is little room for these groups to influence national policy-making (Su and Yang, 2000) and transforming the structure and practices of local companies (Shen, 2005).

To better understand organisations' involvement in internet control measures, we can draw a parallel with Chinese nationalism. Pei (2003) has identified several dimensions of nationalism, including source and bases. In terms of source, he argues that some nationalism is a product of grassroots voluntarism (such as US nationalism) while some is fostered by government elites and promoted by the apparatus of the state (police, military, state-run media). Chinese nationalism is viewed as state-sponsored and an attempt to fill an 'ideological vacuum' left by weakening socialism (Christensen, 1996; Oksenberg, 1997; Sautman, 2001). To put things in context, organisations involved in cyber-control measures do not manifest 'self directed sets of motivations', as was the case with the US chemical industry's response to environmentalism (Hoffman, 1999).

Three other key factors strongly suggest that normative institutions related to cyber-control in China tend to be engaged in government-centric activities rather operating from 'self directed sets of motivations' (Hoffman, 1999). First, as noted above, the state's deep entrenchment in the economy raises the interesting possibility that the government can play a critical role (Pei, 2006).

Second, to succeed in China, e-business organisations need to make room for civil servants and high-ranking government officials (Sikorski and Menkhoff, 1999; Einhorn, Webb and Engardio, 2000). It is important to note that most internet regulations are guidelines only and do not represent formal laws (Shie, 2004). The regulatory vacuum thus makes it important to have good relationships with government officials. For organisations involved in e-business, it is thus important to consider the government's interest.

Third, the concepts of customer service and privacy, which are the focus of most e-business related professional associations in democratic regimes, are not well developed in authoritarian regimes. Terrill (2005) goes even further, arguing that 'because China remains an authoritarian state, we cannot know what the Chinese people want'. Organisations involved in e-business in China thus face little or no pressure related to customer service and privacy.

Formation of cognitive institutions

Note that cognitive components represent culturally supported habits that influence behaviours related to cyber-control. Cognitive feedback loops are associated with cognitive programmes that are built on the mental maps of individual decision-makers and thus function primarily at the individual level (Huff, 1990).

More recently, while regulative and normative pillars related to cyber-control as discussed above are still active, a cognitive institutional pillar appears to be forming. For instance, Yahoo defended its decision to hand over information on a journalist, Shi Tao, claiming that the company was following Chinese 'customs' (Stevenson-Yang, 2006). These decision-makers are thus giving culture-related arguments (Jepperson, 1991) to justify their internet control measures.

Before we proceed, it is important to note that the idea of internet control severely lacks face validity. Put differently, unlike some fields such as environmentalism in the US chemical industry (Hoffman, 1999), the face validity of internet control is highly disputed. The adoption of an idea such as internet control tends to be less problematic when its elements possess a high 'face validity', which results in common agreement among diverse institutional actors concerning the overall utility (Tolbert and Zucker, 1983).

Because of the low face validity, the level of taken-for-grantedness is not highly established for cyber-control and associated activities. As discussed earlier, there are arguments for as well as against Western companies' involvement in networking authoritarian regimes. A related point is that cyber-control's taken-for-grantedness is not homogeneous. For instance, while statements from Yahoo chief Jerry Yang and chair of the ISC, Hu Qiheng, reflect a belief that internet control is consistent with culture, others disagree. Many foreign observers, for instance, were not convinced that Yahoo's decision to hand over information on the journalist had any thing to do with Chinese 'customs' (Stevenson-Yang, 2006).

Evolution of competing institutions

Up to this point, we have concentrated on cyber-control-led institutional evolution in China. As noted above, organisations and populations that inhabit China's cyber-control field are also affected by other sets of formal and informal rules. For instance, anti-censorship companies such as Freegate (Fowler, 2006) and technology multinationals such as Google, Yahoo and Microsoft, which are major participants in China's cyber-control field, are based in the USA. China's cyber-control field is thus influenced by competing US institutions, a situation which is worth examining further. Hoffman notes that:

> to fully appreciate the complexity of institutional dynamics, one must analyze specific institutions at the center of an issue-based field and the competing institutions that may lie within the populations (or classes of constituencies) that inhabit that field. (Hoffman 1999: 352)

China's cyber-control-led institutional evolutions have also triggered changes in competing institutions. For instance, Human Rights in China, a New York-based nonprofit group, has provided financial support to anticensorship companies such as Freegate (Fowler, 2006). Similarly, Amnesty International has accused US-based internet companies such as Google, Microsoft and Yahoo of violating the Universal Declaration of Human Rights in their agreement with the Chinese government to censor internet use in China (US Fed News Service, 2006b). In the USA, the enactment of the Global Online Freedom Act 2006 can be considered as a part of Washington's 'controlling tactics' to 'establish power and dominance over the external constituents' (Oliver, 1991: 157). The act has increased the US government's power over American technology companies. The evolution of formal and informal institutions in Western countries has thus reconstructed the institutional field of cyber-control in China.

Discussion

The nature of the Chinese government's internet control measures

Economic performance is the single most important basis on which many authoritarian regimes seek legitimacy (Naughton, 1993). The Chinese Communist Party's (CCP) resilience can arguably be attributed to a combination of political reforms and economic development (Marsh and Dreyer, 2003; Guoguang, 2006). More to the point, the CCP is counting on the information economy to enhance its image. The CCP expects that a richer and more technology-oriented economy might help increase respect for it (Kshetri and Cheung, 2002). In short, China is interested in fostering a network economy to capitalise on the internet's economic benefits. Hachigian (2001) quotes a senior government official: 'we in the government think we missed a lot of the industrial revolution. And we don't want to miss this revolution'.

Even by the late 1990s, internet control techniques in China were 'blunt'. Internet subscribers were required to register with local security bureaus, enabling government officials to ascertain who was visiting which websites (Rodan, 1998). In an attempt to foster a network economy, this strategy changed subsequently. According to a report from Reporters Without Borders, internet control in China is conducted through 'a clever mix of investment, technology and diplomacy' (McLaughlin, 2005). The government-backed ISC is an example of such diplomacy.

It is also important to note that authoritarian regimes' measures to control the internet are unlikely to catch up with advancements in technologies. On other words, as authoritarian regimes build a better mousetrap, technology companies and savvy internet users come up with better mice. Several examples illustrate this point. For instance, inside many authoritarian regimes, black marketers in cybercafés, universities and private homes are providing access to blocked websites. They typically exploit 'technological loopholes to circumvent government filters and charge fees for access' (Palmer, 2005). To take another example, in Vietnam,[2] activists make extensive use of voice-over internet protocol (VoIP) to contact each other, take part in conference calls and live debates, and post recorded voice messages via online forums available on the websites of VoIP providers such as PalTalk, Yahoo! Messenger and Skype (Johnson, 2006). Importantly, the manner in which VoIP converts conversation to digital bits makes it harder to search for offensive words, compared, for example, with e-mail.

As is the case of China, authoritarian regimes thus need to seek the help of multinationals from the developed world. In China's case, foreign multinationals are co-opting with the Chinese government in exchange for access to the US$125 billion e-business market.

Technology companies' response to the government

Various institutional actors differ in terms of their powers to affect an organisation's outcome. The Chinese government, for instance, can exert a higher level of control on China-based technology companies. Their responses can be described as a habit – 'unconscious or blind adherence to preconscious or taken for granted rules or values' (Oliver, 1991: 152).

Western technology companies, however, are required to appease a diverse set of competing institutions. These companies thus are required to decouple their responses.

Different theoretical contributions and various empirical studies have led to the accepted view that the exact nature of decoupling is a function of the perception of relative powers of competing organisational and institutional interests (Pfeffer, 1981a, 1981b; March and Olsen, 1989; Oliver, 1991; Zajac and Westphal, 1995; Westphal and Zajac, 1998, 2001). These studies also provide support for the notion that substantial responses cannot be made to appease actors that diametrically oppose one another. More to the point, the substantive response relates to the threat or opportunity associated with the actor that is perceived to be more powerful and the symbolic response relates to the threat or opportunity associated with the actor perceived to possess less power (George et al., 2006).

In this regard, it is important to note that China's online transactions were estimated at US$125 billion in 2006 (*Asia Pulse*, 2006). The Chinese government thus possesses enormous power over these multinationals. To gain access to the huge Chinese e-commerce market, Western organisations appear willing to take actions that are non-isomorphic with respect to their home country institutions.

To substantiate this claim, we began by arguing that the Chinese government has built 'The Great Firewall of China' with the help of foreign companies such as Cisco Systems (Gutmann, 2002; Shie, 2004).

Cisco also provided China with hardware specifically designed to assist China's cyber police to conduct surveillance of electronic communications (Jasper, 2006). About a quarter of the vendors at the Security China 2000 trade show, many of them foreign firms, were marketing products aimed at enhancing China's 'Golden Shield' (Fackler, 2000). In short, many Western technology companies have chosen 'to co-opt the source of the pressure' related to cyber-control (Pfeffer and Salancik, 1978; Burt, 1983; Oliver, 1991: 157).

An important point to bear in mind is that organisational decision-makers may differ in terms of their perception of the relative powers of different organisational and institutional interests. Accordingly, their responses to institutional pressures vary. Among foreign technology companies, Yahoo's strategy can be described as compliance – 'conscious obedience to or incorporation of values, norms, or institutional requirements' (Oliver, 1991: 152). Yahoo chief, Jerry Yang, said that he had to make the decision to help Chinese authorities arrest a journalist in order to do business in China (McLaughlin, 2005). It can, however, be argued that unlike many Chinese technology companies, which may have unconsciously adhered to local rules (Oliver, 1991: 152), the strategy of compliance of foreign technology companies such as Yahoo is consciously and strategically chosen to comply with institutional pressure in anticipation of self-serving benefits or access to resources (Pfeffer and Salancik, 1978; Meyer and Rowan, 1983; DiMaggio, 1988; Oliver, 1991: 153).

In the early, 2000s, other foreign companies such as Google and AltaVista responded differently. These companies' response to institutional pressure related to Chinese cyber-control can be described as 'avoidance' (Meyer and Rowan, 1977; Pfeffer and Salancik, 1978; Meyer, Scott and Deal, 1983; Powell, 1988). This can be defined as an 'attempt to preclude the necessity of conformity' (Oliver, 1991: 154). Alternatively, the strategy could also be 'escape', which entails exiting 'the domain within which pressure is exerted' (Oliver, 1991: 155).

It should also be noted that the responses of Yahoo, Google and AltaVista are not socially acceptable in their home country. Threats to resources motivate organisational leaders to conduct broader searches for alternatives that may exist beyond the bounds of social acceptability (March and Simon, 1958; George et al., 2006: 354). Theorists argue that these organisations are likely to lose resources if they adhere to current practices. For this reason, they may undervalue the risks associated with departing from established ways of doing things and with challenging the legitimacy of established ways, and, as a result, attempt to create the framework for new legitimate forms through non-isomorphic change (George et al., 2006: 354).

Conclusions and implications

The foregoing discussion provides a framework for understanding institutional processes associated with internet control measures in authoritarian regimes. The findings are broadly consistent with existing theories on field formation. Nonetheless, this chapter has revealed unique processes and mechanisms associated with internet control measures in China. The theory presented here has thus filled in some of the gaps in macromarketing literature and other related areas of scholarship.

At this point, we should emphasise that there are some problems associated with the reliance on cognitive and normative institutions. Some institutionalists argue that the distinction between normative and cognitive institutions tends to be 'blurred' (Campbell, 2004) and it is unclear when taken-for-grantedness of social norms becomes high enough to become part of cognitive institutions (Clemens and Cook, 1999). Other have pointed out that there is a poor specification of mechanisms by which normative and cognitive structures affect behaviours of various actors (Hirsch, 1997).

The image that a company presents to external parties is a function of the power those parties have over the organisation (Basu, Dirsmith and Gupta, 1999). With new regulative institutions such as the Global Online Freedom Act 2006, the power that the US government has over US-based technology companies has strengthened. This means that these companies have to work to change their images.

Notes

1. At present, there are more than 150 members, most of which are organisations (see *http://www.isc.org.cn/English/*).
2. Authorities in Vietnam have blocked access to pro-democracy blogs and websites, and it is believed that e-mails are scanned.

4

The Chinese government's influence on the third-generation cellular standard: motivators, enabling factors and mechanisms

After postponing several times, China's Ministry of Information Industry is expected to issue third-generation (3G) cellular licences in 2008. Analysts suggest that cellular networks based on the three 3G standards approved by the International Telecommunications Union – China's homegrown Time Division – Synchronous Code Division Multiple Access (TD-SCDMA), the US-based CDMA 2000 and Europe's Wideband CDMA (W-CDMA) – will be established by the 2008 Beijing Olympics. Chinese policy-makers are understandably enthusiastic about TD-SCDMA's preliminary results. Critics, however, have been concerned that Beijing has demonstrated a clear bias towards the TD-SCDMA. The explanations offered in this chapter shed some light on motivators, enabling factors and mechanisms associated with the Chinese government's influence on 3G cellular standardisation.

Introduction

Among the many visible examples of China's evolution as an innovation powerhouse, one is particularly telling: the International Telecommunications Union's acceptance of China's Time Division – Synchronous Code Division Multiple Access (TD-SCDMA) as a global third-generation (3G) cellular standard. Among 16 proposals submitted for IMT-2000[1] standards, the International Telecommunications Union (ITU) selected TD-SCDMA as one of the three global 3G standards. The other two standards are the US-based CDMA 2000 and Europe's Wideband

CDMA (W-CDMA). Indeed, TD-SCDMA is the first ITU standard originated in a developing country (*Intermedia*, 2006).

The question of when China would issue 3G cellular licences had been anybody's guess for the past several years. After postponing several times, China's Ministry of Information Industry issued cellular 3G licences in April 2008. The goal is to enable providers and handset vendors to be fully operational by the 2008 Beijing Olympics (Dano, 2005; Bremner, 2006). Beijing is expected to award 3G licences to both the current carriers China Mobile and China Unicom and also to fixed-line operators China Netcom and China Telecom (Roberts 2007). It is expected that cellular networks based on all three types of 3G standards will be built by the 2008 Beijing Olympics (Silva, 2006). Critics have, however, been concerned about Beijing's 'asymmetric regulation', which arguably tends to treat telecom companies differently (Ogden, 2006a). The 2004 annual US Trade Representative report, for instance, argued that manipulation of telecom standards and secretive regulations in China favour domestic and/or government-controlled firms (Clark, 2004). More to the point, in the 3G arena, it is suggested that Chinese policy-makers have demonstrated a clear bias towards the TD-SCDMA (Clark, 2004; Dean 2005, 2006a; Kong, 2006).

Although China's attempts to develop Chinese standards and set standards for the world in prior product domains have been largely unsuccessful, there is growing recognition that China may have a higher chance of success with its 3G cellular technology. Vogelstein et al. consider the issue from a US perspective:

> For the tech industry, it's China – not Europe, or Japan, or other Asian countries – that will soon be its [USA's] main rival. The implications are profound. No longer content to cheaply make other people's products, a task it has clearly mastered, China wants to be a global standards setter ... One place to watch the flexing of power is in mobile phones. (Vogelstein et al., 2004)

Being a combination of an empire and a modern nation, today's China is 'unusual' (Terrill, 2005). A deeper understanding of how China's idiosyncratic features can influence the government's standardisation measures will provide academics, managers and policy-makers with valuable insights. In this chapter, we investigate unique enabling factors, motivators and mechanisms associated with the Chinese government's influence on 3G cellular standardisation. To achieve this goal, we draw on the conceptual foundation provided by standardisation theories and China's formal and informal institutions.

In the remainder of the chapter, we first briefly review the emergence of TD-SCDMA. Next, we discuss motivations, enabling factors and mechanisms associated with the Chinese government's influence on 3G standardisation activities. The final section provides discussion, implications and conclusion.

The emergence of TD-SCDMA

Key events associated with the evolution of TD-SCDMA are presented in Table 4.1. From the mid-1990s, the Chinese government made research and development (R&D) in mobile telephony a top priority (Niitamo, 2000). Heavy investment in the cellular sector was accompanied by a series of programmes designed to accelerate telecom development, including extensive re-engineering of and intense competition in the mobile sector (Kshetri and Cheung, 2002). China Mobile, China Unicom and other players were forced to adapt to the rigorous disclosure requirements of the NYSE, NASDAQ and Hong Kong's Growth Enterprise Market, and to cut off redundant workers (McDaniels and Waterman, 2000). The state also set up an array of institutes in the mid-1990s for mobile telephony research.

As a component of the government's overhaul of the telecom industry, Datang Telecom Technology Co. Ltd., a telecom infrastructure systems group, was separated from the China Academy of Telecommunications Technology in 1998. The company developed TD-SCDMA in 1998. In 1999, Siemens teamed up with Datang for the further development of TD-SCDMA (Dano, 2005).

The Chinese government has helped drive significant research in 3G technology, more specifically in TD-SCDMA. As of 1999, the Chinese government had invested over US$500 million in 3G research (Liu, 1999). Likewise, as of 2005, the government had invested more than US$123 million in R&D related to TD-SCDMA (Dano, 2005).

The TD-SCDMA proposal was submitted to IMT-2000 for consideration as a global 3G mobile communications standard, and was accepted by the ITU in May 2000 and by the Third Generation Partnership Project (3GPP) in March 2001 (Table 4.1). Notwithstanding the ITU's attempt to create a single global standard, regulators, vendors and carriers could not reach a unanimous agreement (untius.com, 2008) and thus three 3G standards were selected.

In November 2003, Datang developed the first 3G handset based on TD-SCDMA (Wieland, 2004). According to the secretary-general of TD-SCDMA Industry Alliance, about 20 manufacturers had developed

Table 4.1 Emergence of TD-SCDMA: a timeline

Date	Event	Remarks
1992	The '3G' idea was first conceived at the 1992 World Administrative Radio Conference	
1997	The International Telecommunications Union (ITU) called for proposals for IMT-2000 to define an 'anywhere, any time' standard for the future of universal personal communications	16 proposals were submitted to ITU for consideration
1998	Datang was separated from the China Academy of Telecommunications Technology	
29 June 1998	China sent its TD-SCDMA proposal to the IMT-2000; the proposal was signed by the minister and two vice-ministers of the MII	It was the last day set by the ITU for the submission of IMT-2000 standards
July 1998	Datang and Siemens signed an agreement for the joint development of TD-SCDMA	
December 1998	The Third Generation Partnership Project (3GPP) was accepted by six standards-setting bodies: ETSI, ARIB and TIC of Japan, ANSI of the USA, and the TTA of Korea (3gamericas, 2008)	
5 November 1999	The ITU accepted five IMT-2000 standards as radio interface standards: TD-SCDMA, together with SC-TDMA (UMC-136), MC-TDMA (EP-DECT), CDMA 2000 and W-CDMA	
May 2000	At the World Radio Conference, the ITU accepted TD-SCDMA as one of the three 3G standards	CDMA 2000 and W-CDMA were the other two standards
Early 2001	Siemens' cumulative investment for the TD-SCDMA project amounted to US$1 billion	The company had established joint laboratories with Datang in Beijing, Munich, Berlin, Milan, Vienna and London
16 March 2001	TD-SCDMA was accepted by the 3GPP	

The Chinese government's influence on the 3G cellular standard

Table 4.1   Emergence of TD-SCDMA: a timeline (*Cont'd*)

Date	Event	Remarks
11 April 2001	First test of TD-SCDMA was successfully conducted in Beijing	
23 October 2002	Chinese MII allocated 155 MHz for TD-SCDMA and 60 MHz for both CDMA 2000 and W-CDMA	
30 October 2002	TD-SCDMA alliance was established by eight domestic vendors	
November 2003	China developed the world's first 3G handset based on TD-SCDMA	
November 2003	Motorola released a TD-SCDMA module library for the MRC6011 Reconfigurable Compute Fabric device	
January 2006	The MII formally endorsed TD-SCDMA as a Chinese national standard	
August 2006	SK Telecom signed a memorandum of understanding with China's National Development and Reform Committee to help China develop TD-SCDMA	
2008	TD-SCDMA scheduled for launch	

Source: 3gamericas (2008); 3g.co.uk (2003); Yan (2003, 2004).

more than 100 cellphone models based on the TD-SCDMA standard by the end of 2006 (*China Daily*, 2006).

TD-SCDMA is expected to capture a significant share of the 3G cellular markets both abroad and at home (Einhorn, 2003a). One estimate has suggested that 50 million TD-SCDMA-based mobile phones will be sold in 2008, and 85–120 million in 2010 (*SinoCast China*, 2007a). Another report released by CCID Consulting has predicted that there will be over 51 million TD-SCDMA subscribers by 2011 (PR Newswire, 2008a).

The Chinese government's motivations for promoting TD-SCDMA

An examination of formal and informal Chinese institutions helps us to understand the sources of regulative, normative and cognitive legitimacy for TD-SCDMA. In a larger sense, motivations associated with the support for TD-SCDMA can be explained in terms of political cognitive and political normative factors. Mental maps of political elites or 'persons who by virtue of their institutional positions have a high potential to influence national policy making' (Moore 1979: 674) determine a nation's technological landscape. Political elites include legislators, governmental officials, political party officials, leaders of various interest groups, military leaders, etc.

Defence against economic threats

David and Steinmueller (1994: 29), like many standardisation theorists, argue that one of the major motivations behind technology policies and standard-setting for regulatory bodies is to achieve national goals such as protection of domestic employment and reduction in foreign dependence. Governments tend to use standards as instruments to promote such objectives even if they are in conflict with securing a Pareto-optimal[2] outcome at the global level (Bar and Borus, 1987). The government can take standardisation and related measures to overcome consumer abuse arising from businesses' myopia, greed and economic power (Hart, 1998).

Chinese policy-makers' mental maps in relation to foreign technology products and foreign multinationals have led to clear bias towards domestically-developed technologies. Chinese policy-makers think that developed countries tend to see developing countries like China as 'markets' for the transnational corporations' products, thereby threatening their economic security (*China Economic Times*, 2000). The China, Japan and South Korea (CJK) open-source alliance helps us to understand the attitude towards dependence on foreign ICT products. A spokesperson for the CJK initiative has argued that the three countries consider it unacceptable to depend on technology over which they have no price control (Krikke, 2003).

Central to this perspective is a high level of advocacy for national self-reliance and domestic development of technology among Chinese policy-makers, researchers, scientists and military leaders (Simon, 2001). In China, foreign technology imports and the outflow of royalties have been a focus of concerns (Einhorn, 2004a). They want to reverse the flow of fees[3] by exporting Chinese standards.

Success with TD-SCDMA would help China achieve various economic goals. For one thing, China is trying to slip into a higher gear and move

The Chinese government's influence on the 3G cellular standard

beyond low-margin contract manufacturing. TD-SCDMA is part of China's broad efforts to cut spending on technology imports (Dean 2006a) and advance its technological gear. Chinese companies own most of the TD-SCDMA intellectual property (Shannon, 2006). In addition to the fact that most of the core TD-SCDMA intellectual property rights are held by Chinese companies, the standard's Chinese origins mean that most of the major players in the TD-SCDMA value chain are likely to be Chinese (Bremner, 2006). For instance, most of the TD-SCDMA equipment is likely to be supplied by the Chinese vendors such as ZTE and Huawei (Silva, 2006; Yuan, 2006).

Defence against military threats

Another motivation behind technology policies and standard-setting for regulatory bodies is to enhance defence capabilities (David and Steinmueller, 1994: 29). In some countries, national security has been an important element in the equation of choice function that involves different technological standards.

Among Chinese leaders there are suspicions that the country is under cyber-attack from the USA. There has been a deep-rooted perception among Chinese policy-makers that Microsoft and the US government spy on Chinese computer users through secret 'back doors' in Microsoft products.

According to a *China Economic Times* (2000) article, Xu Guanhua, then Chinese vice minister of science and technology, perceived there to be three aspects of national security affected by high-technology: military security, economic security and cultural security. Regarding military security, Guanhua forcefully argued that developed countries have put many high-tech arms into actual battles and discussed the likelihood of ICT-exporting countries installing software for 'coercing, attacking or sabotage'. A spokesperson for the CJK initiative has continued the discussion on the potential military threat by arguing that from a defensive standpoint, the three countries find it unacceptable to depend on technology for which they have no control over the source code (Krikke, 2003).

To capture the feelings that accompany China's dependence on foreign technology, it is important to note that computer hardware and software imported from the USA and its allies are subject to detailed inspection in the country. Chinese technicians take control of such imports and resist or closely monitor if Western experts install them (Adams, 2001). Several years ago, Chinese cryptographers reportedly found an 'NSA Key' in Microsoft products, which was interpreted as pertaining to the National Security Agency. The key allegedly provided the US government backdoor access to Microsoft Windows 95, 98, NT4 and 2000. Although Microsoft has denied this allegation and even issued a patch to fix the problem, the Chinese government remains unconvinced.

The Chinese government has also expressed concerns over security issues in cellular technologies. For instance, in 2004, Beijing raised the ire of US manufacturers by forcing them to embed a Chinese-designed data encryption technology known as WAPI on all WiFi equipment sold in the country. Foreign companies selling wireless local area network equipment in China were required to work with local companies with access to the Chinese-developed encryption software (Einhorn, 2004b). Although critics labelled this measure as a trade barrier to favour local businesses, Beijing maintained that the reason for the encryption standard was 'a matter of making China's wireless networks safe for users'. As China is increasingly relying on wireless data in its communications infrastructure, WiFi insecurity is of more concern to the country than in most other nations (Gillmor, 2003).

To put things in context, the national security concern is also reflected in the 3G industry. In July 2004, Datang Telecom confirmed that it had selected the open source Linux as the operating system for its 3G TD-SCDMA handset (SinoCast, 2004). In addition to the low cost, Linux has a clear security advantage as it has no undetectable vulnerabilities. A vice director of software research at the Chinese Academy of Sciences said Linux would protect the government from attacks by foreign hackers (Computing Canada, 2000).

National representation in ICT standards

There are persuasive arguments for thinking that political normative factors such as desire for the representation of Chineseness in ICT also have led to Chinese policy-makers' bias towards TD-SCDMA. In the Chinese policy landscape, there has been a strongly expressed desire for the representation of Chineseness in ICT. A key point is the matter of national pride in having domestically developed global technological standards.

Part of Beijing's deep thirst for domestic standards comes from its largely unsuccessful previous attempts to create Chinese standards and set standards for the world. Since the 1980s, China has made several attempts to develop a Chinese computer operating system but has failed because of the rapid developments in the global software industry (Goad and Holland, 2000).[4] The Chinese are working very hard to create their own standards in many ICT industries, such as computer operating systems, audio-video compression and 3G data standards (ZDNet Asia, 2003), as they believe they have the ability to set their own standards in these areas (Koprowski, 2004).

TD-SCDMA is touted as a sign of China's growing technological sophistication (Bremner, 2006) and arguably symbolises its 'industrial ambitions' (Morse, 2006a, 2006b). Chinese policy-makers have envisioned TD-SCDMA as a means to help China leapfrog ahead in the global technological race. They hope the success of TD-SCDMA will boost the country's representation in global ICT standards and will further burnish its image worldwide. The chief technology officer of Commit, a multicompany Sino-foreign joint venture in Shanghai that is developing TD-SCDMA chip sets and handset reference designs put the issue this way:

> If they miss 3G, they will have to wait 15 more years to get 4G. So they have to be there, and this is why they have invested in TD-SCDMA. I am completely convinced that they want to give it a go. (Clendenin and Mannion, 2002)

In short, TD-SCDMA is a part of a broad effort by Beijing to create Chinese-made standards to bolster domestic know-how (Dean, 2006a).

A showcase of national achievement in technology use

The Chinese government is counting on the information economy to enhance its image. The Communist Party expects that a richer and more technology-oriented economy might help increase respect for the country. International mega-events such as the Olympic Games and the World Cup provide significant opportunities for demonstrating national achievements in technology fields. For instance, Many South Koreans believed that the country's national image improved significantly around the world following the 2002 World Cup (Kima, Gursoya and Leeb, 2006).

The 2008 Beijing Olympics deserve special attention. This mega-event is a tremendous opportunity for China to internationalise a positive national image. The Olympics can help promote a positive image worldwide using 3G as a showcase of China's economic and technological success. A critical question is thus what Beijing would like the world to see during the 2008 Olympics as a showcase of its economic and technological progress (Morse, 2006a, 2006b).

Ideally, China wants to delay 3G licences until TD-SCDMA can compete with CDMA 2000 and W-CDMA (Kong, 2006). During the Olympics, however, the Chinese government would also like to show China's technological achievement to the world. China thus does not want to miss this opportunity by delaying the issue of 3G licences because of further TD-SCDMA development issues. For this reason, Beijing does not have much time to gaze at the walls. Thus, while it would be ideal to have a working 3G network based on TD-SCDMA by the time of the Olympics, it would still be better to have other 3G networks running – irrespective of any standards – than none at all.

Thus, while Beijing favours TD-SCDMA, it also wants to have 3G services operational before the 2008 Olympics (Dano, 2005; Ogden, 2006b). Some analysts have argued that 3G will be ready only in the cities that are hosting Olympic events and may be postponed in the rest of the country (Clendenin, 2007a). In short, to jazz up its image by showing the world its economic progress and technological achievement, China should deploy 3G sufficiently before the mega-event rather than waiting for TD-SCDMA to mature (Einhorn, 2004a). In April 2008, China Mobile Communications launched a third-generation trial network based on the TD-SCDMA standard in eight Chinese cities (Scent, 2008).

Enabling factors

Up to this point we have devoted the chapter to understanding the motivations of Chinese policy-makers that are influencing the success of TD-SCDMA. It is now pertinent to explore what enabling factors and mechanisms are available for the Chinese government to influence 3G standardisation activities.

The state's deep entrenchment in the economy

China's most notable feature is perhaps the state's deep entrenchment in the economy, which means that several mechanisms are available for the government to influence standardisation activities. As of 2001, 70 per cent of large and medium-sized 'corporatised' enterprises had Communist Party members on their boards of directors (Pei, 2006).

There are thus persuasive arguments for thinking that state priorities come ahead of shareholders' profits for Chinese telecom companies. Furthermore, the government wants TD-SCDMA to be successful (Balfour, 2006a). Telecom companies are thus moving rapidly towards TD-SCDMA, despite being unsure how such moves will affect their economic health. For example, in February 2006, Citigroup downgraded China Telecom from 'buy' to 'hold', arguing that the company was 'stuck in a regulatory whirlpool' (Ogden, 2006a).

Analysts suspect that Chinese carriers do not want the TD-SCDMA licence because the technology is unproven and its deployment may well isolate them from the rest of the world of equipment (Yuan, 2006). China Mobile, the strongest carrier, is expected to be awarded a licence based on TD-SCDMA because the government wants to give it to the carrier with the most resources to support it (Balfour, 2006a). Ironically, China Mobile's dominance in the Chinese cellular market could prove to be a handicap for the company. Thus, compared with firms in the West, Chinese companies are more government-centric and less consumer-centric. Terrill (2005) goes even further, arguing that 'because China remains an authoritarian state, we cannot know what the Chinese people want'.

A weak rule of law

To further explore why the government can play a critical role in 3G standardisation, it is helpful to consider the Chinese legal system. As noted in Chapter 2, the law is arguably 'marginalised' in China (Myers, 1996: 188). It is important to note that most regulations are guidelines only and do not represent formal laws (Shie, 2004). The tricky political landscape and regulatory vacuum thus make it important for firms to have good relationships with the government (Sikorski and Menkhoff, 1999; Einhorn, Webb and Engardio, 2000). In short, to succeed in China, it is critical for telecom companies, especially foreign ones, to take measures to appease the government, irrespective of whether or not such measures are in the companies' best interest.

Mechanisms

Initial conditions favourable to TD-SCDMA

Arthur (1994) has provided evidence which indicates that in markets with two or more incompatible competing technologies, small changes in initial conditions may shape the competitive landscape. The changes in initial conditions arise due to chance or strategy, which may result in one technology gaining a sufficient lead and even dominating the market to become the de facto standard while competing technologies are locked out. Several economists have argued that this can occur even when the dominant technology is distinctly inferior to the alternatives (Lee and O'Connor, 2003).

By issuing TD-SCDMA licences first, the Chinese government hopes TD-SCDMA will have more room to grow and compete with foreign 3G standards (*SinoCast China*, 2007b). TD-SCDMA will thus have a time advantage to grab a larger 3G market share (Financial Times Information, 2006). Analysts expect that China Mobile may be awarded a 3G licence based on TD-SCDMA and licences based on the other two standards could be issued later (Financial Times Information, 2006). Just as important are some of China's formal rules that favour TD-SCDMA. In October 2002, the Ministry of Information Industry's (MII) document No. 479 allocated 155 MHz for TD-SCDMA compared with 60 MHz for both CDMA 2000 and W-CDMA.

Bessen and Farrel (1994: 118) observe that an inferior product may defeat a superior one if members of different stakeholder groups widely expect it to do so. A related point is that a standard's chance of dominating the market increases if it is supported by a reputable governmental or non-governmental organisation (Teece, 1998). In this connection, it is worth noting that in January 2006, the Ministry of Information Industry formally endorsed TD-SCDMA as a national standard, saying it was ready for commercial use. TD-SCDMA was the first of the 3G standards to be formally approved (Dean, 2006a). The other two standards were under review (Dean, 2006b).

Licence timing to overcome 'excess inertia' associated with competing standards

A standard is less likely to be successful if it is launched without having developed technology and a greater body of knowledge about the standard (Krueger, 1991). Thus, were commercial 3G to be launched early in China, the market may lock into proven standards (CDMA 2000 and W-CDMA) because of 'excess inertia' (Farrell and Saloner, 1986). For instance, Westinghouse Electric Company followed this strategy in the development of the electricity standard, and its alternating current became superior to Thomas Edison's direct current (David, 1991).

TD-SCDMA's progress has been a key factor in determining licence timing (Dean, 2006b). The government wants TD-SCDMA to be mature and commercially viable so it can be used by at least one of the networks (Dean 2005, 2006a). However, the reason why Beijing is postponing 3G licensing is mainly because testing the TD-SCDMA standards has taken longer than expected (Yuan 2006). For example, initial trials revealed a number of weaknesses in TD-SCDMA, such as high call drop-out rates (Ogden, 2006b) and failure to support advanced services such as downloading video clips (*Wall Street Journal,* 2006).

Influencing vertical relations

Theoretical and empirical evidence indicate that development of an ICT standard is influenced by the degree of orientation towards the standards of firms in vertical relations (David and Greenstein, 1990). For example, in Japan, expectations and incentives of computer hardware companies influenced standards in the software industry (Cottrell, 1994). Without support from leading handset manufacturers and other firms in vertical relations, TD-SCDMA operators could thus find it hard to attract consumers (*China Daily,* 2006). In this connection, it is worth noting that the state's deep entrenchment in the economy (Pei, 2006) enables Chinese policy-makers to take a number of measures to influence the 3G landscape. To take one example, partly because of Chinese government pressure, firms in vertical relations with TD-SCDMA are increasing their degree of orientation toward this technology.

Global cellular players such as Motorola and Nokia, which had previously betted against TD-SCDMA (Clendenin, 2003), subsequently revised their strategies and are moving toward the incorporation of this technology. Motorola offers a case in point. In November 2003, Motorola released a TD-SCDMA module library for the MRC6011 Reconfigurable Compute Fabric (RCF) device. Motorola's RCF library code kernels for TD-SCDMA have been designed to handle the complex processing step of TD-SDCMA development (tdscdma-forum.org, 2003).

In mid-2003, China Mobile, the former state monopoly, reportedly persuaded major system producers to invest in TD-SCDMA (Ericsson, 2003a). Eight companies – China Mobile, China Telecom, China Unicom, Datang Group, Huawei, Motorola, Nortel and Siemens established the TD-SCDMA Forum to promote the technology with the China Mobile Association Committee as the governing body. According to the secretary-general of the TD-SCDMA Industry Alliance, as of the end of 2006, about 20 manufacturers had developed more than 100 cellphone models based on the TD-SCDMA standard (*China Daily*, 2006).

There are also a number of partnerships related to TD-SCDMA. For instance, Siemens has teamed up with Datang since 1999 (Dano, 2005). In 2003, Datang Mobile, Royal Philips Electronics and Samsung Electronics formed a joint venture company known as T3G to offer complete technology to handset manufacturers to accelerate the production of TD-SCDMA handsets (RCR Wireless News, 2003). Ericsson and ZTE, Nortel Networks and China Putian and Alcatel Alsthom and Datang have also formed infrastructure-related partnerships (Dano, 2005).

Western vendors can also make profits by offering design and management expertise related to TD-SCDMA (Silva, 2006). Qualcomm, for instance, was on the verge of filing patent violation lawsuits against TD-SCDMA-handlers. Nonetheless, the company announced in December 2006 that it had developed TD-SCDMA chips for handsets and was working with Huawei, ZTE and others (Morse, 2006a, 2006b).

Further support for TD-SCDMA came from South Korea's SK Telecom. In August 2006, SK Telecom signed a memorandum of understanding with China's National Development and Reform Committee to develop TD-SCDMA (*SinoCast China*, 2006a). SK Telecom also announced its plans to deploy a TD-SCDMA experimental station in Korea's Bundang region to carry out TD-SCDMA trials by 2007 (*SinoCast China*, 2006a). Based on the experimental station, SK Telecom plans to develop TD-SCDMA related products and services (*SinoCast China*, 2006a).

Conclusion

We provided an explanation as to why and how Chinese policy-makers are influencing the Chinese 3G standardisation landscape. This chapter has thus contributed to the conceptual and empirical understanding of factors influencing the success of standards originating from developing countries such as China. We analysed the nature of regulative, normative and cognitive legitimacy to the domestically developed standards in China. From a theoretical perspective, our framework helps further explain clear contexts and attendant mechanisms associated with the government's influence on technological standardisation.

The basic idea behind the Chinese government's support for the home-grown TD-SCDMA is simple: it helps to achieve various economic, political and military goals. The above discussion indicated that with regard to standardisation activities, the only developing country to create an ITU-accepted standard has many idiosyncratic features. The essential point made in this chapter is that the Chinese government seems to have a notoriously high ability to influence standardisation activities. For TD-SCDMA, however, the battle to overtake competing standards concerns more than just the Chinese government's support.

The discussion in this chapter contains some practical and policy implications.

Customer heterogeneity and TD-SCDMA

Even if the 'initial conditions' are not favourable enough for the TD-SCDMA to become a dominant 3G standard in the market, the technology has some distinct characteristics that are likely to attract certain operators and consumers. Katz and Shapiro have noted:

> Customer heterogeneity and product differentiation tend to limit tipping and sustain multiple networks. If the rival systems have distinct features sought by certain customers, two or more systems may be able to survive by catering to consumers who care more about product attributes than network size. Here, market equilibrium with multiple incompatible products reflects the social value of variety. (Katz and Shapiro, 1994)

TD-SCDMA performs better than its rivals on at least some dimensions and thus has the potential to attract some segments of operators and customers. On the bright side, TD-SCDMA combines CDMA's spectrum efficiency and the asymmetric data-transfer capability of the global system for mobile communications (GSM). Compared with W-CDMA and CDMA 2000, TD-SCDMA uses a relatively small amount of frequency resources. For this reason, TD-SCDMA is more suited to deployment in dense cities than the competing standards (Clendenin, 2002). Some argue that TD-SCDMA's efficient usage of network resources leads to savings of as much 30 per cent over W-CDMA (Lynch and Chau, 2000). Thus, TD-SCDMA can also be deployed as a complementary technology rather than a competitor to W-CDMA and CDMA 2000 (Lynch and Chau, 2000).

As noted above, mobile phones and base stations installed with TD-SCDMA chips are more likely to originate from China than those based on competing standards and thus are likely to be cheaper (Business CustomWire, 2003; Clendenin, 2003). Moreover, intellectual property licensing fees are also likely to be lower for TD-SCDMA than for competing standards (Bremner, 2006).

TD-SCDMA in developing economies

We noted earlier that because of the cost factor, developing economies are potentially good candidates for TD-SCDMA. In addition to the cost factor, because of cultural and political proximity, third-world multinationals tend to do better in developing countries than in the developed world (*Economist*, 2005b). Experts say that China's current technology, know-how and capabilities are more effective for competing in developing countries compared with industrialised countries (*Economist*, 2005b). Chinese technology companies Huawei, ZTE, Haier and TCL already have significant operations in developing economies. Some analysts argue that if TD-SCDMA is successful in China, carriers in Africa and Latin America will use TD-SCDMA to launch 3G services within two to three years (Morse, 2006a, 2006b) due mainly to its potentially lower cost compared with competing standards (*Electronics Weekly*, 2006).

China's influence in developing economies and TD-SCDMA

Success in developing countries is also about having good relationships with policy-makers and operators in those countries. The Chinese government provides some developing countries, especially those in Africa, with soft loans, extremely generous credit and financial support. For instance, in April 2005, China Development Bank provided Nigeria with a loan of US$200 million to deploy a nationwide CDMA450 wireless access technology to be built by Huawei. Analysts argue that in return for such support, China may require African carriers to adopt TD-SCDMA as they upgrade their wireless networks (Morse, 2006a, 2006b).

Quality issues

TD-SCDMA is a good six years behind its competitors, and still has much to do. Until the autumn of 2006, while data links for TD-SCDMA handsets reached a 3G speed of 384 kbits/second, voice connection reliability was only in the 80–95 per cent range (Clendenin, 2007a) – the Chinese government has a target of at least 95 per cent reliability. As of early 2007, China Mobile, China Netcom and China Telecom were running TD-SCDMA trials in Beijing, Qingdao and Xiamen respectively (Chau, 2007; Roberts, 2007). Furthermore, the inescapable fact is that Chinese firms such as Datang do not have the innovation DNA of many innovative Western firms. Past experiences in the telecom industry have indicated that market entries of companies with little or no experience may create risks for the reliability and quality of the network (David and Steinmueller, 1994: 230). Datang, the TD-SCDMA's sponsor, has relatively little experience with cellular technology compared with its rivals. If TD-SCDMA proves to be inferior, it is likely to increase quality-related risks for users.

A lack of TD-SCDMA handsets also remains a problem. There are fewer models handset that run TD-SCDMA compared with those that run the competing standards, W-CDMA and CDMA 2000 (Clendenin, 2007a). It is also important to note that some renowned handset suppliers such as Motorola declined to bid for 3G handsets based on TD-SCDMA (Tanner, 2007).

Notes

1. International Mobile Telecommunications (IMT-2000) is a general term for technologies planned to be included in the ITU's world standards for 3G mobile communication.
2. A feasible allocation is said to be Pareto optimal if there are no other feasible allocations that make at least one person better off without making any individual worse off.
3. For instance, China is the world's largest maker of DVD players. By adopting its own technology, it could save US$2 billion a year in royalties paid to an 18-company consortium (Calbreath, 2004).
4. China's attempt in the mid-1990s to introduce its CD standard, Super Video CD to the world also faced foreign market resistance as well as a lack of strong consumer support within the country.

5

Drivers of foreign multinationals' research and development activities in China

Developing economies such as China are rapidly moving from the periphery to the centre of the global economic system. Among many indicators that illustrate this trend, one is particularly telling: a rapid increase in foreign multinationals' research and development (R&D) activities in these economies. No other developing economies come even close to China in terms of the amount and quality of such activities. Some multinationals are performing cutting-edge R&D in China. Moreover, some products developed in Chinese R&D labs are being first introduced in China and other Asian countries before being launched in Europe and North America. This chapter examines the characteristics that affect the amount and quality of multinational R&D activities in a host country. Building on the model, we analyse foreign technology firms' R&D operations in China.

Introduction

Developing Asian economies such as China are rapidly moving from the periphery to the centre of the global economic system (Wells, 1983; Amsden, 2001; Mathews, 2006a, 2006b). Among the many indicators that illustrate this trend, the rapid increase in activities related to research and development (R&D) by foreign multinational corporations (MNCs) in these economies has commanded a great deal of attention. No other developing economies come even close to China in terms of the amount and quality of foreign MNCs' R&D activities. Some MNCs are performing cutting-edge R&D in China. Moreover, some products developed in Chinese R&D labs, such as new cellphone models, are first introduced in China and other Asian countries before being launched in Europe and North America (Buckley, 2004; Schwartz, 2005).

Researchers have just begun to grapple with the nascent trend of developing economies' increasing centrality in the global economic system, but these early studies have mainly dealt with the internationalisation of successful firms based in these economies (Wells, 1983, Amsden, 2001; Mathews, 2006a, 2006b). As firms based in the developing world benefit tremendously from foreign MNCs' R&D activities through spillover mechanisms such as reverse-engineering, skilled labour turnovers, enterprise spin-offs, demonstration effects and supplier–customer relationships (Cheung and Lin, 2004; UNCTAD, 2005), studies focusing on drivers of MNCs' R&D in developing economies could help our understanding of the emergence of challenger firms from these economies.

From a Western perspective, China, which received cumulative foreign R&D investment of US$4 billion by mid-2004 (UNCTAD, 2005), has been a highly visible example of a challenger developing economy. According to a 2004 global study by the Economist Intelligence Unit, 39 per cent of respondent firms favoured China as the site for future overseas R&D investments compared with 29 per cent favouring the USA (*R&D*, 2004). In another survey conducted by the UN Conference on Trade and Development (UNCTAD) in 2005, 62 per cent of firms named China as the world's most attractive prospective R&D location for the next five years (EIU ViewsWire, 2005a). Vin Caraher, president and CEO of Thomson Scientific explains: 'it is clear that China is rapidly shifting from a low-cost manufacturing base to a low-cost innovation base, where complex R&D processes, experimentation and exploration are conducted' (PR Newswire Europe, 2006). A *Business Week* article echoes this view: 'China's lab may soon rival its powerhouse factories' (Einhorn, 2006a). Faced with examples and evidence like the above, the 2004 report of the US President's Council of Advisers on Science and Technology argued that China's progress on the R&D and high-technology fronts has 'created a new level of nervousness on the part of many industry and academic professionals' (Stokes, 2005). China's emergence as a serious challenger to the West can be attributed to MNCs' R&D investments in the country.

Before proceeding, we offer some clarifying definitions. Adaptive R&D is related to production and entails adapting imported technologies. Innovative R&D, on the other hand, involves 'the development of new products or processes' (UNCTAD, 2005).

Foreign firms' R&D in China: a brief survey

The accumulated R&D investment of foreign MNCs in China reached US$4 billion by mid-2004 (UNCTAD, 2005). By mid-2005, there were an estimated 750 foreign R&D centres in China, growing at a rate of 200 new centres every year (*R&D*, 2004; Walsh, 2005; Yuan, 2005; Asia Africa Intelligence Wire, 2006). It is important to note that there were fewer than 50 such centres in 1998 (*Newsweek*, 2006). Likewise, more than two-fifths of the 210,000 invention patent applications in 2006 came from foreign MNCs (*China Daily*, 2008).

A 2005 report by China's Ministry of Commerce revealed that about half of the MNCs in China planned to establish their own R&D centres within the country over the next two years, and over 60 per cent of those based in the country planned to increase R&D investments through mergers and acquisitions (*R&D*, 2005). The same study found that one-third planned to increase their R&D employees in China over the next two years. Another study conducted by the Economist Intelligence Unit in 2004 found that the majority of respondent firms (39 per cent) favoured China as the site for future overseas R&D investments. This compares favourably with 29 per cent for the USA and 28 per cent for India (*R&D*, 2004). Similarly, 62 per cent of the firms responding to a 2005 UNCTAD survey named China as the world's most attractive prospective R&D location for the next five years (EIU ViewsWire, 2005a). In a further survey conducted by *R&D Magazine* of more than 400 R&D scientists, engineers, managers, directors and executives in the USA, one-third of the respondents planned to outsource at least some of their R&D activities to other countries; 61 per cent of those outsourcing said they would outsource to China (Studt, 2006). Likewise, according to a survey conducted in 19 countries by Booz Allen Hamilton and INSEAD, 31 per cent of R&D employees worldwide will work in China and India by 2007, compared with 19 per cent in 2004 (Rajagopalan, 2006).

For economies in the developing world looking to benefit from foreign investment in R&D, what matters is not simply the volume of R&D generated in the economy or the access to foreign technology and know-how, but also the sophistication of the R&D activities and the strength of the linkages between the multinationals and domestic firms. Traditionally, the most innovative R&D activities were performed in technologically advanced countries. In recent years, contrary to the widely-held belief that MNCs perform 'less cutting-edge R&D' in developing countries (Economist Intelligence Unit, 2005), many foreign companies have built state-of-the-art R&D facilities and are generating globally-oriented innovations in the developing world. While MNCs conduct mostly adaptive R&D in Latin American and African economies (UNCTAD, 2005), an increasing proportion of R&D conducted in China is innovative in nature.

Thus, whereas MNCs' R&D investment in China was initially adaptive, companies such as Motorola and IBM have been investing heavily to expand their Chinese R&D centres to develop products for the global market (Chen and Dean, 2006). In 1993, Motorola established its first R&D lab in China. By September 2005, the company had 16 Chinese R&D centres and had invested over US$500 million (*People's Daily*, 2005). In 2005, the firm had 2,000 R&D personnel, a figure expected to increase to 3,000 by 2006 (Asia Africa Intelligence Wire, 2006). Likewise, Dell opened its R&D centre in Shanghai in 2002, which employed 500 people in early 2007 (Einhorn, 2007).

Consider the following examples:

- The Nokia 3610 model, introduced to the Asia-Pacific market in 2002, was the first product developed entirely by the Nokia Product Development Centre in Beijing (UNCTAD, 2005). By, 2004, 10 per cent of the company's mobile handsets sold globally were designed in the Nokia Product Development Centre in Beijing (UNCTAD, 2005). By 2005, the company conducted 40 per cent of its global R&D for the cellphone business in its Beijing research centre (Powell and Steptoe, 2005) and 40 per cent of handsets were designed and developed in the country (Economist Intelligence Unit, 2005). In April 2005, Nokia's researchers in the Beijing centre developed a new system that significantly enhanced the speed of the company's network and its capacity to handle huge volumes of data (Powell and Steptoe, 2005).

- As of 2005, Microsoft Research Asia, in Beijing's Haidian district, transferred over 100 technologies from product teams for the development of products such as Office XP, Office System 2003, Windows XP, Windows Server 2003, XP Media Center Edition, Windows XP Tablet PC Edition, Xbox, MSN, Vista and the MPEG-4 video-streaming standard (Kanellos, 2002b; Friedman, 2005; Taft, 2006).

- In 2000, a group of engineers at Motorola's Beijing R&D centre started working on cellphones based on the Linux operating system. Since then, Linux has been an integral part of the company's software strategy. As of mid-2005, Motorola offered five Linux-based phones globally

(Mehta, 2005). The A780 model, which is now available in the USA and Europe, was developed in China, and lets users write on the screen with a finger rather than a stylus (Chen and Dean, 2006).
- Dell's engineers in its Shanghai lab developed low-cost PC models. The company is working on adapting those models for Brazil, India and other emerging markets (Einhorn, 2007).

Host country characteristics affecting foreign MNCs' R&D activities

Figure 5.1 presents a preliminary model of a theory of MNCs' R&D activities in the developing world. Each arrow in the model represents

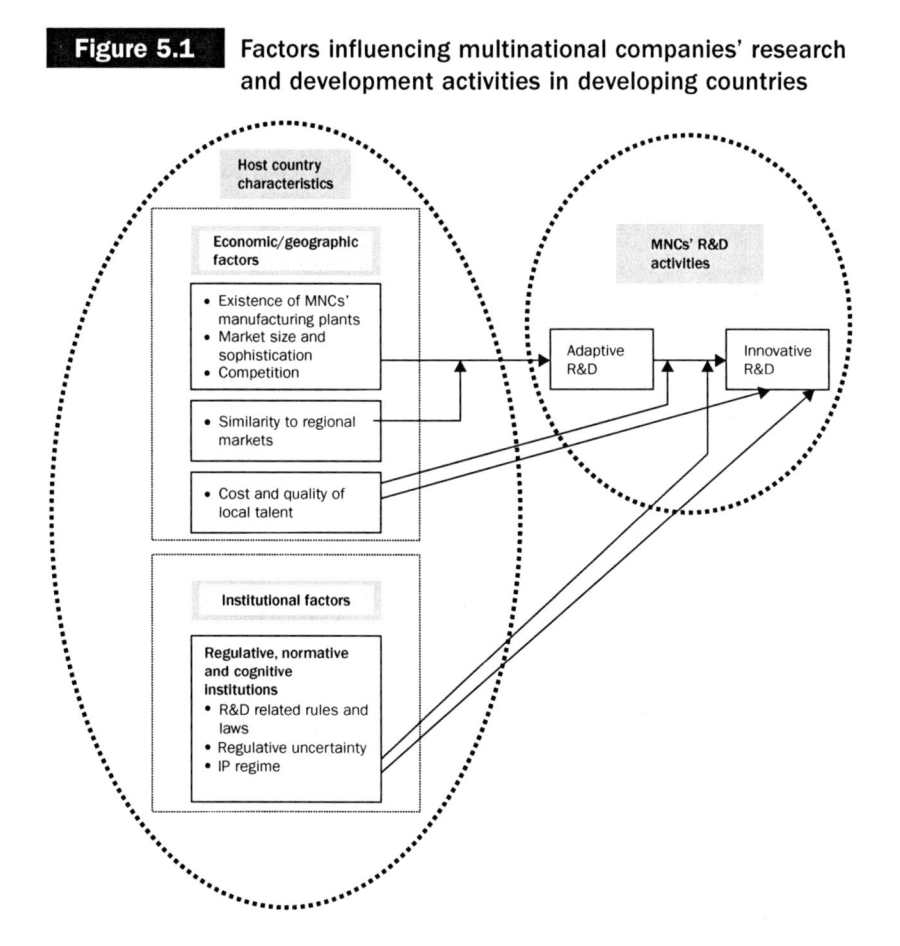

Figure 5.1 Factors influencing multinational companies' research and development activities in developing countries

- Host country characteristics
  - Economic/geographic factors
    - Existence of MNCs' manufacturing plants
    - Market size and sophistication
    - Competition
  - Similarity to regional markets
  - Cost and quality of local talent
  - Institutional factors
    - Regulative, normative and cognitive institutions
      - R&D related rules and laws
      - Regulative uncertainty
      - IP regime

- MNCs' R&D activities
  - Adaptive R&D
  - Innovative R&D

proposition(s) related to motivations behind a firm's adaptive and innovative R&D activities in a developing country.

Economic and geographic factors

Existence of foreign manufacturing plants leading to adaptive R&D

Theoretical and empirical evidence indicates that overseas production acts centrifugally and thus decentralises a firm's R&D functions. Production activities require a range of competencies including adapting products to manufacturing conditions (Blanc and Sierra, 1999). Some R&D thus needs to be done close to where it is implemented (Reich, 2005). Put differently, companies' have an inherent propensity to seek and maintain R&D activities near to manufacturing plants. Internationalisation of production thus leads to R&D internationalisation (Pearce and Singh, 1992). As Blanc and Sierra (1999: 190) observe, 'R&D tends to follow establishment of manufacturing abroad with a certain time lag'.

These theoretical insights are verified by observations and surveys. Florida and Kenney (1994) observed a number of Japanese applied R&D facilities in the USA close to manufacturing plants, providing technical support and helping to strengthen technological capabilities. For example, Fujitsu established a telecommunications research facility on the site of its Richardson, Texas plant to assist with the development of new products and customisation of products for its US customers (Florida and Kenney, 1994).

The findings of some surveys are also worth considering. A study conducted among UK-based pharmaceutical companies indicated proximity to manufacturing plants as one of the top two factors for deciding R&D location (Howells, 1984). Similarly, R&D executives responding to a survey conducted by *R&D Magazine* ranked proximity to the company's manufacturing plant as an important consideration in planning a new R&D facility (Goldner, 1992). Still another survey conducted among Japanese biotechnology and electronics firms found that desire to be close to the firm's offshore manufacturing plants was among the top three factors influencing the location of R&D activities (Florida and Kenney, 1994).

The rapid growth of MNCs' R&D in China, as noted above, can be partly attributed to its strong manufacturing base. Commenting on the rapid flow of R&D investment in China, the managing director for General Atlantic Partners, a US private-equity fund said: 'Whenever manufacturing is located in a country, innovation always follows ... Manufacturing has migrated to China and is there to stay' (Einhorn, 2006a). As noted in the last section, although most MNCs initially manufactured low and mid-tech goods in China, they subsequently brought R&D activities to support manufacturing activities. In other words, the need to maintain a close tab on production facilities has led to MNCs shifting R&D operations to China. Motorola, Siemens, IBM, Intel, Nokia and GE, among others, are all conducting manufacturing-related R&D in China (Reich, 2005).

Market size and sophistication leading to adaptive R&D

R&D activities are also needed for tailoring products to the local market (Blanc and Sierra, 1999; Moris, 2004; Normile, 2005). A study by Fors and Zejan (1996) found that the need to adapt products to foreign markets is a major motivation for establishing R&D centres abroad. Another survey conducted among Japanese biotechnology and electronics firms indicated that proximity to markets was the most important reason for establishing R&D facilities abroad (Florida and Kenney, 1994).

Benefits associated with establishing and maintaining proximity to China's huge and rapidly growing consumer base has helped to attract MNCs' R&D activities to the country. A Boston Consulting Group analyst notes: 'investing in research in China at this stage is as much of a commercial as a scientific decision' (*Economist*, 2005c). It is important to note that although China is an emerging market, its consumers want the latest innovations (O'Driscoll, 2006).

In the healthcare industry, for example, the pharmaceuticals market in China is growing rapidly and an increasing number of Chinese people are seeking Western-style treatments (O'Driscoll, 2006). Factors contributing to such changes include urbanisation, lifestyle changes and associated chronic diseases (Comtex News Network, 2006). AstraZeneca and Novartis have each announced plans to invest US$100 million in medicine-related R&D in China (Comtex News Network, 2006).

There is thus a tremendous need to conduct market-specific research in China (Economist Intelligence Unit, 2005). First, the Chinese market possesses many characteristics that differentiate it from Western markets. This is particularly true for sectors such as pharmaceuticals and software.

In pharmaceuticals, treatments for ailments that are more common to Chinese consumers can be better developed locally (Adcock, 2005). Likewise, early Chinese versions of Microsoft Windows lacked ease of use; for example, they required many keystrokes to create a single Chinese character. Chinese consumers also believed that Microsoft's Chinese products were technically inferior to the English versions. Microsoft was thus forced to do more local R&D in China. Chinese-language Live Search Map services launched by Microsoft's Virtual Earth team, for instance, focus on Chinese customers (China.org, 2008a).

Chinese consumers also have different product usage habits and lifestyles which make it necessary to conduct local R&D. P&G offers another case in point. P&G has invested over $1 billion in China since 1988 in four factories and a Beijing R&D centre. Hundreds of P&G research managers live with Chinese families in cities and on farms to learn how they use everything from detergent to toothpaste (Roberts, Arndt and Engardio, 2005).

Second, many companies consider their Chinese R&D centres as critical to their success. According to the American Chamber of Commerce, the profits of US companies working in China averaged 42 per cent in 2002 – significantly higher than the worldwide average (Liu, 2003). For instance, the chemical company, DuPont realised the necessity of a major R&D facility in China to develop new products to meet the rapidly changing needs of the country's customers (Tremblay, 2006). China is the company's fastest-growing market. Likewise, Dow Chemical Co.'s chief technology officer and corporate vice president of R&D recently put the issue this way: 'To send over an English-speaking, jet-lagged engineer to help a Chinese guy figure out why his product isn't working just isn't cost-effective' (Rajagopalan, 2006). In a similar fashion, Motorola, Siemens, IBM, Intel, Nokia and GE, among others, are conducting the necessary R&D for tailoring their products to the Chinese market (Normile, 2005).

Competition leading to adaptive R&D

For some MNCs operating in the developing world, R&D has become a major competitive tool to maintain an edge over their rivals. Consider, for instance, the competition between Motorola and Nokia in the Chinese cellular market. In 2003, Nokia introduced 15 new cellphone models in the Chinese market (CNETAsia, 2003b). Motorola's chief operating officer Mike Zafirovski said Motorola would respond by accelerating its plan to introduce 14 new phones designed for the

Chinese market and by increasing the R&D budget by 30 per cent (Charny, 2003).

Similarity to regional markets influencing adaptive R&D

MNCs can also use a R&D facility in an economy as a launch pad to develop products for the regional market. Some R&D centres in China are used to develop products for the Asian market in general (UNCTAD, 2005). For instance, Oracle's Beijing lab develops products related to the open source Linux operating system for its Asian customers (Buckley, 2004). Likewise, Nortel Networks is planning to employ its R&D centre in Beijing to offer services to the Asian region (*SinoCast China*, 2006b). Quoting a World Technology Evaluation Center report, the Bureau of Industry and Security (1998a) argues that the primary motivation for entering the Chinese market is 'to be nearer to the fastest growing electronics markets, which are now in Asia, and where the market demand ... for electronics is significant'.

Similarly, in 2004, the US-based direct sales firm Amway announced its plan to open a new R&D centre in Shanghai. The company aimed to use the centre to develop skincare as well as nutritional products for the Asian market (SPC Asia, 2004). In November 2006, Novartis announced plans to build an integrated biomedical R&D centre in Shanghai. R&D activities at its Shanghai centre will initially focus on addressing medical needs unique to Asia, such as infectious causes of cancer endemic to the region (Comtex News Network, 2006).

The above paragraphs provide support for the notion that a large and sophisticated regional market provides an additional incentive for developing globally-oriented innovations. It is important to note that for some products, adaptive R&D conducted in China, which initially focused on local and regional needs, subsequently evolved into global innovative R&D (UNCTAD, 2005). This is especially true for highly standardised products like those related to cellular telecommunications for which the size and sophistication of the Asian market have been rapidly increasing. Handset manufacturers such as Motorola, Nokia and Siemens are developing their latest cellphone models, based mainly on the Linux operating system, in China. In addition, they are introducing some of these models to the Chinese market before releasing them in Europe and North America (Buckley, 2004; Schwartz, 2005). Other companies such as IBM and Intel are also running full-scale labs that are globally-oriented and work on their companies' most advanced products (Buckley, 2004).

R&D costs leading to innovative R&D

While costs are of secondary importance for some firms' R&D decisions, they are more important for others (Moris, 2004). According to Studt (2006), about half of the respondents in a recent survey said that lower costs were important or very important reasons for outsourcing R&D in developing economies such as China. In another survey conducted by Booz Allen, 30 per cent of the responding companies named cost as their primary motive in choosing an R&D location (Rajagopalan, 2006). Indeed, increased R&D costs have forced many MNCs to locate R&D activities in developing countries such as China (Miroux, 2005).

China's cheap R&D personnel and low cost of materials, facilities and clinical experiments are attracting MNCs (O'Driscoll, 2006; Studt, 2006). Estimates suggest that R&D centres in China cost about one-tenth to one-half of their counterparts in the USA or Europe (Adcock, 2005; Laudicina and White, 2005). For instance, the annual cost of employing a chip design engineer (salary, benefits, equipment, office space and other infrastructure) in 2002 was US$28,000 in Shanghai and US$24,000 in Suzhou, compared with US$30,000 in India and US$300,000 in the USA (Ernst, 2005). Similarly, for work related to drug development, the annual cost of employing a chemist in the USA is about US$250,000 a year compared with $25,000 a year in China (Zamiska, 2006).

For this reason, pharmaceutical giants such as GlaxoSmithKline and Eli Lilly are collaborating with Asian biotech research companies in an attempt to cut the cost of developing and bringing new drugs to market, which is estimated to average US$500 million (Engardio et al., 2005). While many Western MNCs are thus cutting R&D at home, they are rapidly expanding R&D operations in developing countries (Engardio et al., 2005).

Size and quality of local R&D talent leading to innovative R&D

The availability and quality of local R&D talent and work ethic play a critical role in locating an MNC's R&D works (Goldner, 1992; Florida and Kenney, 1994). A study sponsored by the Kauffman Foundation has indicated that collaboration with universities and intellectual capital are key factors in deciding an MNC's R&D operations in developing countries (CNN Money, 2006). Intel CEO Craig Barrett (2005) has said: 'If the world's best engineers are produced in India ... that is where our companies will go. We locate facilities where we can find or import talent to produce our products'.

It would be obviously erroneous to suggest that low costs are the only motivators for MNCs conducting innovative R&D in China. Indeed, China's labour pool comprises locally trained scientists and engineers as well as returning Chinese with Western training and experience (Boswell and Baker, 2006). China thus has a large number of highly trained and highly motivated scientists performing high-quality research (Yegin 1995; Erickson, 2001; O'Driscoll, 2006). MNCs are expanding R&D activities in China to tap the huge pool of local talent and ideas and to broaden their technology base (UNCTAD, 2005; Studt, 2006).

Intel CEO Craig Barrett, for instance, has argued that the Chinese are 'capable of doing any engineering, any software job, any managerial job that people in the United States are capable of' (Segal, 2004). Indeed, more scientists and engineers are staying in or returning to China with graduate degrees from foreign countries to perform R&D work for foreign affiliates or local firms, or to start their own businesses (UNCTAD, 2005). In 2002, over 18,000 Chinese people with degrees from foreign universities returned to China. This figure was 47 per cent more than in 2001, double that in 2000, and over three times the figure for 1995 (Lynch, 2003). By 2005, 170,000 Chinese with graduate degrees overseas returned to China (EIU ViewsWire, 2005b). MNCs such as AstraZeneca want to capitalise on the huge and growing Chinese research pool to launch products that would help them compete globally (O'Driscoll, 2006).

A study on China's electronics sector observes:

> In fact, all US electronics companies are increasing their Asian investments in R&D to take advantage of favourable industrial-government partnerships and engineering workforces that are highly motivated and well trained (frequently in the United States). (Bureau of Industry and Security, 1998a)

Among many examples of globally oriented R&D centres to be established in China are the Nokia China R&D Centre (1998), the Motorola China Research Institute (1999), the Nortel China R&D Centre (2001), the Ericsson China Central R&D Institute (2002) and Sony Ericsson's global R&D centre in Beijing (2004) (UNCTAD, 2005).

Table 5.1 A comparison of China with other developing and developed economies in terms of R&D related indicators, 2004

Country	Population (million)	Patents granted to residents (per million people)	Receipts of royalties and licence fees (US$ per person)	R&D expenditures (% of GDP)	Researchers in R&D (per million people)	Total no. of researchers in R&D
China	1,308	NA	0.2	1.3	663	867,204
India	1,087	1	NA	0.8	119	129,364
USA	295	281	178.2	2.6	4,484	1,324,574
UK	60	64	202.1	1.9	2,706	161,007
Japan	128	874	122.7	3.1	5,287	676,207
Germany	83	156	617.0	2.5	3,261	269,358

Source: UNDP (2006).

As noted above, MNCs are also benefiting tremendously by conducting globally-oriented R&D activities in China. Economists such as Jagdish Bhagwati argue that it is more useful to regard industry in terms of comparative advantage or occupation-specific industry rather than in terms of high-skilled and low-skilled (Bhagwati, 1998). Some developing countries are thus specialising within certain industries requiring a highly-skilled workforce (Bernstein, 2004). For instance, China ranks second in the world, only after the USA, in terms of total number of researchers in R&D (Table 5.1). Although China may fall behind the USA in terms of the overall quality of researchers, Chinese researchers have demonstrated competence to perform high-quality R&D in areas such as open source software, nanotechnology and cellular telecoms. As noted above, MNCs are capitalising on China's strengths in these areas to generate globally-oriented innovations.

Institutional factors

R&D related rules and laws influencing innovative R&D

Local rules and laws determine incentives and disincentives associated with R&D. Significant government support has been a major force behind the increasing volume of MNC R&D activities in China (Bureau of Industry and Security, 1998a). The Chinese government ties incentives to MNCs' commitment to R&D activities and technology transfer (Moris, 2004). For instance, in the mid-1990s, foreign investment approval authorities encouraged more R&D activities. These were designed to show foreign firms' commitment to China's technological development. The state encouraged MNCs to compete against each other in terms of the amount and quality of R&D activities conducted in China (Adcock, 2005). A number of surveys have found that government support in the form of tax concessions and other incentives has been among the key drivers in the inflows of R&D in China (*Health Insurance Law Weekly*, 2006; O'Driscoll, 2006; Studt, 2006).

Indeed, in addition to tax concessions and incentives, some components of China's regulative environments are more R&D friendly than in the West. In this way, some R&D activities are moving to China to benefit from regulative and jurisdictional arbitrage. Animal testing is a good example. US laws generally require that all drugs must be tested on at least two animal species before being submitted for approval by the Food and Drug Administration (FDA) (Pocha and Brown, 2006). While R&D involving animal testing faces strong resistance because of political issues in the USA and Europe, and religious issues in India (Pocha and Brown, 2006), such resistance is almost non-existent in China. Drug development projects that involve animal testing are thus rapidly moving to China.

Regulative uncertainty influencing innovative R&D

A lack of clear policies has been a barrier to firms' operations in China (*Asia Pulse*, 2006b). Some scholars go even further, arguing that in China:

> the law is marginalized and the legal system relegated to a lowly position in a spectrum of meditative mechanisms, while at the same time available for manipulation by powerful sectors within the state and the society at large. (Myers, 1996: 188)

The legal uncertainty created by the regulatory vacuum makes it important to have good relationships with government officials. Some MNCs have received access to the giant market by gaining favour with influential government officials (O'Driscoll, 2006). Conducting R&D in

China is one way to influence the government officials. R&D activities and the transfer of technologies are critical to succeed in China. A US Department of Commerce study found that transferring important technologies and next-generation scientific research to Chinese companies is required for any access to China's low-cost labour force or the market. Many high-tech firms are willing to pay the price by transferring technology and management expertise in exchange for market access (Bureau of Industry and Security, 1998a). For this reason, MNCs such as AstraZeneca are supporting China's focus on innovation and are substantially increasing R&D investments in the country (O'Driscoll, 2006).

The intellectual property regime influencing innovative R&D

Control over strategic R&D activities has been one of the important centripetal forces. This is especially important for innovative R&D activities. Because of this, theorists have suggested that a firm's decision to decentralise R&D activities is a function of the amount and nature of knowledge it can access in relation to what it can control (Loasby, 1998; Blanc and Sierra, 1999). Firms are therefore creating internal and external organisations to optimise the combination of capabilities they can control versus those they can access (Loasby, 1996). In developing countries with weak intellectual property regimes, the creation of such organisations to protect intellectual property results in a lower volume of R&D activities.

Despite the abundant and growing market, some MNCs hesitate to do conduct R&D in China because of the weak protection of intellectual property rights (Watson, 2005). In an attempt to minimise the loss of important intellectual property, MNCs have maintained 'hierarchical modular R&D structures' (Quan, 2005) at different levels. The existence of these hierarchical modular R&D structures is inherently negatively correlated with the amount and quality of R&D conducted in the host country.

First, 'core' R&D is developed in the MNC home country, while 'peripheral' or 'non-core' R&D' is conducted in developing countries such as China (Teece and Chesbrough, 2005). For instance, the Japanese multinational, Matsushita conducts all of its basic research as well as the development and production of technologically-sensitive products in Japan, while conducting simpler processes in other Asian countries, especially China (Parker, 2004). Similarly, Kyocera, the Japanese semiconductor and electronic component manufacturer, produces its most complex materials in Japan. Workers in its overseas plants, such as those in China, are told precisely how to mix materials sent from Japan with local materials, but are given no information about what the ingredients are. If some parts or components break on a machine in its overseas plant, Kyocera's Japanese engineers disassemble and repair it in privacy (*Economist*, 2004a). Local workers thus have no access to Japanese technology and know-how.

Such modular structures exist across firm boundaries as well as within firms. In their interviews among Chinese semiconductor firms, Teece and Chesbrough (2005) found evidence of hierarchical modular approaches. One firm, a joint venture between a Japanese MNC and a local Chinese partner, separated its production into three distinct processes. The joint venture did not rotate people across the three processes, thus, no departing employee would have knowledge of more than a single process (Teece and Chesbrough, 2005). Without the knowledge required to put the pieces together, any employee leaving that company can take little or no intellectual property with them. Moreover, Japanese companies are increasing their use of 'black-box' technologies. These are embedded in advanced manufacturing processes (*Economist*, 2004a) and cannot be disassembled and analysed, thus making them difficult to reverse-engineer, thereby protecting their patents (Parker, 2004).

Notwithstanding these measures, a certain amount of intellectual property leakage cannot be avoided (Parker, 2004). A weak intellectual property regime worsens the situation. Japanese technology firms have found that their intellectual property protection measures have made it difficult to foster loyalty in their Chinese operations (*Economist*, 2004a). Loyalty of local employees is becoming more important as many foreign MNCs are increasing the ranks of local Chinese engineers in the R&D centres. For instance, Matsushita planned to increase the number of local general managers in its China operation from 40 per cent in 2004 to 70 per cent by 2006 (Parker, 2004).

Spillover effects

The fact that foreign firms are active in technology and R&D-intensive sectors means that a significant amount of spillover is likely to take place in the local economy. Technology spillover takes place through a number of channels such as reverse-engineering, skilled labour turnovers, enterprise spin-offs, demonstration effects, and supplier–customer relationships (Cheung and Lin, 2004; UNCTAD, 2005). Studies have found that the amount of technology transfer is positively related to the amount of spillover (Blomstrom and Zejan, 1992). Obviously, innovative R&D activities tend to produce greater spillover effects than adaptive R&D.

Reverse engineering

In the early days of China's market opening, the government's IT spending mostly went to activities geared towards the reverse-engineering of hardware technologies (Brisendine, 2002). For a large number of Chinese firms, this is still the method most often employed to gain access to foreign technologies (Sanford, 2004; Liu, 2007). IT products reverse-engineered in China range from simple hardware to sophisticated technology products such as Pentium processors and the CDMA mobile chipset. It is worth noting that China has a huge pool of talented engineers that understand circuit design, which is important in the reverse-engineering process (*Electronic News*, 2002). A survey conducted by *EE Times-China*, for instance, revealed that about 35 per cent of engineering teams in China copy core designs from other companies (Zhao, 2004). A major motivation behind reverse-engineering is to cut royalties paid to foreign intellectual property owners (Stevenson-Yang and DeWoskin, 2005).

Mobility of labour

Employee turnover is a major way through which technology and knowledge spillover to the domestic economy takes place. This is especially important for firms in developing countries for two reasons (UNCTAD, 2005). First, tacit knowledge associated with R&D is embedded in individuals' knowledge and experience rather than hardware or capital equipment. Second, such knowledge is difficult to gain in other ways (UNCTAD, 2005). An experienced researcher joining a local company with all his/her knowledge is probably the strongest form of technology transfer.

Training offered by foreign multinationals has offered Chinese firms an unprecedented opportunity to attract qualified R&D personnel. Chinese companies' rapid growth and global aspirations have made them attractive employers (Roberts, Balfour and Engardio, 2005). They have been able to lure employers from multinationals. To take some examples, the president of Shanda Interactive Entertainment is a former president of Microsoft's Chinese operations, while Lenovo's head of corporate communications has previously worked for Ogilvy & Mather and General Electric. Likewise, Huawei has hired people away from Motorola and Nokia, while Haier and China Netcom have attracted people from McKinsey, A.T. Kearney and the Boston Consulting Group (Balfour and Roberts, 2005). The Beijing-based Sohu.com, one of China's main internet portals, has recruited hundreds of its employees from multinationals, including a group leader in human resources who had previously worked at McDonald's (Hymowitz, 2005). Similarly, a research director of a multinational's R&D centre in China attracted a whole team of researchers back to the Chinese Academy of Science by offering them the opportunity of doing independent research as well as other incentives (UNCTAD, 2005). This represents a drastic change compared with some years ago, when no 'self-respecting' white-collar employee of a foreign company would have joined a local one (Balfour and Roberts, 2005).

Enterprise spin-offs

Another channel for spillovers is 'spin-off' firms or innovations from foreign affiliates. A number of trained researchers working in multinationals have become new entrepreneurs. For instance, China Techfaith Wireless was formed by a 14-person team that left Motorola China in 2002 (UNCTAD, 2005). The company filed for a $150 million initial public offering in the USA in May 2005 and was listed on the NASDAQ (Clendenin, 2005a). Photonic Bridge, another R&D firm in China, was also founded by a team of engineers and researchers from Lucent.

Demonstration effect

The presence of foreign companies or products can also stimulate innovations among local firms (Cheung and Lin 2004). Other indirect benefits include encouragement of commercial culture among scientists and engineers, and implantation of an R&D and innovation culture among local companies (Miroux, 2005). Using provincial data from 1995 to 2000, Cheung and Lin (2004) found positive effects of foreign direct investment (FDI) on the number of domestic patent applications in

China. They found the strongest spillover effect on minor innovation, such as external design patent; this points to the existence of the 'demonstration effect' of FDI. Liu's (2002) study based on 29 manufacturing industries in the Shenzhen Special Economic Zone for the period from 1993 to 1998 indicated similar effects.

Conclusion

Our purpose in this chapter was to develop a framework for studying foreign MNCs' R&D activities in China. We identified and analysed a host of economic, geographic and institutional factors influencing the amount and quality of MNCs' R&D activities in the country.

Some of the drivers of R&D are changing rapidly in China. For instance, the Chinese market is growing very rapidly in terms of size and sophistication. Such growth is likely to further accelerate MNCs' R&D activities in China.

For one thing, rapidly improving the intellectual property regime will significantly alter China's attractiveness as an R&D location. For instance, China has recently enacted new piracy laws. Under these laws, buyers of pirated goods can be fined 5–10 times the value of the goods, while manufacturers face jail time and confiscation of their equipment (Kanellos, 2002a). The government has provided significant empowerment to regulatory agencies involved in intellectual property rights issues such as the State Administration of Industry and Commerce, the State Administration of Press and Publications, the Intellectual Property Rights Office and the State Pharmaceutical Administration (Yang, 2002). Similarly, plans were announced to open special centres in 50 Chinese cities by 2006 to handle complaints regarding infringement of intellectual property rights as well as to provide consulting services (MacLeod, 2006). As noted in Chapter 1, Pfizer was recently successful in defending its Viagra patent against a major Chinese ministry-level government agency (Boswell and Baker, 2006).

MNCs can capitalise on China's strengths in some technology sectors more than others to develop products for the global market. For instance, although China's overall technology posture lags behind that of industrialised countries, some Chinese high-tech sectors are very strong and sophisticated. In particular, Chinese companies have developed sufficient prowess to compete globally in areas related to nanotechnology, open source software, cloning technology and cellular telecommunications.

6

Forces shaping the development of the Chinese technology workforce

Among global IT analysts there is a lack of consensus regarding the high-tech manpower that China will require to support its emergence as a global IT leader. This chapter investigates the lack of agreement in the literature regarding the size and quality of the Chinese technological workforce. We examine how four mega-effects have shaped the trajectory of the Chinese technology workforce.

Introduction

The most profitable niches in international IT markets are heavily skill-dependent (Campbell, 2001). Although China's scientific and technological progress so far can be arguably attributed to its 'smart and dedicated people rather than to the purchase of expensive scientific instrumentation' (Frieman, 1999), it is not clear whether the country will have sufficient manpower to accommodate its emergence as a global IT leader.

Global IT analysts disagree as to their assessments of China's high-tech manpower. Intel CEO Craig Barrett, for instance, argues that the Chinese are 'capable of doing any engineering, any software job, any managerial job that people in the United States are capable of' (Segal, 2004). Others maintain that the size as well as quality of the Chinese scientific workforce has been a key factor in attracting multinational companies' R&D centres to China (Normile, 2005). Yegin (1995) has suggested that China has a large number of highly-trained scientists performing high-quality research. Sender (2003) notes that 'China is no longer just about vast pools of cheap labor, it is increasingly about the combination of that with skilled human capital'. Moreover, it is widely believed that high-tech workers in China are highly motivated (Erickson, 2001). Nils Newman, a co-author of the National Science Foundation-supported study on worldwide technological competitiveness notes:

> For scientists and engineers, China now has less than half as many as we do, but they have a lot of growing room ... It would be difficult for the United States to get much better in this area, and it would be very easy for us to get worse. It would be very easy for the Chinese to get better because they have more room to maneuver. (Toon, 2008)

Farrell and Grant's (2005) study, however, has indicated a serious shortage of high-tech labour in China. They argue that failure to address this shortage would hinder the development of China's IT industry. They further maintain that unlike in manufacturing industries, if employees lack required skills, their willingness to work for long hours makes only a marginal difference in the IT industry, especially in IT-enabled services (Farrell and Grant, 2005). Similarly, a *Business Week* (2005) article has noted a lack of managerial expertise in Chinese workforce. Others argue that, in China, 'a good IT professional is hard to find, and good IT managers are even scarcer' (Trombly and Marcus, 2006). Likewise, some foreign investors have rated India higher than China in terms of the availability of a highly-educated workforce and management talent (Laudicina and White, 2005).

In the literature, there has thus been a surprising lack of agreement as to the size and quality of the Chinese technological workforce. This chapter investigates the inconsistent findings noted above. The remainder of the chapter is structured as follows. The next section provides a brief survey of the Chinese technology workforce. Then, we discuss four mega-effects that have shaped the Chinese technology workforce landscape. Finally, we provide discussions and implications.

The Chinese technology workforce: a brief survey

Table 6.1 compares China and India in terms of indicators related to the technology workforce. In this section, we elaborate some of the indicators:

- *College/university enrolments*: College and university enrolment in China's grew by 132 per cent during 1980–90 (Watanabe, 1994) and

Table 6.1 China and India in terms of major indicators related to the high-technology workforce

Indicator	China	India	Remarks
Adult literacy rate (%, 2002)*	90.9	61.3	
Percentage of tertiary students in science, maths and engineering*	53	25	
Researchers in R&D per million people (1999–2001)*	584	157	
Patents granted to residents (per million people, 2000)	5	0	During 1996–2001, the number of US patent applications by Chinese entities more than tripled to 1,252 (Normile, 2005)
Human development index (2002) (rank out of 177 countries)*	0.745 (94)	0.595 (127)	
No. of college graduates (2005)[†]	3.3 million	3.1 million	US: 1.3 million
Percentage of university students studying engineering[†]	33	4	
No. of engineering graduates (2005)[†]	600,000	350,000	US: 70,000
Percentage of engineering graduates 'qualified' to work at global companies[†]	10	25	
Annual cost of employing a chip design engineer (salary, benefits, equipment, office space and other infrastructure)(2002)[¶]	US$28,000 in Shanghai, US$24,000 in Suzhou	US$30,000	US$300,000 in the USA
Annual earning of an entry-level programmer (2004)[§]	US$1,704–1,760	US$1,500	
Employee turnover rate (2004)**	14.4%	13.1%	The Philippines: 18.1%, Taiwan: 17.7%

Sources: *United Nations Development Program (2004a); [†]Farrell and Grant (2005); [‡]Colvin (2005); [¶]Ernst (2005); [§]Rai (2004); **Knight Ridder Tribune Business News (2005). Another report in the Economist (2005e) indicated a nationwide employee turnover rate of 11.3 per cent in 2004 vs. 8.3 per cent in 2001.

by 300 per cent during 1980–2004 (Barshefsky and Gresser, 2005). The number of college and university students increased from 6 million in 1996 (Parker, 2004) to 11 million in 2000 (Dillon, 2004), further increasing to 14 million in 2003 (Parker, 2004). In 2005, the number of students enrolled in Chinese universities was expected to be between 16 million (Dillon, 2004) and 20 million (Barshefsky and Gresser, 2005). Likewise, 330,000 Chinese students intended to enrol in graduate programmes in China in 2004, which is 20 per cent higher than in 2003 (Mooney et al., 2004). These growths are the result of the rapid expansion of Chinese educational infrastructure. For instance, during 1999–2004, China opened 67 new private colleges and universities, and targeted the top nine for extra funding so as to make them 'the world's best' (Parker, 2004). Reflecting these trends, in 2004, 2.4 million new employees with undergraduate degrees and 150,000 with master's or doctoral degrees joined China's workforce (Tong, 2005). It is worth noting, however, that only 5.2 per cent of Chinese adults have a college degree compared with 25 per cent in the USA (Sovich, 2006).

- *IT enrolments*: More important, perhaps, is China's high proportion of students in science, maths and engineering fields. By the end of 2002, 70 per cent of Chinese universities had set IT specialities (*SinoCast China*, 2003a). In 2003, about 1.3 million students specialised in IT fields (*SinoCast China*, 2003a). An estimate by the research firm Gartner suggests that, in 2002, China had about 400,000 IT professionals and that this number was increasing by 50,000 each year (Ulfelder, 2003). Likewise, another estimate suggests that the number of computer science graduates produced by China's universities will soon reach 200,000 annually (Flanigan, 2003). It is, however, suggested that only a small proportion of China's workforce is considered to be of world-class standards. One estimate suggests that there are only 160,000 engineers suitable for work in multinationals (Farrell and Grant, 2005) – this is about the same as in the UK.

- *Chinese science and technology workers abroad*: In 2004, over 15 per cent of students studying outside their home country were from China, which compares with 5 per cent from India (Dahlman, 2007). Similarly, among foreign-born scientists and engineers working in the USA, more originate from China than any other country. One estimate suggests that over 150,000 Chinese engineers and scientists work in the USA (Miscevic and Kwong, 2002). Likewise, of the 8,504 PhD degrees granted by US universities in science and engineering in 1991, over one-third went to Chinese students (Alarcon, 1999).

- *Chinese graduate returnees:* An increasing number of Chinese educated abroad are returning home. Most of these returnees are university graduates and possess international business and technical expertise (Martinsons, 2005). In 2002, over 18,000 Chinese graduated from foreign universities returned to China. This figure was 47 per cent more than in 2001, double that in 2000, and over three times the figure for 1995 (Lynch, 2003). In the early 2000s, graduate returnees and scholars accounted for 81 per cent of the academicians of the Chinese Academy of Sciences, 54 per cent of the members of the Chinese Academy of Engineering, and 72 per cent of the leading scientists of the country's national research projects (*China Staff*, 2003). Likewise, in some departments of the University of Peking, up to one-third of the faculty members have PhD degrees from US universities (*Economist*, 2005d).

- *Turnover rates*: Fierce competition and a limited supply of talent have led to high turnover rates in the Chinese high-tech workforce (Longo, 2004). In 2004, 10 per cent of executives in Shenzhen and 8.5 per cent in Beijing changed job (*Economist*, 2005e). It is suggested that Chinese IT companies lose one-third of their workforce in a year and managers change their jobs every 15 months (Sovich, 2006). Chinese and multinational companies are taking measures to combat high turnover rates. Some companies, for instance, offer training as an extra incentive to reduce employee turnover (*Economist*, 2005e). To prevent workers from using this training as a means to move to other companies, however, some firms do not provide certificates after the completion of the training (Lipper, 1997). Although China's overall high-tech employee turnover rate is higher than India's, the reverse has been the case for some companies that have operations in both countries. For instance, Tata Consultancy's staff turnover in China is less than 6 per cent, compared with 15 per cent in its Indian operations (Rai, 2004).

Some qualitative indicators

Table 6.2 provides some qualitative indicators of the Chinese high technology workforce. These are explored in further detail below:

- *Ability to achieve practical solution*: Zita (1991) has described a shift in the emphasis of the Chinese workforce from 'quest for knowledge'

Table 6.2 Some qualitative indicators related to the Chinese high-tech workforce

Indicator	Explanation
Ability to achieve practical solution	Chinese students get little practical experience in projects or teamwork A high degree of bias toward theory
Hard vs. soft skills	Emphasis on hard subjects 'Soft' concepts of management
Language proficiency	A large population with good command of Asian languages Poor English has been a major weakness
Composition of the IT talent pool	High-tech workforce in China has a tendency to move to senior technological or managerial position after some years of work experience
Collaboration with the outside world	Chinese academics, engineers and scientists get exposure to the outside world

to 'search for practical applications'. More recent studies, however, have indicated China's limited progress on this front. Farrell and Grant's (2005) study, for instance, indicates that compared with engineering graduates in Europe and North America, who work in teams to achieve practical solutions, Chinese students get little practical experience of projects or teamwork. Other studies have found that Chinese students rely heavily on lectures or materials copied from textbooks and lack the initiative to develop original thinking as demanded by the situation (Branine, 1996). It is argued that they allocate more labour and capital to resolve problems, rather than thinking about new processes (*Economist*, 2005f). In sum, a high degree of bias toward theory has been a major difficulty for the Chinese high-tech workforce (Farrell and Grant, 2005).

- *Hard vs. soft skills*: As noted above, an increasing proportion of Chinese students are taking the 'hard' subjects (Barshefsky and Gresser, 2005). China's import of Western or other modern management techniques has also concentrated on the tangible and quantitative approach. A study conducted in the early 1990s has indicated that 'soft' concepts of management, such as marketing and consumer behaviour have not been integrated into Chinese thinking (Borgonjon and Vanhonacker, 1992). Such concepts are perceived by the Chinese Communist Party as a threat to the communist ideology.

Moreover, the driving force of national production has traditionally been a central plan, rather than the consumer (Borgonjon and Vanhonacker, 1992). China has, however, made some progress in this direction. In 1991, the Ministry of Education authorised nine universities to award master's degrees in business, and under a joint agreement between the Chinese government and the European Union, the China Europe International Business School opened in 1994 (Ridgway, 2005). As of 2005, some 95 universities were allowed to offer MBA degrees (Ridgway, 2005). It is, however, suggested that Chinese managers still lack modern management thinking and skills such as negotiating international partnerships and exiting loss-making activities (*Economist*, 2005f).

- *Linguistic skills*: In terms of linguistic skills, China has a large population with good command of Asian languages such as Japanese and Korean (Einhorn and Kripalani, 2003a). Although poor English has traditionally been a major weakness of the Chinese workforce, the government is taking a number of measures to address this. For example, since 2001, the Chinese Ministry of Education has required students to start learning English as early as the third grade (Farrell and Grant, 2005). Similarly, college students must pass English proficiency exams to graduate. Thanks to these measures, the number of English-speaking graduates in the workforce doubled during 2000–04 to surpass 24 million (Filippo, Hou and Ip, 2005). Nonetheless, because of the short supply of teachers and professors, English classes are very large, even at universities (Farrell and Grant, 2005).

- *Composition of the IT talent pool*: Compared with India, the high-tech workforce in China has a tendency to move to senior technological positions after some years of work experience. In the software industry, for instance, China's major challenge has been a programmer-heavy skill-profile that constrains it from moving beyond coding contracts. About 90 per cent of Indian IT professionals are blue-collar. It is suggested that 'elite cultivation' in China's IT education results in over half of IT graduates pursuing senior technological or managerial positions after some years of work experience (Kharbanda and Suman, 2002).

- *Connection with the outside world*: An increasing number of Chinese academics, engineers and scientists are participating in international scholarly meetings and workshops that provide exposure to global standards and practices (Bureau of Industry and Security, 1998b). By the mid-1990s, China was cooperating with at least 83 countries on science and technology-related projects (Bureau of Industry and Security, 1998b). China is also aggressively pursuing relationships with foreign universities (Martinsons, 2005). For instance, as of 2004, Yale University had partnerships with 45 institutions in 16 Chinese cities (Parker, 2004).

Four mega-effects in the Chinese technology workforce landscape

The Chinese high-tech workforce landscape is arguably shaped by Confucianism, socialism and capitalism (Yang, 2002; Yang, Zhang and Zhang, 2004). The last two effects – socialism and capitalism can be explained in terms of four mega-effects (Table 6.3).

The Cultural Revolution effect

The 1966–76 Cultural Revolution severely damaged the Chinese educational infrastructure (Yegin, 1995: 567). Chinese engineers and technical workers lost significant research and practical experience, pushing the Chinese IT sector two to three generations behind that of the West (Zita, 1991). Many were taught that 'capitalism was evil' (*Economist*, 2005e).

Advanced research was brought to a standstill. Following the Cultural Revolution, the concentration of scientific and technical workers in China was one of the lowest in the world. For instance, in 1985, only five of every 1,000 workers in China were scientific or technical workers; at that time, the comparable figure was 97 for the Soviet Union, 60 for the USA and 49 for Japan (US Embassy Beijing, 1996).

In 1966, China stopped publishing scientific journals and discontinued subscriptions to foreign journals (Library of Congress, 1987a). In 1967, the Chinese Academy of Sciences – the scientific research base of China, founded in 1949 and modelled on the Soviet Science Academy – had its budget reduced to less than one-sixth of what it had been in 1965 (Di Capua, 1998; Yongxiang, 2001). Scientists and technical workers were paid less than manual labourers, severely damaging their productivity and morale (US Embassy Beijing, 1996). Western scientists visiting China in the mid and late 1970s noted that the Chinese scientific landscape was characterised by a lack of theory and an extreme emphasis on short-term problems (Library of Congress, 1987a).

Table 6.3 Four mega-effects in the development of China's high-tech workforce

Effect	Explanation	Major features/effect on high-tech workforce	Remarks
The Cultural Revolution effect	Science and technology industry was a major victim of the 1966–76 Cultural Revolution	1966: discontinuation of scientific journal publication and subscriptions to foreign journals; scientists and technical workers were paid less than manual labourers	
The market track (open-door) effect	Effects of the post-1978 economic and political reforms	In 1978, China allowed its students to go overseas for study; some went overseas with the sponsorship and financial backing of the government and many others were permitted to study abroad with own financing	
The Tiananmen effect	The government took measures to re-establish control over students (in China and abroad) following the Tiananmen events	Students abroad were afraid of punishment for political reasons if they went home 1990: China required university graduates to work for five years before becoming eligible to study abroad August 1992: US Congress passed a bill that allowing Chinese studying in the USA to apply for permanent residency	
The prosperity effect and the government's incentives	China's economic prosperity and government incentives attracted overseas Chinese scholars and students	The government called for nation-building and provided a number of incentives; cities and provinces started providing their own incentives for returnees	2003: over 4,000 companies run by 15,000 graduate returnees in 76+ industrial parks

Chinese technological development was, however, relatively advanced in strategic areas related to national defence, such as nuclear technology, missile delivery capabilities and air force. The development of these sectors can be attributed to an 'ideologically and functionally protected community of sophisticated Chinese scientists and engineers' (Barnett, 1972). For instance, it took only three years for China to move from its first atomic test (16 October 1964) to the first thermonuclear test (17 June 1967); this compares with eight years taken by the USA, five years by the UK and four years by the Soviet Union (Garden, 2001). By this time, China had developed missiles that were capable of reaching parts of Japan and the eastern Soviet Union (Lewis and Litai, 1987). China achieved its first successful satellite launch on 24 April 1970, using a CSS-3 intercontinental ballistic missile, which was followed by the second successful launch on 3 March 1971 and the third on 26 July 1975 (Cooper and Contant, 2001). Except for the nuclear weapon and the missiles to deliver them, Chinese military research made virtually no progress beyond copying the Soviet technology of the 1950s (*Asia Times*, 2000).

The market track (open-door) effect

Deng Xiaoping introduced economic and political reforms in 1978.[1] A notable feature of his open-door policy was the 'dual-track approach' (La, Qian and Roland, 2001). The 'market track' of this approach played an especially important role in development of high-tech manpower.

Deng and China's other leaders made a strategic decision to send Chinese scholars overseas for academic and scientific training.[2] A major goal of the reform was to compensate for the decade lost due to the Cultural Revolution by training a new generation and retraining an older generation of researchers, scientists and teachers (Zweig, 1997). In 1978, Deng said:

> Any nation or country must learn from the strong points of other nations and countries ... We must actively develop international academic exchanges and step up our friendly contacts with scientific circles of other countries. (Mann, 1986)

Following the, 1978 political and economic reforms, Chinese students started going overseas in substantial numbers (Table 6.4). Chinese leaders knew that many students would not return, but they were ready

Table 6.4 Cumulative number of Chinese leaving overseas for study

Year	Cumulative no. of Chinese leaving for overseas for study	Cumulative no. returning to China	Remarks
1985*	37,000		15,000 in the USA
1992†	170,000	60,000	The rate of return for graduate students was only about 10 per cent
1995†	210,000	70,000	Return rate from the USA was estimated at 5–10%
1996¶	250,000–270,000		There were over 100,000 Chinese students in the USA in 1996 (Rubin, 1996)
2002 end§	580,000	160,000	400,000 Chinese students studied abroad during 1983–2002 and 130,000 of them returned back to China (*SinoCast China Business Daily News*, 2003a); during 1978–2002, 63,533 were financed by the government and 77% of them returned home (*China Staff*, 2003)
2005**	700,000	170,000	

Sources: *Mann (1986), †Sun (1992), †Tempest (1995), ¶*South China Morning Post* (1997), §*China Staff* (2003), **EIU ViewsWire (2005b).

to pay this price to get well-trained personnel (Mann, 1986). Although some of them were sponsored by the government, many were self-financed.

In the early years of the open-door policy in education, the return rate of Chinese students studying abroad was high. Sending primarily advanced or mature scholars, with established careers and families in China, universities and research labs gained significant benefits from the educational exchanges (Lampton, 1986; Chen and Zweig, 1993). Before the open-door policy, Chinese technology workers were typically educated in the then USSR. Gradually, a high-tech workforce educated in Europe or North America started to grow (Bureau of Industry and Security, 1998b).

Another mechanism by which the post-1978 reforms contributed to China's technological workforce was through Deng's emphasis on institutional linkages to foster innovations. This led to the formation of a large number of research institutes. By 1987, there were over 5,000 state-operated research units employing more than 230,000 scientists (Baark, 1988). By the late 1980s, China had 1,000 tertiary education institutions (*People's Republic of China Yearbook 1990*, cf. Yegin, 1995). Similarly, in the mid-1990s, China's 15,000 non-government scientific research institutions had 400,000 scientists.

The effect of the Tiananmen riots

Students had led the Tiananmen Square riots in June 1989. Following these events, re-establishing control over students became the primary objective for China's leadership (Southerland, 1990). Arrests of students and democrats continued for months after the riots (Rosett, 1989).

Events on the home-front led to strong anti-government sentiments among overseas Chinese students and scholars, resulting in strong attitudes against returning home (Zhang, 1992). A survey found that while only 6.1 per cent of the respondents had planned to seek US residency before the Tiananmen events, 31 per cent planned to do so after the events (Zhang, 1992). Another study found that while 58 per cent of the respondents had definitely planned to return upon graduation, following the Tiananmen events, a majority planned to wait for political developments (Zweig, 1997). Some were afraid of punishment for political reasons if they went home (Sun, 1992; Smart and Grace, 1993). Family members also discouraged students from returning home (Chang and Deng, 1992). The Tiananmen incident thus accelerated the already high rate of brain drain.

What is more, a number of Western governments refused to cooperate with the Chinese government and allowed Chinese students to stay in their countries. In August 1992, US Congress passed a bill that allowed Chinese students who had entered the USA between 5 June 1989 and 11 April 1990, to apply for permanent residency beginning July 1993, unless the president declared it safe for them to return to China (Sun, 1992; Hertling, 1997). Over 50,000 Chinese students and scholars became permanent US residents (Zweig, 1997). Likewise, over 10,000 secured working rights in Canada; and in Australia, over 20,000 Chinese students were accorded an opportunity to stay (Zweig, 1997).

The Chinese government also restricted overseas study by requiring university graduates to work for five years before they become eligible to study abroad (Southerland, 1990). Those going abroad without working for five years were required to repay substantial amounts to the state for their education (Southerland, 1990). The Chinese Communist Party also started promoting the political slogan, 'Contribute more to the four modernizations' (Rosett, 1989). The state also reportedly ordered some students to cease their studies and begin agricultural work instead (Rosett, 1989).

In a circular made public in August 1992, the State Council encouraged all students abroad to return to China for employment after receiving their degrees (*Beijing Review*, 1992). The new policy outlined in a State Council circular was the most detailed public formulation until that time (Sun, 1992). Chinese students and scholars abroad, however, were sceptical of the government's measures. For instance, in a survey of 273 Chinese students in the USA and Canada, more than half 'did not trust that the Chinese government would keep its word about allowing people who returned to go out of China in the future' (Tempest, 1995; Zweig, 1997).

The prosperity effect and the government's incentives

In addition to the Tiananmen effects discussed above, traditionally low salaries, inadequate or nonexistent research facilities and scarce housing discouraged Chinese scientists and scholars abroad to return home (Sun, 1992). The trend has, however, been reversed due to government policies and China's rapid economic growth, which together have encouraged a large number of Chinese to return home (Martinsons, 2005). Many have also been encouraged to return because of the record salaries that investment banks, brokerage houses and other companies have begun to offer (Huus, 1994).

The government's call for nation-building and incentives for returnees

China arguably wanted its students to return with expertise in foreign technology, so that they could contribute to economic development and nation-building (Skeldon, 1996). Due to the low and even decelerating trends of returning Chinese, especially following the Tiananmen events, Chinese officials feared that there would be a severe shortage of qualified people to achieve economic growth (Sun, 1992). The Chinese Communist Party also acknowledged the loss of the middle generation of academics caused by the Cultural Revolution (Hertling, 1997). Chinese officials were further concerned by the US green-card phenomenon (Tempest, 1995).

The Chinese government formulated policies to attract Chinese science and technical workers living abroad and use them as a national competitive advantage (Einhorn, Carey and Gross, 2005). It launched a media campaign praising students and scholars abroad and outlined preferential treatment for returnees (Sun, 1992). China encouraged even short visits by its overseas students (Tempest, 1995) and in April 1992 helped pay for 39 Chinese students studying abroad to come to Beijing to attend a scientific meeting (Sun, 1992). The government also matched them with jobs in government research facilities (Tempest, 1995).[3] Those returning home were no longer required to go through the cumbersome process of getting a separate Chinese exit permit (Sun, 1992).

Gradually, a higher proportion of Chinese students and scholars abroad started to return to China. Analysts argue that the higher rate of return can be attributed to increased domestic opportunities rather than patriotism and the government's call for nation-building. In a survey of Chinese students in America, Chen and Zweig (1993) found that patriotism was a factor for only 19 per cent of students planning to return.

Central and local governments launched dozens of programmes and incentives to lure overseas Chinese scientists, researchers and students back to China (Tempest, 1995). These included a large numbers of jobs allocated for those educated abroad, provision of research grants, preferential hiring policies and specially designated institutes such as the Qinghua University Higher Research Center (Bureau of Industry and Security, 1998b). A number of technology parks, including Beijing's Zhongguancun, were established exclusively for returnees. By 2003, 4,000 companies run by 15,000 graduate returnees were spread over 76 industrial parks established across the country (*China Staff*, 2003). Chinese graduate returnees were also offered tax breaks, cheap office space, start-up loans and advice on how to navigate the local bureaucracy (EIU ViewsWire, 2003a).

Delegations of Chinese officials also started travelling to US universities to meet with Chinese students. They talked about opportunities in China and assured students that they would be allowed to leave the country if they were not happy (Tempest, 1995).

Major cities devised their own special policies to attract returning students. Returnees were also allowed to earn a proportion of profits made on products they developed (Sun, 1992). To attract IT talent, Shanghai started to issue blue identity cards to non-Shanghai residents. Shenzhen sent its own recruiting delegation to several US universities in 1992 and issued its own set of preferential regulations to attract students from abroad (Sun, 1992).

In 1997, the provincial government of Fujian announced its plan to recruit a number of deputy departmental heads through open selection (*China Business Information Network*, 1997). Returned students with a PhD degree and over three years of experience were eligible to apply. Similarly, during 1992–99, Henan Province offered returned students financial assistance for research projects worth a total of US$980,000, attracting about 2,000 returnees, including 200 with doctoral degrees (*Asiainfo Daily China News*, 1999a). Likewise, in May 2003, Nanjing University advertised internationally to fill 300 new professorial slots (Parker, 2004).

Attention to human resource development

Contrary to previous five-year plans' emphasis on 'growth-first' priorities, China's current five-year plan has paid more attention to human resources (EIU ViewsWire, 2005c). The plan aims to strengthen compulsory education for a minimum of nine years and promote key specialities and talents among the workforce, including administration, enterprise management and technology (EIU ViewsWire, 2005c).

Increased returnee rate

The experience of other East Asian countries such as Japan and South Korea indicates that the proportion of returnees increases over time (Skeldon, 1992). Similar trends are observed in China. In the early 1980s, a survey of privately financed Chinese students at American universities found that about 85 per cent 'would probably not return' to China (Mann, 1986). A similar survey conducted more recently has indicated that about 50 per cent of Chinese studying in foreign universities plan to return to China (*China Staff*, 2003). The proportion of returnees is even higher for government-sponsored students – over 90 per cent during 1996–99 (*China Business Information Network*, 1999).

As noted above, following the Tiananmen events, Chinese students studying overseas were reluctant to return to China, but the number of returning students increased over time (Table 6.4). In the early 2000s, the number of returned overseas students increased at 13 per cent every year (*SinoCast China*, 2003b). In the late, 1990s, foreign-educated Chinese scholars accounted for over half of the top scientific researchers working on key research projects (Bureau of Industry and Security, 1998b).

Foreign companies' research and training

Foreign companies' training facilities and R&D laboratories have also played a critical role in the development of Chinese technology manpower. Some companies, such as GE and Motorola, have their own in-house 'university' (*Economist*, 2005e). Motorola's in-house university teaches language and computer courses to support staff, technical courses to engineers and management to its managers (Lipper, 1997). Ericsson China opened a Beijing training centre in mid-1995 to provide managerial and technical skills to employees (Lipper, 1997). The Finnish company Nokia's Creative Thinking Corner workshops, for instance, encourage creativity and innovation. In 2002, Microsoft pledged US$750 million to help train software and hardware developers, in part by providing free access to Microsoft products and to develop computer courses in 36 universities (Schafer, 2004).

By mid-2005, there were over 600 R&D laboratories affiliated to non-Chinese multinationals in China (Normile, 2005). Some R&D needs to be done close to where it is implemented and for this reason China's huge manufacturing base has triggered R&D activities in China (Reich, 2005). Other R&D activities, on the other hand, are needed for tailoring products to the local market (Normile, 2005). China's huge consumer base has also helped the development of R&D activities. Motorola, Siemens, IBM, Intel, Nokia and GE, among others, are conducting manufacturing-related R&D in China (Reich, 2005) as well as the work needed for tailoring products to the Chinese market (Normile, 2005).

To take one example, Intel opened a US$50 million research lab in Beijing in 1998 (Dean, 1999). In June 2005, the company also set up a US$200 million Intel Capital China Technology Fund to invest in Chinese innovation in computers, electronics and semiconductors. One of the major goals of the project is to develop Chinese talent (Wallace, 2005).

Some R&D centres, on the other hand, have broader missions. For instance, Microsoft Research Asia[4] is one of the company's four global research centres (Normile, 2005). Motorola's 19 research centres in China employ more than 1,600 R&D engineers. Similarly, in October 2003, Oracle opened its China Development Centre in Beijing, which concentrates on Linux and e-government solutions (Tan, 2003).

Market forces in the high-tech workforce

Market forces began to prevail in the Chinese technology workforce in the late 1980s (Maurer-Fazio, 1995). Starting the early 1990s, technology companies started advertising jobs offering record salaries (Maurer-Fazio, 1995). In 1992, the mayor of Shenzhen reportedly went to Xian and offered high salaries to senior technologists willing to move to his city. The mayor of Xian countered with similarly attractive deals to keep his technologists (*Economist*, 1992). Recruiters also offered spacious accommodation, commissions and bonuses (Maurer-Fazio, 1995). The affected cities countered by making even more attractive offers to lure back their workers (Huang, 1993). It is suggested that in China's major cities, salaries for mid-level IT managers are increasing by 8–10 per cent per year due to the shortage of such employees (Trombly and Marcus, 2006).

Discussion and implications

The four mega-effects discussed above and their somewhat conflicting effects help us understand the size and nature of the Chinese high-tech workforce. For instance, the Chinese workforce consists of an interesting combination of managers from state firms that are too bureaucratic as well as entrepreneurs with a private sector background who are 'unconstrained by capital or the law' (*Economist*, 2005e). The effect of the Cultural Revolution is still felt today. After a lifetime under socialism, many high-tech employees lack the mindset to adopt Western-style working practices (*Economist*, 2005d). Chinese technology workers who returned home after experiencing the personal freedoms of the West (Hertling, 1997), on the other hand, are uncomfortable in the bureaucratic environment.

Nationalism and patriotism have also played important roles in attracting overseas Chinese back home. An increasing number of Chinese working for Silicon Valley companies also 'want to work for American companies but still do something for their countries' (Hull, 1997).

The attractions created by economic opportunity and government incentives have also led to saturation in the foreign-educated high-tech workforce market. Media reports have indicated that Chinese returning home with graduate degrees are finding that their foreign credentials are no longer as valuable on the Chinese job market as they used to be in the 1990s (*SinoCast China*, 2003a; Mooney et al., 2004).

Where is the future leading?

China's weakening socialism is expected to affect the high-tech workforce as well. The size and power of bureaucratic managers and those believing in socialism can be expected to decline over time. The virtuous circle created by domestic opportunities in China and an increasing number of foreign-educated Chinese joining the country's workforce will further enhance China's technology workforce. Even more important perhaps is the current five-year plan's emphasis on human resource development and the shift from the 'growth-first' priority. If the 'growth-first' priority has helped China achieve record economic growth, it is reasonable to expect that current emphasis on human resource development will significantly increase the size and quality of China's technology workforce.

Notes

1. Deng, who became a vice premier in April 1973, was 'unsympathetic' to Mao's policies. Together with Premier Zhou Enlai, Deng exerted strong influence on the policy landscape. They favoured the modernisation of the Chinese economy, which was formally confirmed at the Tenth National Party Congress in August 1973 (see Library of Congress, 1987b). At the Fourth National People's Congress in January 1975, Zhou Enlai outlined a programme identifying science and technology as one of the four sectors (along with agriculture, industry, national defence) prioritised for modernisation (Library of Congress, 1987b). Deng also initiated a significant transformation in the Chinese Academy of Sciences. In 1975, Hu Yao-bang was appointed as the head of the CAS with the Party leadership instruction, 'CAS should be rectified and its leadership be reinforced' (Yongxiang, 2001).
2. Some argue that Chinese students and scholars were traditionally sent abroad to pursue critical civil and military dual-use technologies and that this practice continues today (Political Transcript Wire, 2005).
3. The Chinese government, however, demanded that those who had joined organisations 'hostile to China' should resign their positions and 'no longer take part in any activities to violate the Chinese constitution and law and oppose the Chinese government' (Sun, 1992).
4. In 1998, Microsoft opened its research facility and committed US$80 million in six years (Dean, 1999).

7

The Chinese software industry: structural shifts

Although China is far ahead of India in terms of most economic and technological indicators, its commercial software industry has failed to catch up with India. Nevertheless, the Chinese software industry has been undergoing major structural shifts in terms of market, technology, product, actors and focus. This chapter examines such shifts. First, Chinese players are likely to benefit from the domestic software market demand, which is higher and increasing more rapidly than in India. Second, China has put greater emphasis on the development and deployment of open source software, which is likely to be a disruptive innovation in the software industry. Third, new and powerful actors such as Red Flag Linux and Beijing Software Industry Production Center have emerged in the Chinese software landscape. Fourth, Chinese players have actively engaged in collaboration with global software giants and regional economies. Such collaboration has promoted synergistic advantage and allowed them to focus on core competences. Fifth, China's leadership in the global electronics industry combined with collaboration with global giants allows its players to add value by bundling their software with domestic hardware and software of multinational corporations. In addition to these structural shifts, China is also finding ways to overcome its competitive disadvantage by removing the skills bottleneck.

Introduction

Following the post-1978 economic and political reforms, China has left India far behind in terms of almost all economic and technological indicators. During 1978–2003, China's inward foreign direct investment

(FDI) exceeded US$400 billion (Farrell, Gao, and Orr, 2004). In the 1990s, India's share in world trade grew from 0.5 to 0.7 per cent, while China's share rose steeply from 1.8 to 4.0 per cent. In fiscal year 2004–05, China's total exports of electronic information-related products reached US$268 billion (Jin, 2006) compared with India's US$18 billion (Hindu Business Line, 2006). Other indicators, such as the penetration rates of fixed and cellular telephones and PCs also point to the fact that the Chinese high-technology industry is more sophisticated and developed than India's (Table 7.1).

Yet there is one area where China has failed to catch up with India: the commercial software industry. This industry is very important for developing countries such as China and India as it requires much less initial investment than manufacturing industries.

Although China is the world's second-largest software-outsourcing destination after India (Forrest, 2007), the gap between the software industries of the two economies is substantial. India's software exports exceeded US$12 billion in 2003 compared with China's US$2 billion. Likewise, the Chinese domestic software market is dominated by foreign companies (Goodman, 2002). India's National Association of Software and Service Companies aims to export US$60 billion worth of software by 2010, while the Chinese government's target has been US$10 billion for the same year (Einhorn, 2006b). Perhaps even more importantly, low-value application development and maintenance services account for a large proportion of Chinese software exports (Verma, 2005).

There is a lack of consensus among global software industry analysts as to whether China can close this gap in the future. Some, including an ex-president of India's National Association of Software and Service Companies, believe that China severely lacks the skills and manpower necessary to catch up in the race. Another school of thought maintains that China's status as a regional and a potential global economic superpower combined with its leadership in the electronics industry will help to close the gap. A study by Gartner Group, for instance, has suggested that by shifting the technological gears, China could catch up with India in commercial software (Einhorn et al., 2002).

What will be the future gap between the software industries of the two nations? Given that these two economies account for two-fifths of the world population and are among the few developing countries that have established good track records in the export of high-technology products, this is a very important question. Mindful that a definitive answer requires further research, this chapter outlines the major structural shifts in the Chinese software industry that will influence the

The Chinese software industry

Table 7.1 A comparison of China and India in terms of major indicators related to the software industry

	China	India
Export of software and back-office services (US$)	3.6 billion (2005)[†] 10 billion (Chinese government's target for 2010)[¶¶]	39.6 billion (FY 2006–07)[*] 60 billion (NASSCOM's target for 2010)[¶¶]
Major software export destination	60% Japan (2004)[†]	63% USA 24% Europe 3.6% Japan (2002)[*]
High-tech export as a percentage of manufacturing exports[††]	19	4
PCs per 1,000 people (2000)[¶¶]	28	7
Annual PC sales (2007)[**]	28 million	6.5 million
Telephone mainlines (2004, per 1,000 people)[§§]	241	41
Cellular subscribers (2004, per 1,000 people)[§§]	258	44
Internet users (2003, per 1,000 people)[§§]	73	32
No. of internet users (mid-2007)[§]	162 million (220 million in February 2008)	42 million
Consumer expenditure on hardware (US$, 2003)[†††]	1,004 million	260 million
Technology achievement index (Max possible: 1)[††]	0.299	0.201
E-government score (2002, out of 100) (rank out of 198 countries)[***]	56.3 (7)	45.1 (59)
No. of software enterprises (2002)	6,000+	3,000+
Average size of software firms (no. of employees)	25	174
No. of CMM level 5 companies (2004)	2	60
GNP per capita (US$)[¶¶] 2004 1980	 1,490 134	 640 262
Total ICT expenditure (US$, 2001)[†††]	66.6 billion	19.6 billion
ICT expenditure as % of GDP[†††]	5.7	3.9

Table 7.1 A comparison of China and India in terms of major indicators related to the software industry (*Cont'd*)

	China	India
R&D expenditure (2000–03, % of GDP)[§§]	1.3	0.8
Total population (2004) [§§]	1,308 million	1,087 million
Adult literacy rate (%, 2004)[§§]	90.9	61.0
Tertiary students in science, maths and engineering (% of all tertiary students)[§§]	53	25
Researchers in R&D per million people (1999–2003)[§§]	663	119

Sources: [*]itworld.com (2007); [†]*People's Daily* (2006); [‡]Xinhuanet (2007); [¶]rediff.com (2004); [§]InternetWorldStats.com (2008); China Economic Net (2008) for February 2008 data for China; [**]David (2008) for India; Liang (2008) for China; [††]UNDP (2002); [‡‡]UNDP (2001); [¶¶]UNDP (1997, 2006); [§§]UNDP (2006); [***]West (2002); [†††]Calculated from *Euromonitor* (2007); [‡‡‡]UNDP (2004b); [¶¶¶]Einhorn (2006b)

locus of the future gap. While China has been simultaneously working to improve its skill base – its major competitive disadvantage compared with India – the structural shifts in terms of market, technology, product, actors and focus are likely to have a significant influence on the magnitude and direction of the future gap. Analyses that do not consider such factors are likely to be too simplistic and cannot make a meaningful prediction.

Forces of structural changes in an industry

Different theoretical contributions and various empirical studies have led to the accepted view that dynamic and cumulative processes lead to an industry's structural changes and evolution (Hamilton, 1919: 315; Veblen, 1959: 132–3; Siu, 2006). A range of institutional actors play a prominent role in the evolution process. As to the dynamic nature of ingredients related to the evolution, Hamilton (1919: 315) has observed that 'not only are things happening to them, but changes are going on within them'.

A second point to bear in mind is that the processes influencing structural change and evolution within an industry are cumulative. This is because:

> the situation of today shapes the institutions[1] of tomorrow through a selective, coercive process, by acting upon men's habitual view of things, and so altering or fortifying a point of view or a mental attitude handed down from the past. (Veblen, 1959: 132–3)

Put differently, in a given period, the structure and performance of an industry and its interrelationship with its superset – the economy – are the results of the complex interactions of social, economic and political factors of the preceding periods (Schmid, 1987).

The industrial transformation processes are associated with 'persistent external attractors and responsive internal actors' which are interdependent (Schmid, 1987; Parto, 2005: 41). In this regard, economic agents – the internal actors – are 'creative interpreters of messages' from the environment, which is composed of external actors. The economic agents' creativity is 'deeply intertwined with the processes of investment and innovation' (Metcalfe, Foster and Ramlogan, 2006: 29).

Prior researchers have noted the important roles of social, economic and political factors in bringing about structural transformation in industry. Kalafsky (2006) has documented how factors such as demographic changes and availability of workforce brought structural changes to Japanese manufacturing. Similarly, in China, the complex interactions of the newly emerging market forces and the vestiges of the planned economy have brought organisational and managerial changes (Chan, 2006).

Some researchers (e.g. David, 2000; Malerba, 2006) have introduced the concept of co-evolution to conceptualise and explain structural changes, transformation and evolution within industries. Co-evolutionary processes entail knowledge, technology, actors, demand and institutions, and tend to be path-dependent (David, 2000). To understand and predict structural changes in any industry, we thus need to take into account factors such as the co-evolution of products and processes, inter-firm cooperation and other organisational dynamics (Malerba and Orsenigo, 1996; Sornn-Friese, 2005).

Structural shifts in the Chinese software industry

The above observations point to the fact that like other economic sectors, the software industry realigns and adapts to major shifts in technology, markets and other external factors (Kumerasan and Miyazaki, 2001). The Chinese software industry is undergoing adaptation and realignment in response to structural changes in the environment. We analyse five major structural shifts in the Chinese software industry. They are related to the market (deepening the penetration in the domestic market); technology (with an emphasis on open source software); product (increased bundling of Chinese software with hardware and foreign corporations' software); focus (increasing collaboration with multinational corporations and regional economies and focus on core competence); and actors (emergence of new powerful players).

Market shift: deepening penetration in the domestic market

A fundamental structural shift in the Chinese software industry can be explained in terms of co-evolutionary processes (David, 2000) related to the stimulation of domestic demand and degree of orientation of Chinese software firms towards the domestic market.

Estimates of the sizes of the two economies' domestic software markets vary widely across different sources. Based on various proxies, such as electronic industries and penetration rates of major technologies including fixed and mobile telephone lines, PC and internet, however, it can be assumed that domestic software demand in China is much higher and growing at a much faster rate than India (Table 7.1). In 2002, the Chinese PC business software market grew by 20 per cent to reach US$2 billion. The Chinese electronic industry exceeded US$130 billion in 2002 (20 per cent of the world market) (EIU ViewsWire, 2003b) and was estimated to exceed US$300 billion in 2006 (In-Stat, 2006a). India's electronics industry, on the other hand, amounted US$12.3 billion in 2002 and is estimated to reach US$35.6 billion in 2010 (Carbone, 2004). China is already the world's second biggest PC market and is expected to become the biggest by 2010.

Recent institutional changes are stimulating domestic software demand in China. China's new piracy law has increased domestic opportunities for software companies. A recent study estimated that 82 per cent of software used in China is pirated (Fenton, 2008). As a World Trade Organization (WTO) member, China is compelled to adhere to the Trade-Related Aspects of Intellectual Property Rights (TRIPS) agreement. The TRIPS agreement requires China to provide adequate legal and enforcement tools to prevent piracy. Unlike other new members, China is subject to an annual review of its WTO obligations for the first 10 years of its membership. Many countries are likely to put pressure on China to act on intellectual property infringement. Under new laws, buyers of pirated goods are fined and manufacturers are likely to face jail time and confiscation of their equipment.

More importantly, Chinese policies are increasingly gravitated towards domestic firms. Although China's domestic software market has been so far dominated by foreign companies, domestic vendors are rapidly capturing market share. In 2003, domestic companies captured 30 per cent of the Chinese software market (EIU ViewsWire, 2003b). By 2010, China hopes that 60 per cent of the domestic software market will be captured by Chinese firms (Hale and Hale, 2003).

The size of the government's software consumption and recently enacted local procurement laws favour domestic players. In all developing countries, the government is the single biggest user of IT products (Nidumolu et al., 1996) and the Chinese government is much ahead of other developing country governments in the adoption of modern IT. A significant part of China's informatisation strategy is the implementation of e-government programmes. For instance, China ranked seventh among 198 countries in terms of e-government performance in 2002. In terms of IT adoption, the Chinese government is rated ahead of global high-tech leaders such as Switzerland, the UK, Singapore and Germany (West, 2002).

The Government Procurement Law enacted in January 2003 requires government departments to procure domestic goods and services where possible. In early 2002, even before the existence of such a law, out of seven government software contracts, six went to Chinese vendors. Despite China's accession to the WTO, government procurement is excluded from the scope of the WTO's multilateral trade rules, meaning that China is not obliged to open government procurement to foreigners.

Chinese policies are thus geared towards favouring domestic software manufacturers (Ackerman, 2006). Fear about dependence on foreign countries, combined with a sense of national pride resulted in the principle of national self-reliance during the Mao Tse Tung era. Since then, this principle has been guiding the Chinese economic system. Although attitudes toward technology imports and foreign investment have changed drastically, a high level of advocacy for national self-reliance and domestic development of technology still exists among Chinese policy makers, researchers and scientists, and more so among military leaders. For instance, China is the world's largest manufacturer of DVD players. By adopting its own technology, it can save $2 billion a year in royalties that would otherwise be paid to an 18-company consortium (Calbreath, 2004). There is a strong determination to achieve self-reliance and reverse the flow of such fees by exporting Chinese standards.

The developed domestic market has also provided additional benefits to the players in the Chinese software industry. In contrast, the major disadvantage faced by Indian IT companies is soft demand from local customers.

Technology shift: investment in disruptive innovation

Another distinguishing feature of the structural shift of the Chinese software industry is the increased emphasis on the development and deployment of open source software (OSS). OSS development was the only software project on the government's list of top technology priorities in 1999. Similarly, in 2004, Linux internet server software and Linux mobile phone software were among the 19 projects given financial support from the fund established by the State Council in 1986 to encourage research and development in IT.

No other countries come even close to the level of advancement China has achieved in the deployment of OSS, particularly Linux. OSS satisfies various characteristics of discontinuous or disruptive innovation in the software industry. Having the edge in OSS is thus likely to result in Chinese firms emerging as powerful players against the incumbents in the global software industry.

Despite initial inferior performance, disruptive innovations tend to be 'cheaper, simpler, smaller, and more convenient to use' (Christensen, Raynor and Anthony, 2003). These characteristics seem to be exactly what OSS possesses. While OSS has experienced some initial problems related to compatibility issues with user skill and business partner technology, it has many disruptive features. For instance, OSS has a lower initial cost and slower obsolescence of basic infrastructure features and thus a lower total cost of ownership (TCO). Second, the amenability to modification makes localised customisation easier. Third, OSS communities provide supportive environments for transition. Fourth, OSS has high levels of compatibility and portability for old and used hardware (Kshetri, 2004). Such characteristics are likely to establish an entirely new performance trajectory for OSS.

The scale of OSS development and deployment in China is large enough to be noticed at the global level. Red Flag's Linux applications are already used in China's aircraft, weapons systems, vehicles, industrial equipment and consumer devices (Calbreath, 2004). Thanks to OSS, the Chinese military has been able to use supercomputers that have been domestically developed. Dawning, a spin-off from the Chinese Academy of Sciences, has developed Dawning 4000L, a supercomputer with over 3 tera floating operations per second (TFLOPs) or 3 trillion calculations per second computing power. Dawning 4000L is based on a Chinese-designed Linux operating system.

Nearly all major global OSS players including HP, Intel, Sun, Oracle and IBM have significant operations in the Chinese Linux industry. In November 2003, Sun announced a deal to sell 200 million copies of a Linux-based Java Desktop System to the government. Multinational corporations (MNCs) have collaborated with Chinese partners to get a slice of the exponentially growing Chinese Linux market.

Disruptive innovations either create new markets by targeting non-consumers or compete in the low end of an established market. Given limited computerisation in developing countries, most firms and individuals are currently non-consumers of software. Developing countries have historically been unable to access the high-technological products because they lacked skills and wealth. Software demands in these markets are typically small and hence not attractive for large companies such as Microsoft, IBM, Oracle and SAP. The disruptive and discontinuous characteristics of OSS as noted above are likely to make it attractive for developing countries. As in other disruptive innovations, the incumbents (e.g. Indian software firms and software giants from developed countries) may lack the ability to play the new game in the field of OSS (Christensen, Raynor and Anthony, 2003). What is more, with the creation of new consumption and the improvement in innovation, perpetrators of OSS may also attract incumbents' customers.

Actor shift: emergence of new players

Chinese software companies are generally much smaller than Indian firms (Table 7.1). For instance, China's largest company, Oriental Software, has about 1,300 employees compared with 26,000 at India's Infosys and 24,000 at TCS. Over half of China's software houses have fewer than 50 employees (Leung, 2003). Typical players in the Chinese software industry thus suffer from an absence of management skills and a lack of experience working on sophisticated large-scale projects (EIU ViewsWire, 2003b).

An important structural shift in the Chinese software landscape has been the emergence of new and powerful players. In August 1999, the Chinese government established Red Flag Linux with the backing of the Chinese Academy of Science. Beijing Software Industry Production Center was similarly a group established by the government to organise OSS development. Building on Red Flag and Cosix Linux, and coordinating the efforts of 100 software engineers across 18 organisations, Yangfan Linux was launched by the centre in early 2002 and was installed on 2,800 government computers in Beijing during its first six months. Yangfan is based on versions of Linux developed by Red Flag and the China Computer Software Corporation. The centre also developed office applications and other Linux-based software. In early 2004, the government selected 29 domestic firms and encouraged them to export software to USA and Europe (Baijia, 2004).

Focus shift: collaboration and concentration on core competence

Chinese companies have collaborated with MNCs and regional economies in various software projects. Such collaboration has promoted synergistic advantages and allowed Chinese firms to focus on core competence. For instance, Red Flag has collaborated with TurboLinux in the introduction of Chinese-language versions of Linux; with Miracle Linux to develop Linux versions of Oracle's database software and Asianux (a just-for-Asia Linux); with HP importing HP's Integrity and Proliant servers to Red Flag's Linux; with LinuxLab of the USA in the development and marketing of software; with Eforce, Culturecom and Mobile Telecom in the development of the Chinese 2000 Mobile Linux Operating System for handheld devices; and with Oracle in the development of Red Flag Data Center OS4.0.

Similarly, the Chinese Academy of Sciences, Chinese Software Institute launched Hongqi Linux – an all-Chinese operating system – and its application solutions with Compaq in August 1999; and Chinese Penguin64 with Singapore's Donovan Systems in 2000. Likewise, Motorola collaborated with Lineo, a developer of embedded Linux, and Caldera, a provider of Linux for commercial application. IBM has an alliance with Kingsoft, China's biggest office-automation software vendor, to develop desktop Linux applications. Red Flag is also working with companies like HP for the development of OSS products for international markets.

China has also identified regional economies which have similar missions and has collaborated with them in a number of software projects. The Japan-China-Korea (JCK) open-source alliance, announced in November 2003, is such an initiative to promote OSS by co-sourcing. The JCK initiative is guided by the Japanese IT Services Industry Association, the Chinese Software Industry Association and the Federation of Korean Information Industries. The three associations have over 1,000 corporations as their members, including nearly all the major players in the Japanese and Korean IT industry. The JCK partnership is working on open source business models, standardisation of software, and training software engineers. There has been a division of labour in the partnership: China will develop PC operating systems, Japan will focus on software development and security, and Korea will develop software for PDAs (Krikke, 2003). They are setting up a database to coordinate efforts and avoid duplication.

Product shift: bundling with hardware and software

As noted above, China's high-tech sector is far ahead of India's and has a wider global presence. However, while India's software exports represent 70 per cent of the country's overall IT revenue, Chinese software exports represent only 10 per cent (Verma, 2005). China's high-technology exports as a proportion of manufactured exports totalled 19 per cent in 2001 compared with India's 4 per cent. In 2001, China exported over US$24 billion of IT hardware compared with India's US$1 billion. China's high-tech exports grew by 25.4 per cent in 2001 to reach US$46.5 billion. China can capitalise on its leadership in the electronic industry and collaboration with global software giants to add value by bundling its software with hardware and with software manufactured by foreign companies. Given that OSS is being widely used in devices ranging from cars to coffee pots and supercomputers to cellphones, value addition by bundling seems promising for the Chinese software industry. The following are some examples:

- *Bundling with cellphones*: For instance, the Chinese TD-SCDMA mobile standard is the only 3G standard of the Time Division Duplex (TDD) mode approved by the International Telecommunications Union (ITU), and it uses the 3G frequencies allocated to TDD by ITU.

TD-SCDMA is more suited for use in dense cities, such as in developing countries. Estimates suggest that it will capture a significant share of the global market following its launch in 2005. In July 2004, Datang Telecom Technology, the intellectual property owner of the TD-SCDMA, confirmed that it had selected Linux as the operating system for its 3G handsets.

- *Bundling with PCs*: The Chinese desktop PC industry is advanced compared with that in other developing countries. Legend (now Lenovo), a Chinese firm, is the market leader in the Chinese desktop PC market. According to International Data Corporation, in 2003, Lenovo's market share in China was about 30 per cent. In the last quarter of 2007, Lenovo's share in China's PC market was 29 per cent (Business Wire, 2008). Players like Lenovo can enter into the international computing market and sell PCs that use OSS operating systems and applications. LG, a Korean MNC, has successfully employed such a strategy. LG is exporting Pentium 4 PCs to India that are low-priced because it uses Red Hat 8 Linux as the operating system. Low price has boosted the demand for LG's PCs in India.

- *Bundling with other software*: Bundling software manufactured by Chinese companies with that of MNCs will be beneficial to both parties. In OSS, the demands of some segments are boosted by the success of complementary programmes; the expertise of MNCs in such segments boosts demand for their offerings (Lerner and Tirole, 2002). There has already been such bundling for the domestic market. For instance, IBM signed a marketing agreement with Red Flag to bundle IBM DB2 Express Database Software with Red Flag Linux, which will be used for small and medium-sized businesses.

Overcoming competitive disadvantage

China is also finding ways to overcome its competitive disadvantage by removing the skills bottleneck. Despite its lack of human resources and competence in the software sector, China is considerably ahead of India in terms of basic indicators such as concentration of researchers in R&D, literacy rate, and proportion of tertiary students in maths and engineering (Table 7.1). In addition, the Chinese government has formulated policies to attract Chinese science and technical manpower living abroad and turn it into a national competitive advantage. Among the foreign-born scientists and engineers working in the USA, more originate from China than from any other country. Thanks to increased domestic opportunities, a higher proportion of Chinese engineers trained in advanced countries are returning to China.

On the linguistic front, China has a large population with good command of Asian languages such as Japanese and Korean. In 2006, China was the destination for over 60 per cent of Japan's outsourced software businesses (Xinhuanet, 2007). They can overcome poor English with training and experience (Einhorn and Kripalani, 2003b). English is taught in Chinese schools and college students are required to pass exams on English proficiency in order to graduate. In 2002, 20 million people were undergoing English language training in China.

To combat its lack of high-level international quality certification, China is also making attempts to enhance its technological skills. For instance, the Chinese government is providing incentives for software firms that attain Capability Maturity Model (CMM) level 3 or higher (Leung, 2003). Likewise, in an attempt to foster the OSS skill base, China has recently made OSS a required course in 35 universities and 35 prevocational schools.

Conclusion: where is the future leading?

India has outperformed China in terms of skill base and has an established track record and a solid customer base. For instance, over one-fifth of Fortune 1000 companies outsource their software requirement to India. Moreover, the USA and EU, India's major software destinations, have a much higher level of software demand than Japan, China's major software destination (Table 7.2).

In contrast to India's one-dimensional focusing on developing its skill base (e.g. number of CMM level 5 companies and the size of the technical manpower pool), China has been undergoing major structural changes and a multidimensional development. The structural shifts have the potential to dramatically reshape the Chinese software industry and change the competitive landscape.

The magnitude and direction of the future gap between the performances of the software industries of the two nations are thus functions of the degree of disruptiveness of OSS; the success of collaborative efforts with MNCs and at the regional level; the capacity to develop a domestic technological base (e.g. success of TD-SCDMA); the international performance of the Chinese electronic industry; and domestic growth of the Chinese software market.

Table 7.2 Market sizes of major software destinations of China and India

Economy	Software category	Market size (US$m)	Remarks
USA (2004)	PC business Networking Home, leisure and entertainment PC Multimedia Pre-recorded music and video	9,010 4,300 9,200 7,700 34,100	The USA is the world's largest software market and accounted for half the world software sales in the last decade. Moreover, US IT and financial services companies have geared up much faster than their European counterparts to exploit the advantages of offshore programming. Chinese firms are rapidly making inroads to the USA.
UK (2004)	PC business Home, leisure and entertainment PC Multimedia Networking Pre-recorded music and video	1,690 757 825 1,022 4,900	Among the Western European countries, the UK is India's largest software destination. Software demand from the UK has been facilitated by the historical links between the two countries in addition to India's proficiency in the English language.
Japan (2003)	PC business	7,700	Language and cultural barriers are hindering the growth of India's exports to Japan.
South Korea (2003)	PC business	189	

Source: Calculated from *Euromonitor* (2007).

Attacking competitors' markets

The major software destinations for China and India are culturally dependent. The USA and the UK are India's major markets, thanks to its English language competence (Tables 7.1 and 7.2). On the other hand, China's cultural similarity with Japan and South Korea has facilitated its software export business. Hong Kong based enterprises are also outsourcing software to China thanks to geographic proximity and political links. China's market (Japan), albeit small, is much safer for Chinese software firms. China's engagement in regional collaborations such as the JCK initiatives will also allow it to work with its customers and further strengthen the market.

Compared with Indian software giants such as Infosys and Tata Consulting Systems, Chinese companies are much smaller and less known in the global market. Language and cultural barriers compound the problem for Chinese firms in the US and EU market. As mentioned earlier, only a few firms selected by the government are focusing on the US and EU market.

Although China, according to an IDC study, is the fourth-favourite software outsourcing location for US companies after India, Canada and Ireland (Baijia, 2004), software sales by Chinese firms to the US market are a fraction of those by Indian firms. In 2003, only 18 per cent of China's software outsourcing sales came from the USA (Baijia, 2004). Chinese companies are focusing on US-invested companies in China rather than directly competing with Indian firms (Baijia, 2004). Nonetheless, as the USA and EU countries are major buyers of China's hardware products, the structural shifts discussed above can facilitate the penetration of software in these economies.

The developing world markets

The software industry is growing rapidly in developing countries. Although Indian software firms export to over 95 countries, less than 10 per cent of these exports go outside the USA, EU and Japan (Tables 7.1 and 7.2). India thus has a very limited presence in developing countries. China's electronic exports in developing countries and increasing bundling with hardware will help boost growth. Moreover, almost all major developing countries in Asia, Africa and Latin America are rapidly gravitating towards OSS and have high-profile national OSS projects. They are thus likely to favour OSS over proprietary software. Superimposition of lower TCO of OSS onto China's increasing dominance in the global IT industry therefore allows China to significantly alter the dynamics of competition in these economies.

Note

1. As noted earlier, 'institutions' are defined as macro-level 'rules of the game in a society' (North, 1990: 3), or 'a set of socially prescribed patterns of correlated behavior' (Bush, 1987: 1076).

8

Diffusion of open source software: institutional and economic feedback

The scale of open source software (OSS) development and deployment in China is large enough to be noticed at the global level. Nearly all major global OSS players including HP, Intel, Sun, Oracle and IBM have significant operations in China and have collaborated with Chinese partners to get a slice of the exponentially growing market. In June 2004, China's Linux-based 'Dawning 4000A' was the world's tenth fastest supercomputer. In 2003, 89 per cent of Linux's Asia-Pacific revenue outside Japan came from China. OSS arguably exhibits a 'private-collective' innovation model, a hybrid between the 'private investment' model and the 'collective action' model. The 'private' component entails availability of opportunities for private investments and returns in innovations related to OSS. The 'collective' component is related to the public good nature of OSS. Opportunities for such innovations vary across economies. This chapter examines how the two components of the OSS innovation model fit with unique institutions, enterprises and the markets of socialist-capitalist China. The Chinese way of transition to market economy is significantly different from those of Eastern Europe and the Soviet Union. While its radically reformed enterprises and markets resemble those of capitalist countries, China has not dismantled its political and social institutions. The newly created industries and markets under the Chinese dual-track approach of reform favour the 'private' component of the OSS innovation model. Yet the unaltered political institutions and the Chinese Confucian heritage have provided fertile ground for the 'collective' component. The chapter argues that the rules of the game defined by institutions in China and the newly created enterprises and markets have provided near optimum incentives for the development of OSS.

Introduction

China has become a global economic and technological powerhouse. It is fast emerging as a main rival for industrialised countries in high-technology industry and has the potential to set global standards in its fields of expertise (Vogelstein et al., 2004). The Chinese open source software (OSS) industry is one of the highly visible examples to illustrate this trend. The scale of development and deployment of OSS in China is large enough to be noticed at the global level. Nearly all major global OSS players, including HP, Intel, Sun, Oracle and IBM, have significant operations in the Chinese OSS industry and have collaborated with Chinese partners to get a slice of the exponentially growing OSS market.

There are many types of OSS available. From the Chinese perspective, Linux, the flagship of OSS (Applewhite, 2003), deserves special attention. Indeed, Linux has been a cornerstone of the Chinese software industry. In November 2003, Sun announced a deal to sell 200 million copies of its Linux-based Java Desktop System to the Chinese government. In the second half of 2003 alone, Linux sales in the Chinese PC market exceeded 800,000 copies (*SinoCast China*, 2004b). In June 2004, China's Linux-based Dawning 4000A was the world's tenth fastest supercomputer. In 2003, 89 per cent of Linux's Asia-Pacific revenue outside Japan came from China (Andrews, 2003). China's Linux market grew by an annual rate of 27.1 per cent in 2005 (Global News Wire – Asia Africa Intelligence Wire, 2006) and the Chinese IT research firm, CCID Consulting, predicted a compound annual growth rate (CAGR) of 49.3 per cent during 2005–09 for the Chinese Linux server software market. Likewise, according to International Data Corporation (IDC), the CAGR of the Chinese Linux market for 2006–10 will be 34 per cent.

Many Chinese technology players are embracing OSS enthusiastically. According to CCID Consulting, the Chinese Linux market reached RMB200 million (US$28 million) in 2007, which was 20 per cent higher than in 2006 (PR Newswire, 2008b). China Mobile is among the 33 technology companies led by Google supporting the development of a standardised open source mobile operating system (Palenchar, 2007). Likewise, in 2006, Chinese telecom manufacturer, ZTE announced an advanced Linux-based smart-phone project in China (Schillings, 2006).

OSS possesses many unique characteristics. Programmers and companies engaged in the development of open source processes have exhibited 'startling' behaviour, at least to economists (Lerner and Tirole, 2002: 2). Of the two models of innovation available in the literature – the 'private investment' model (PIM) and the 'collective action' model (CAM) – the OSS innovation model is not conceptually akin to any of them. Von Hippel and von Krogh (2003) have documented how the OSS innovation model needs to be seen as a hybrid between the PIM and the CAM. They show that the OSS innovation model entails two components: the private component (e.g. availability of incentives for private investment in OSS-related products) and the collective component (e.g. OSS has several characteristics of a public good).

Institutionalists have recognised that the success of an innovation is tightly linked to the context provided by institutions in an economy (Storper and Walker, 1989; Sabel and Zeitlin, 1997). Institutions provide the 'cognitive, normative, and regulative structures' (Scott, 1995: 33) that determine institutional preference for an innovation. The rapid diffusion of OSS in China raises an important question: how have Chinese institutional and market conditions contributed to that growth?

Theoretical models of the effect of institutions, industries and markets on the unique OSS innovation model are not well developed. Building on the literatures on OSS and institutional theory, we investigate OSS diffusion in China and propose a model that explains why China has moved quickly and massively to OSS deployment. The proposed model combines mainly the hybrid characteristics of OSS innovation with the Chinese institutional, industry and market contexts.

This chapter proceeds as follows: the next section examines the innovation model associated with OSS. This is followed by a discussion of China's unique transition to market economy and incentives for various types of innovations. We then propose a model that explains institutional, industry and market preferences for OSS in China. Finally, we provide discussions and implications.

OSS as a hybrid between the private investment model and the collective action model

The existing literature has two mutually exclusive models of innovation – the 'private investment' model' and the 'collective action' model. The former assumes that innovations are supported by private investment with expectations of private returns (Demsetz, 1967; von Hippel and von Krogh, 2003). Private returns typically in the forms of patents, copyrights and trade secrets are considered to be a rational justification for offering incentives to investors (Arrow, 1962; von Hippel and von Krogh, 2003).

The CAM applies to public goods. A public good is an economic good characterised by non-rivalry (zero marginal cost for additional users) and non-excludability (once created, it is impossible or at least prohibitively difficult to exclude a certain group of users from gaining access to it) (Olson, 1967; von Hippel and von Krogh, 2003). Under such a model, contributors do not control the knowledge they have developed and supply it to the public unconditionally (von Hippel and von Krogh, 2003). The CAM thus avoids the social loss problem associated with the restricted access to knowledge in the PIM. A key concern for such a model, however, relates to motivating potential contributors to contribute to the project (von Hippel and von Krogh, 2003).

The OSS innovation model deviates from the PIM as well as the CAM (von Hippel and von Krogh, 2003, Table 8.1). The first deviation is that in OSS, users rather than manufacturers are innovators. The typical 'innovators' in OSS are not employees or contract workers and hence do not receive direct compensation for their contributions (Markus, Manville and Agres, 2000). Manufacturers therefore find indirect ways to profit. For instance, LG, a Korean MNC, exports Pentium 4 PCs to India. Because it uses Red Hat 8 Linux as the operating system, it can afford to sell these computers for a low price; in turn, the low price has boosted demand for the product. Second, innovators freely reveal the software that they have developed at their own expense, violating an assumption of the PIM. The OSS innovation model also deviates from the CAM in terms of the nature of project goals; recruitment of contributors; group size; and how users are obliged to contribute to the project (Table 8.1).

Von Hippel and von Krogh (2003) argue that a hybrid model between the PIM and the CAM is ideal for OSS. The innovation model associated with OSS thus has two components. First, many companies in the software industry are making huge profits under a variety of business models, capitalising on the success of OSS (Markus et al., 2000). There are thus incentives for private investment in the model – the private component. Second, just like a voluntary organisation, OSS is 'licensed as a public good' or is given away for free (Markus, Manville and Agres, 2000) and for this reason collective action is needed in the innovation process – the collective component.

Table 8.1 PIM, CAM and characteristics of Linux projects in China

Assumption	Deviation (von Hippel and von Krogh, 2003)	Characteristics of Linux projects in China
PIM		
Innovation takes place by software manufacturers' private investment with expectation of returns (Demsetz, 1967)	Typical OSS innovators are software users rather than manufacturers	In terms of individual users vs. manufacturers, China's tech skill gap is lower than in other developed countries
Free revealing of software code will reduce an innovator's profits (Audretsch and Feldman, 1996)	Open source innovators freely reveal software developed at their private expense	There are reports that Chinese programmers are not revealing the OSS codes they have developed[1]
CAM		
Project goals matter (McPhail and Miller, 1973; Snow et al., 1980; Benford, 1993)	Goal statements of successful OSS projects are varied: technical and narrow to ideological and broad – and from precise to vague	Chinese open source projects have broad goals: to achieve self-reliance and national pride
Effective recruiting of contributors is critical for the project to succeed (Benford, 1993; Taylor and Singleton, 1993)	OSS projects typically do not actively recruit	In China, there have been coordinated government efforts to recruit contributors; for example, 100 software engineers from 18 organisations were recruited to launch Yangfan Linux
Smaller groups are likely to be more effective because incentives can be tailored (Olson, 1967)	OSS projects have become successful even in large groups	
It will be effective to convince participants not to free-ride (Schwartz and Paul, 1992)	OSS projects take no measures to discourage free-riding	

CAM, collective action model; PIM, private investment model; OSS, open source software.

The transformation of institutions, enterprises and markets

Close linkage between and among economic and technological agents (Clark, 2002), especially inter-firm linkages (Cooke, Uranga and Etxebarria, 1997; Nooteboom, 1999; Avermaete et al., 2003) foster innovations. A primary weakness of socialist economies is the lack of such linkages and an innovation system that supports the coordination of firms' economic and technological efforts (Kogut and Zander, 2000). Before 1978, China severely lacked such a system. There was no formal mechanism for diffusion of technology. Moreover, advanced research was brought to a standstill during the Cultural Revolution (1966–76), a critical period in the development of the global electronics industry (Zita, 1991).

The economic and political reforms began in 1978 after Deng Xiaoping came into power, and modernisation of the science and technology sector became one of the dominant goals of the Chinese Communist Party (Wang, 1993). Deng emphasised institutional linkages to foster innovations. He initiated reforms that facilitated the diffusion of technology and endorsed the concept of 'technology as a commodity', which led to the formation of a large number of research institutes. By 1987, there were over 5,000 state-operated research units employing more than 230,000 scientists (Baark, 1988). The transfer of technology from state-operated research units to industrial enterprises fuelled rapid technology diffusion. Under the model, technology was, however, treated like a purely public good and technology modernisation attempted to absorb Western technology without culture (e.g. private investment) (Zita, 1991).

Countries that rely on central planning tend to misallocate investment, hindering innovations (*Economist*, 2002b). Put differently, capitalism and the innovative capability of a nation co-vary positively. Since the early 1980s, China has started moving toward the capitalist end of the socialist–capitalist continuum (Schneider, 1999).

Under the 'market track' of the Chinese 'dual-track approach', economic agents were allowed to participate in the market at free-market prices, provided they fulfilled their social obligations (Lau, Qian and Roland, 2001). The incentives for private investment provided by the 'dual-track approach' also attracted a large number of foreign MNCs. During 1978–2003, multinational companies invested over US$400 billion in China, of which US$53 billion was invested in 2003 alone (Farrell, Gao and Orr, 2004). Favourable government policy in China has attracted a large number of electronic companies. The Bureau of Industry and Security cites a World Technology Evaluation Center report:

> In fact, all US electronics companies are increasing their Asian [especially Chinese] investments in R&D to take advantage of favorable industrial-government partnerships and engineering workforces that are highly motivated and well trained (frequently in the United States). (Bureau of Industry and Security, 1998a)

China's transition to market economy has followed a trajectory significantly different from that in Eastern Europe and the Soviet Union. Reform in China has been described as 'gradual', 'partial' and 'experimental' compared with that in Eastern European countries, which has been described as 'rapid', 'comprehensive' and 'big-bang' (Qian, Roland and Xu, 1999; Kalthil and Boas, 2003; Jones, 2004). China has undergone a transition to market economy without any 'alteration in political structure' (Solinger, 1995: 27). Nee describes:

> China's transition from central planning has assumed a trajectory quite different from that of Eastern Europe and the Soviet Union. Whereas Eastern Europe and the Soviet Union rejected communism for Western-style democracies and initiated rapid state-guided transitions to market economies, China has steadfastly refused to carry out reform of its political institutions and has fixed its course to remake the economic institutions of state socialism not by revolution but by reform. (Nee, 1992: 1)

Similarly, Overholt argues:

> One central difference between China's approach and the East European approach has been construction versus destruction. Instead of focusing on the destruction of socialist institutions, China has concentrated on the construction of market institutions. (Overholt, 1994: 29)

China's unique route to market reform is exactly what makes the country suitable for OSS development. Unique institutional, industry and market characteristics have also separated China from other countries in terms of the characteristics of OSS projects (Table 8.1). The radically reformed industry and market have encouraged private investment in innovations related to OSS. The 'market track' has also resulted in a shift of emphasis from 'quest for knowledge' to 'search for practical applications' (Zita, 1991). Yet policies that are reminiscent of the Chinese communist public goods regime (Solinger, 1995, 127) and the Chinese Confucian heritage are compatible with the public good nature of OSS. In sum, while there was no incentive for any type of innovation in China prior to 1978, its unique locus to capitalism has encouraged private investments as well as collective actions to foster innovations.

Drivers of OSS diffusion in China: institutions, industry and market

Adams (1996) argues, 'like fire, technology depends on its environment to flare or die'. It is thus important to ask what factors have helped OSS to prosper in China? Figure 8.1 presents the proposed model depicting factors influencing a country's propensity to adopt OSS. The left part of

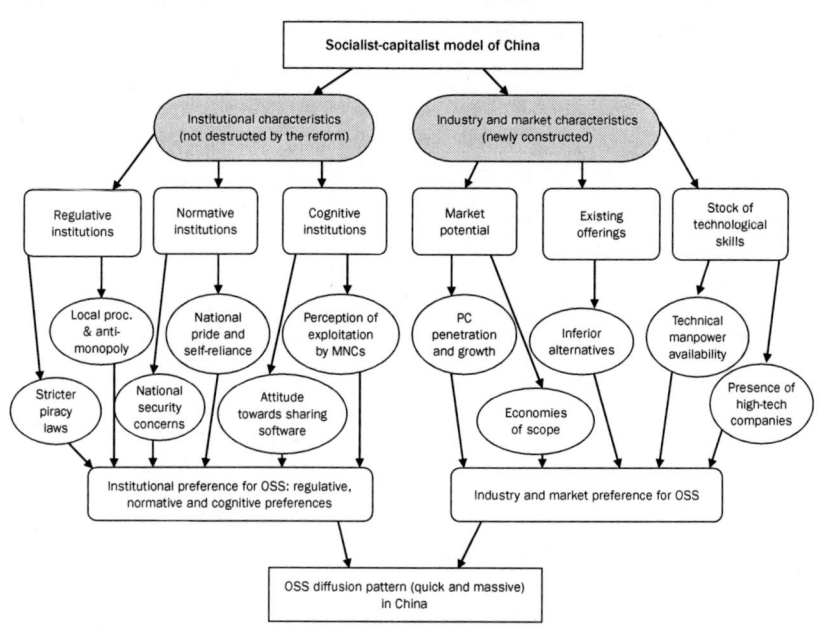

Figure 8.1 Institutions, industry and market related factors influencing OSS diffusion pattern in China

the model includes institutions: the 'cognitive, normative, and regulative structures' (Scott, 1995: 33) that act as the macroeconomic rules of the game for players in the computing industry. Roughly speaking, these are Chinese political and social institutions that have not been destroyed under the 'gradual' and 'partial' approach to market reform (Overholt, 1994; Qian, Roland and Xu, 1999). The right part represents different components of industry and market-related factors impacting the structure of demand and supply. These factors are the outcomes of the 'market track' of the Chinese 'dual-track approach' to reform (Lau, Qian and Roland, 2001).

Institutional factors

The three institutional pillars proposed by Scott (1995, 2001) – regulative, normative and cognitive – are employed to examine institutional processes influencing the diffusion of OSS in China.

Regulative institutions and OSS diffusion

The 2003 local procurement law, in addition to stricter piracy laws and the anti-monopoly and anti-trust laws currently under consideration provide regulative preference for OSS in China.

Local procurement and anti-monopoly laws

The government procurement law enacted in January 2003, requires government departments to procure domestic goods and services where possible (Ebusinessforum, 2004). In early 2002, even before the law was enacted, out of seven government software contracts, six went to Chinese vendors. Although China has now joined the World Trade Organization (WTO), government procurement is excluded from the scope of multilateral trade rules governing the WTO. Thus, China is not obliged to open its government procurement to foreigners. Estimates suggest that the government accounts for 25 per cent[2] of the Chinese software market (*Financial Express*, 2004). The local procurement requirement thus gives a major boost to the growth of the Chinese OSS industry. Moreover, a report by the Chinese State Administration for Industry and Commerce in mid-2004 argued that anti-monopoly and anti-trust laws were urgently needed (*Far Eastern Economic Review*, 2004). The report specifically cited Microsoft's dominance in the Chinese software market.

Stricter piracy laws

Estimates suggest that over 92 per cent of the software used in China was pirated in the early 2000s (Kessler, 2004), and that this had reduced to about 82 per cent by early 2008 (Fenton, 2008). The high rate of piracy has worked against the diffusion of OSS in China. As Tschang (2008) notes regarding the slow diffusion rate of Firefox, the open source browser, in China, 'with pirated copies of Windows XP or Vista selling on the street for less than US$2, there is little economic incentive for Chinese internet users to download Firefox'.

As a WTO member, China is compelled to adhere to the Trade-Related Aspects of Intellectual Property Rights (TRIPS) agreement. The TRIPS agreement requires China to provide adequate legal and enforcement tools to prevent piracy. Unlike other new members, China is subject to an annual review of its WTO obligations for the first 10 years of its membership. Many countries are likely to put pressure on China to act on intellectual property infringement. China has recently enacted new piracy laws. Under the new laws, buyers of pirated goods can be fined 5–10 times the value of the goods and manufacturers face jail time and equipment confiscation (Kanellos, 2002a). An earlier attempt by the Chinese government to mandate a nationwide switch from Windows to Red Flag Linux failed (Raymond, 2004). China's requirement to comply with TRIPS, however, has made the relative value proposition of OSS stronger.

Normative institutions and OSS diffusion

National security concerns and the principles of self-reliance and national pride are common features of Chinese institutions that inform decision-makers' normative preference for OSS, thus leading to pro-OSS policies.

National security concerns

A product of China's transition to capitalism was the evolution of the People's Liberation Army (PLA) as a major player in the Chinese computing industry.[3] For the PLA and the government, maintaining national security is a vital concern. This is highlighted in an editorial on 'information colonialism', published in the *People's Liberation Army Daily* in February 2000:

> Without information security, there is no national security in politics, economics and military affairs. While learning from others, China should not be under their control. (Goad and Holland, 2000)

An article published in *China Economic Times* on 12 June 2000 describes how the then Chinese vice minister of science and technology, Xu Guanhua, considered high-technology to be a key factor in national security (China.org, 2000). Guanhua reportedly said that developed countries have put many high-tech arms into actual battles and discussed the likelihood that technology-exporting countries might have installed software for 'coercing, attacking or sabotage'. In particular, the Chinese government has suggested that Microsoft and the US government spy on Chinese computer users through secret 'back doors' in Microsoft products (Mahlow, 2003).

OSS has no undetectable vulnerabilities and, for this reason, it is perceived as having a clear advantage in terms of military security. A vice director of software research at the Chinese Academy of Sciences said Red Flag would protect the government from attacks by foreign hackers (*Computing Canada*, 2000).

Self-reliance and national pride

As noted earlier, another prescriptive dimension entails the principle of self-reliance and national pride. National pride combined with a fear of dependence on foreign countries resulted in the principle of national self-reliance in the Mao Tse Tung era (Terrill, 1977). Since then, this principle has been guiding the Chinese economic system. Although the attitude toward technology imports and foreign investment has changed drastically (Solinger, 1995: 127), the lack of significant alteration in political structure has led to a high level of advocacy for national self-reliance and domestic development of technology among Chinese policy-makers, researchers, scientists and military leaders (Simon, 2001).

The foreign software used in China's computers has thus long been a focus of concern. Over the past 20 years, the Chinese government's various attempts to develop a Chinese operating system have failed because of the rapid development of the global software industry (Goad and Holland, 2000) and the lack of innovative capabilities fitting the innovation model associated with proprietary software. This has become an uncomfortable reality for Chinese policy-makers. Beijing thus perceived Linux as a leapfrogging technology that would overcome their inability to develop an independent operating system. Chinese scientists and engineers have made several attempts to create Chinese standards in computer operating systems and audio-video compression to 3G data standards (CNETAsia, 2003a). They want to achieve self-reliance and reverse the flow of royalty fees[4] by exporting Chinese standards. Having a custom-made operating system is also a matter of national pride.

In general, the Chinese government sees OSS as a powerful opportunity to catch up and even pull ahead in the global technology race. China's Linux development projects are thus driven by 'ideological and broad' (von Hippel and von Krogh, 2003: 215) goals: to achieve national self-reliance and national pride.

As noted in Chapter 7, OSS represents an important shift in the Chinese government's focus on the Chinese software industry. Beijing started Linux development in 1998, establishing Red Flag Linux in August 1999. Red Flag Linux is backed by the Chinese Academy of Science, and headed by the son of former president, Jiang Zemin. In 1999, Linux development was the only software project on the government's list of top technology priorities. The previously discussed open source alliance between Japan, China and Korea (JCK) is another important initiative to promote Linux by co-sourcing.

Cognitive institutions and OSS diffusion

A deep-rooted perception of exploitation by foreign MNCs combined with the Chinese attitude toward software sharing provide cognitive preference for OSS in China.

Perception of exploitation by foreign MNCs

Although a large proportion of Chinese users do not pay for software, due mainly to widespread piracy, those who do pay feel that the licence fees of Microsoft and other MNCs are expensive. For instance, when Microsoft Windows 98 was launched, its US$241 list price translated to four months' salary for the average Chinese worker. One internet entrepreneur even sued Microsoft for unfair pricing (Smith, 2000). A piracy lawsuit by Microsoft against a small local firm worsened the situation and reinforced the perception of the company as a foreign bully.

Other regional economies also concur with China's view regarding exploitation. A Japanese spokesperson for the JCK initiative argued that the three countries consider it unacceptable to be dependent on software over which they have no control regarding the source code or price (Krikke, 2003).

Attitude toward software sharing

A distinguishing feature of OSS is that unlike proprietary software, it can be 'legally' shared. There are three sources of variation between China and the Western world in terms of attitude toward software sharing: importance of family and society; the perception of the origination of knowledge; and the developmental stage of the computing industry.

In terms of the first difference, Gallaway and Kinnear argue that the contemporary cognitive institutions in Western countries do not encourage sharing:

> Sharing and cooperation are indispensable parts of our history and culture. Yet, in a culture built around an economy based on ideals like acquisitiveness, individualism, competitiveness, and materialism, the open source movement will seem an odd fit with our usual patterns of social articulation. (Gallaway and Kinnear, 2004)

The 'community-based, evolutionary knowledge creation' model of OSS (Lee and Cole, 2003) seems to have a good fit with Chinese society. Confucianism, which emphasises the importance of the family and society over the rights of the individual person (Hofstede and Bond, 1988), has had a strong influence on Chinese politics and culture. Chinese people thus seem to have a higher propensity to share software.

Second, in terms of the perception of how innovations originate, Chinese people differ drastically from Westerners. Intellectual property protection in Europe and the USA is based on the premise that an individual can create and own ideas (Mittelstaedt and Mittelstaedt, 1997). Alford (1995: 25) contends that the different systems of copyrights and attitudes towards piracy can be explained by the conflict between Confucianism and Western concepts of intellectual property, which he argues have not existed in the entire history of China.

The willingness of a society to protect an innovation depends on whether the society perceives the innovation as a unique mental act (invention); an observation about the laws of nature (discovery); or the transmission of ideas from external powers (revelation) (Mittelstaedt and Mittelstaedt, 1997). If a society's culture is based on the premise that knowledge comes through "revelation", the contribution of a software code writer will be seen as almost nonexistent (Mittelstaedt and Mittelstaedt, 1997). In the colonial period, Western countries' attempts to impose intellectual property laws on China failed as the Marxist principle further reinforced Confucianism (Alford, 1995).

Third, the novelty factor related to IT products in developing countries leads to a different attitude towards software sharing. At a 4 per cent penetration rate, a computer is still an exotic product in China. It should be noted that software sharing was more common even in the USA when computers were rare and found only in universities (Gallaway and Kinnear, 2004).

Industry and market-related factors

Various demand and supply-side factors mainly created by the 'market track' of the Chinese 'dual-track approach' to reform (Lau, Qian and Roland, 2001) fall under this category (Figure 8.1). Demand-side factors include existing and new activities that can be performed by OSS; incremental utility of OSS over existing offerings; purchasing power and consumers' propensity to adopt new technologies. Supply-side factors include stock of technological skills and existence of suppliers.

Market potential

China's transition to market economy resulted in rapid economic growth. During 1978–2003, China's per capita income increased fivefold (Farrell, Gao and Orr, 2004). Moreover, Chinese consumers seem to have a higher propensity to adopt high-tech products. For instance, in the mid-1980s, the penetration rates of consumer durables in China were about the same as South Korea, Japan and the then Soviet Union (Sklair, 1994).

Although China had only 36 million PCs in 2002, which translates to less than 3 per cent of its population, its PC market is the world's fastest-growing. According to the International Data Corporation (IDC), 13 million PCs were sold in China in 2003, surpassing Japan (CNN, 2003; Flynn, 2004). China is already the world's second biggest PC market and is expected to become the biggest by 2010. IDC has projected that the packaged software market in China will exceed US$6 billion by 2007 (Ebusinessforum, 2004). The underdeveloped computing market means no or lower switching costs. Individuals and firms in China are more likely to choose Linux because of lower acquisition and installation costs[5] or lower total cost of ownership (TCO) than Windows (Rabano, 2000; Kshetri, 2004).

In developing countries, the government is the single biggest user of IT products and has a greater externality power (Nidumolu et al., 1996). A large part of China's informatisation strategy is the implementation of e-government programmes. In 2002, China ranked seventh among 198 countries in terms of e-government performance, and its e-government performance was rated ahead of global e-commerce leaders such as Switzerland, UK, Singapore and Germany (West, 2002).

Economies of scope

OSS has the additional virtue that it can be used in a number of applications. Economies of scope exist if the cost can be reduced by increasing the variety of activities that can be performed by OSS. The size of the Chinese electronic industry provides a proxy for the potential economies of scope. As noted in Chapter 7, the Chinese electronic industry was estimated to exceed US$300 billion in 2006 (In-Stat, 2006a).

It is worth keeping in mind that the cellular industry is among the many sizable sectors for potential Linux deployment. Given the size and growth of the Chinese mobile market, the development of Linux for mobile devices will have a powerful global impact. The Chinese mobile market is the biggest in the world. Working with Red Flag, Eforce, Culturecom and Mobile Telecom introduced the Chinese 2000 Mobile Linux Operating System for handheld devices in 2002, and the Beijing government adopted it as the country's official operating system. Similarly, in mid-2003, Transmeta Corporation, the developer of Midori Linux, formed an alliance with China 2000 Holdings to develop a Linux operating system for mobile devices in China. In July 2004, Datang Telecom Technology, the intellectual property owner of the TD-SCDMA standard, confirmed that it had selected Linux as its 3G handset's operating system (*SinoCast China*, 2004c). According to research firm Research and Markets, Linux will have gained 34 per cent of the Smartphone Operating System market by 2010.

Foreign multinationals in the cellular telecoms sector are engaged in designing Linux-based cellphones in China. By 2004, 10 per cent of mobile handsets sold globally by Nokia were designed in Nokia's product development centre in Beijing. In May 2005, Nokia announced its plan to expand R&D activities in China so that 40 per cent of handsets produced by its mobile phone business division would be designed and developed in the country. Similarly, in 2000, a group of engineers at Motorola's Beijing R&D centre started working on cellphones based on Linux. Since then, Linux has been an integral part of the company's software strategy. As of 2005, Motorola offered five Linux-based phones globally.

Existing offerings

Foreign software, especially Microsoft Windows, dominates the Chinese computing market. Chinese versions of Windows, however, lack ease of use; for example, they require many keystrokes to create a single Chinese character. Chinese consumers also believe that Microsoft's Chinese products are technically inferior to the English versions.

In addition, the USA and its allies in the Coordinating Committee for Multilateral Export Security have previously restricted high-technology exports to China. Indeed, US law still restricts the sales of computers that exceed specified performance limits, measured in 'millions of theoretical operations per second'. The development of supercomputers is thus arguably driven by the pressure to produce them domestically (Lemon, 2003). For the People's Liberation Army (PLA), a major player and consumer in the Chinese computing market, Linux deployment has been an alternative to conventional supercomputing. The PLA was among the earliest adopters of Linux, using it for large-scale clustering by combining together less-powerful computers to perform complex calculations (over 100 servers are clustered in a typical single system).

Thanks to domestic development of Linux, the Chinese military has been able to use domestically developed supercomputers. Dawning, a spin-off from the Chinese Academy of Sciences, developed Dawning 4000L, a supercomputer with over 3 tera floating operations per second (TFLOPs) or 3 trillion calculations per second computing power. Dawning 4000L is based on a Chinese-designed Linux operating system. The Chinese government has bought one Dawning 4000L for possible military use.

Stocks of technological skills

Technical manpower availability

OSS is arguably the first important computer-industry development that has included the 'third-world hackers' as participants and major contributors (Mann, 1999). China is estimated to account for a large proportion of such hackers.

During the, 1966–76 Cultural Revolution, Chinese engineers and technical workers lost significant research and practical experience, pushing the Chinese IT sector 2–3 generations behind the West (Zita, 1991). Thanks to increased domestic opportunities created by the reform, a higher proportion of Chinese engineers trained in advanced countries are returning to China. Gartner Group estimates that China has 400,000 IT professionals and that the number is increasing by 50,000 annually. Estimates suggest that the number of computer science graduates produced in China will soon exceed 200,000 annually. In an attempt to foster the OSS skill base, China has recently made Linux a required course in 35 universities and 35 prevocational schools (SinoCast, 2004a).

What is more, Chinese IT workers have inclined towards Linux development. A survey conducted in 2002 indicated that 44 per cent of Chinese developers had written one or more application and 65 per cent expected to write one by the following year (Olavsrud, 2003).

Presence of foreign high-tech companies

Users are typical innovators in OSS. Nevertheless, commercial companies find indirect ways of making profits (von Hippel and von Krogh, 2003). The demands of some segments are boosted by the success of a complementary open source program, and the expertise of these companies in such segments boosts the demand for their offerings (Lerner and Tirole, 2002). These segments are not efficiently supplied by the open source community. Such 'reactive strategies' (Lerner and Tirole, 2002) also entail encouraging and subsidising the open source movement through such activities as allocation of a few programmers to open source projects. Oracle and Red Flag, for instance, have set up a joint centre to help solve problems experienced by Red Flag customers.

Thanks to ever-increasing industry–government partnerships and a highly-motivated and well-trained engineering workforce, a large number of high-tech firms have made inroads into China (Bureau of Industry and Security, 1998a). The number of companies involved in developing Linux as well as providing software and services for Chinese Linux users is increasing rapidly. Betts (2005) reports that several big banks had announced plans to switch from Unix to Linux, thanks mainly to the latter's better performance and vendor support.

Researchers have identified three categories of motivations for MNCs to globalise their technological activities: international exploitation of technological capabilities by means of activities such as exports, moving production activities abroad and licensing; collaboration among public and business institutions; and generation of innovations across more than one country (Archibugi and Michie, 1997; Iammarino and Michie, 1998).

International exploitation of technological capabilities

MNCs are providing products and services not generated by Chinese players. In August 1999, in its Linux strategy in China, Motorola announced that it would provide Linux-based platforms and services of support, training and systems integration (*Asiainfo Daily China News*, 1999b). In April 2000, Motorola and San Francisco-based TurboLinux announced their plan to jointly develop a Chinese-language version of an embedded Linux operating system running on the PowerPC 8240 microprocessor. Similarly, Motorola and Picsel Technologies have made Picsel Browser available on Motorola phones in China. Picsel Browser delivers full web and multi-document browsing on a Linux-based Smartphone (Wireless News, 2004). In 2000, IBM reported plans to install Red Flag's Linux operating system on its S/390 computers, which are widely used in China's banking and securities networks (Rabano, 2000). IBM also signed a marketing agreement with Red Flag to bundle IBM DB2 Express Database Software with Red Flag Linux.

Collaboration among public and business institutions

Foreign high-tech companies are collaborating with public and business institutes in China to provide complementary products. In August 1999, the Chinese Academy of Sciences, Chinese Software Institute launched Hongqi Linux, an all-Chinese operating system – and its application solutions – with Compaq (*Asiainfo Daily China News*, 1999c), and Chinese Penguin64 with Singapore's Donovan Systems in 2000. Red Flag has collaborated with a number of foreign companies in a wide range of Linux-related projects (Table 8.2). Similarly, Motorola collaborated with

Table 8.2 Red Flag's collaborations in Linux-related projects: selected examples

Partner	Project	Remarks
TurboLinux	Introduction of Chinese-language versions of Linux	
Miracle Linux	Linux versions of Oracle's database software and Asianux (a just-for-Asia Linux)	Miracle is Oracle's Japanese unit
HP	Porting HP's Integrity and Proliant servers to Red Flag's Linux Development of Linux products for international markets	
LinuxLab of the USA	Development and marketing software	
Oracle	Development of Red Flag Data Center OS4.0	
Eforce, Culturecom, and Mobile Telecom	Chinese 2000 Mobile Linux Operating System for handheld devices	Selected as the official operating system by the Beijing government

Lineo, a developer of embedded Linux, and Caldera, a provider of Linux for commercial applications (*Asiainfo Daily China News*, 1999b). IBM has an alliance with Kingsoft, China's biggest office-automation software vendor, to develop desktop Linux applications. The JCK partnership, meanwhile, is working on open source business models, standardisation of software and training software engineers. There has been a division of labour in the partnership: China will develop PC operating systems, Japan will focus on software development and security, and Korea will develop software for PDAs (Krikke, 2003). The partnership is setting up a database to coordinate efforts and avoid duplication.

Generation of Linux-related innovations in China

MNCs are also generating Linux-related innovations in China. For instance, Motorola has 19 research centres in China, with more than 1,600 R&D engineers. Similarly, in October 2003, Oracle opened its China development centre in Beijing, which concentrates on Linux and e-government solutions (Tan, 2003).

Conclusion

The model presented in Figure 8.1 provides an insight into the determinants of OSS diffusion in China. China's unique assimilation of capitalism and socialism has fostered OSS development. Policies that are reminiscent of the Chinese communist public goods regime (Solinger, 1995: 127) and its Confucian heritage are likely to be compatible with the public good nature of OSS and hence provide a fertile ground for the collective component of the model. The regulatory institutions (e.g. the newly introduced piracy and local procurement laws), the normative institutions (the shared social knowledge that people hold regarding and the importance of national security and domestically developed operating systems), and the cognitive institutions (the beliefs, values and social norms related to software sharing, and the deep-rooted perception of exploitation by foreign MNCs) provide institutional preference for OSS. China's radically transformed industry and market, on the other hand, favour the private component of the OSS innovation model. Low penetration and high growth potential of the computing industry and affordability of Linux boost the demand for OSS. China's market openness, meanwhile, has resulted in the wide availability of technological skills. It is thus fair to say that China's institutions, industry and market have provided near-optimum circumstances for the development and deployment of OSS.

The Chinese Linux industry is mainly the result of initiatives taken by the government. Unlike typical OSS projects that can be conceptualised as 'private provision of a public good' (Johnson, 2002), Linux-related innovations in China represent the public provision of a public good. Future research is needed to examine the nature of the roles of Linux communities in China's quick and massive move to OSS deployment and how Chinese OSS communities differ from the rest of the world. A comparison of the Chinese pattern of Linux development and Linux projects in former Soviet bloc countries, which have followed quite different trajectories to free market transition, will provide further insights into the role of unique institutions, enterprises and the market in China.

Notes

1. In April 2003, Red Hat executives commented that Linux software developers in China acted 'against the spirit of open source programming by not sharing their modified source code' (World IT Report, 2003). Although Chinese commentators dismissed the allegation, 'industry insiders' still support it.
2. In 2001, software purchased by the Chinese government amounted RMB 4 billion (US$500 million) and accounted for 14.1 per cent of the Chinese software market, which was expected to double in 2002 (Ying, 2002)
3. With China's market openness, the huge amount of funds needed to modernise the economy left little resources for defence spending. Deng Xiaoping thus encouraged the PLA to get into business. The PLA exploited money-making opportunities opened up by the reforms. Each of the PLA's central and regional commands owned at least two enterprise groups. By 1998, the more than 15,000 military businesses in various sectors of the economy generated US$18 billion or 2 per cent of the country's GDP (Roberts, Clifford and Crock, 1998).
4. For instance, China is the world's largest maker of DVD players. Adopting its own technology, it could save US$2 billion a year in royalties being paid to an 18-company consortium (Calbreath, 2004).
5. For instance, Sun plans to provide a complete Office clone and Linux operating system for US$50 per employee per year to its business customers; by comparison, Microsoft Office costs over US$400 (Takahashi, 2003). However, Microsoft cannot legally reduce the price of its software in foreign countries as dumping is not allowed.

9

Drivers of broadband diffusion in China

China's telecommunications networks are among the largest in the world. In this chapter, we offer a framework that incorporates factors driving broadband diffusion in a developing economy. The factors are related to demand, cost and input conditions, industry structure, and transfer and export conditions. We then compare and contrast aspects of the framework to explain the different patterns of broadband diffusion in China and India.

Introduction

China's telecommunications networks are among the largest in the world. By the end of 2006, China had 52 million broadband subscribers (Chan, 2007) compared with 1.3 million connections in India at that time (Businessworld , 2006). Clearly, this difference is too great to be explained by the difference in their per capita income. By 2010, China is expected to have 139 million broadband subscribers (Morris, 2006).

This chapter examines the drivers of broadband diffusion in China. In the remainder of the chapter, we first briefly review the diffusion of broadband in China. We then examine factors influencing the development of the broadband industry in a developing economy and translate those considerations into the context of China. The final section provides conclusion and implications.

Broadband diffusion in China: a brief survey

As of 2005, in terms of internet users as well as broadband users, China ranked second in the world – only after the USA. In the first quarter of

2006, China added 3.7 million broadband lines, compared with 3.3 million lines in the USA (ZDNet Research, 2006). By the end of 2006, there were expected to be more broadband lines in China than in the USA (*Business Week*, 2004a; Wagner, 2005). The proportion of Chinese internet users with broadband access increased from 6.6 per cent at the end of 2002 to over 50 per cent by early 2006 (*Economist*, 2006a) and about 66 per cent by July 2006 (msnbc, 2006). Paradoxical as it sounds, in 2006, a Chinese internet user was more likely have a broadband connection than his/her US counterpart (Koprowski, 2006).

While a number of OECD countries have cable television connections to provide internet access (ITU, 2005a), the digital subscriber line (DSL) is the dominant form of broadband in China. Indeed, in 2004, China had 11 million DSL users, the world's highest (*SinoCast China*, 2004d).

With respect to cellular broadband, China introduced wireless fidelity (WiFi) in 2002. By mid-2003, 80 per cent of China's five-star hotels, airports and high-grade office buildings in Beijing, Shanghai, Guangzhou and Shenzhen were connected to China Netcom's WiFi network (*SinoCast China*, 2003c). By the end of 2003, China Telecom, China Netcom and China Mobile had 10,000 hot-spots deployed or planned for rollout (Clark and Harwit, 2004). The Chinese WiFi market exceeded US$24 million in 2003 (Koprowski, 2004) and US$250 million by 2005 (itfacts.biz , 2005). Venture capital companies such as Intel Capital are capitalising on the huge potential of WiFi in China by funding WiFi deployment (Clark and Harwit, 2004). In April 2008, China Mobile Communications launched a third-generation trial network based on the domestically developed Time Division-Synchronous Code Division Multiple Access (TD-SCDMA) standard in eight Chinese cities (Scent, 2008). The research firm, Analysys, estimated the first round of 3G investment to be in the region of US$1.25–2.5 billion and the first phase demand to be for 10–20 million lines (3g.co.uk, 2006). Given that even in the countryside, it is possible to access the internet at fixed line speeds using a mobile handset (*Economist*, 2006b), the potential for cellular broadband in China seems massive.

Regional distribution of broadband users

Rapid broadband growth in China is characterised by a high degree of geographical disparity and regional imbalances (Martinsons, 2005). Broadband networks in some regions such as Shanghai and other coastal towns (Mallaby, 2005) are considered to be more developed than in some parts of the industrialised world (COMMWEB, 2006). Wilson (2005) has observed that it is easier to get broadband connections in some Chinese cities than in many small cities in the USA.

Like the rest of the Chinese economy, the internet and broadband reflect a geographical bias towards cities and the east. For instance, in 2006, broadband penetration was 18 per cent in Chinese cities compared with 3 per cent in rural China (Morris, 2006). Likewise, as of 2006, 40 per cent of internet users were in Beijing, Guangzhou and Shanghai, whereas only 1 per cent lived in western China (*Economist Intelligence Unit Limited*, 2006). According to the China Internet Network Information Center, as of December 2005, only 2.6 per cent of the rural population had internet access compared with 16.9 per cent of the urban population (Wikipedia, 2008). Nevertheless, almost every county has broadband and internet cafés, and high-speed connections are ubiquitous and cheap even in remote towns (*Economist*, 2006b).

Factors driving broadband diffusion

Numerous factors influence the development of the broadband industry in a developing economy (Figure 9.1). Following Beise (2001), and Lehrer, Dholakia and Kshetri (2002), we divide factors driving the growth of the broadband industry into three groups: *nature of domestic demands and inputs* (Linder, 1961; Vernon, 1966), *industry structure* (Porter, 1990), and *export and transfer conditions* (Tilton, 1971; Beise, 2001). The nature of domestic demand and inputs includes factors such as consumer preferences, income, availability and costs of input, infrastructures and government regulations (the roles of technology parks will be covered in detail) and technological economies of scope (a function of prior national experience with previous generations of technology). The importance of industry structure regarding the performance of the broadband industry can be explained in terms of the industrial organisation theory. According to this theory, industry structure determines a firm's behaviour, strategy and performance (Bain, 1956; Porter, 1990). Competition level, size and distribution of broadband suppliers, as well as nature and structure of related industries will be analysed under this category. Factors such as trade policy, export orientation of the firms in the country, strategic regulation, and market size are covered under transfer and export conditions. Figure 9.2 presents a preliminary model relating factors driving broadband and associated mechanisms.

The Rapidly Transforming Chinese High-Technology Industry and Market

Figure 9.1 A comparison of broadband and related technologies in China and India

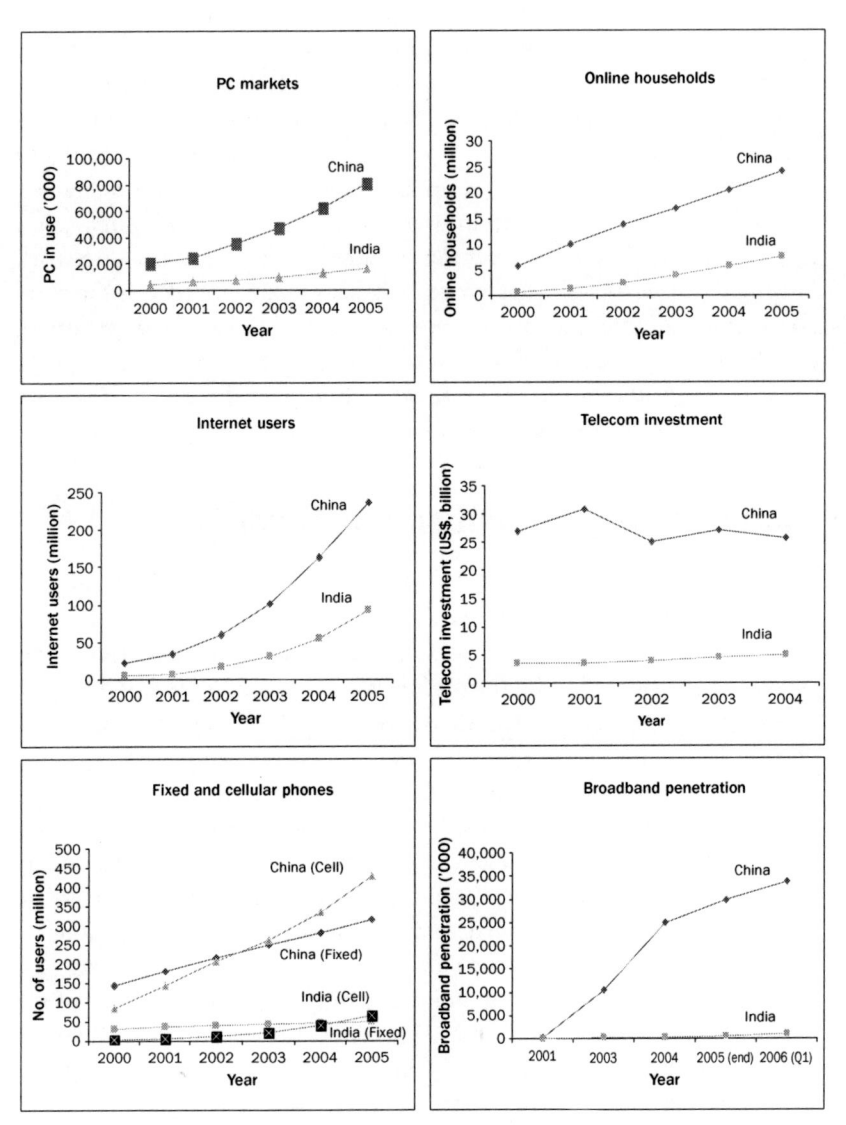

The government's role

Different theoretical contributions and various empirical studies have led to the accepted view that governments can address barriers to technology adoption through new laws, investment incentives, foreign

Figure 9.2 A proposed framework to explain broadband diffusion in developing countries

Supply characteristics of broadband
- Cost/affordability
- Quality
- Content availability
- Marketing/innovativeness

Technological competitiveness of firms
- Production of technologies and technological standards
- Export orientation

Government policy
- Investment
- Market openness
- Licensing and tax

Related infrastructure
- Usability and upgradeability of existing infrastructure

Input characteristic
- Quality of IT workforce
- Cost of hiring

Industry structure
- Competitive rivalry
- Horizontal collaboration

Impact on the economy
- Advancement towards information economy
- International competitiveness

Broadband diffusion
- Diffusion level
- Diffusion pattern
- Width of adoption
- Depth of adoption

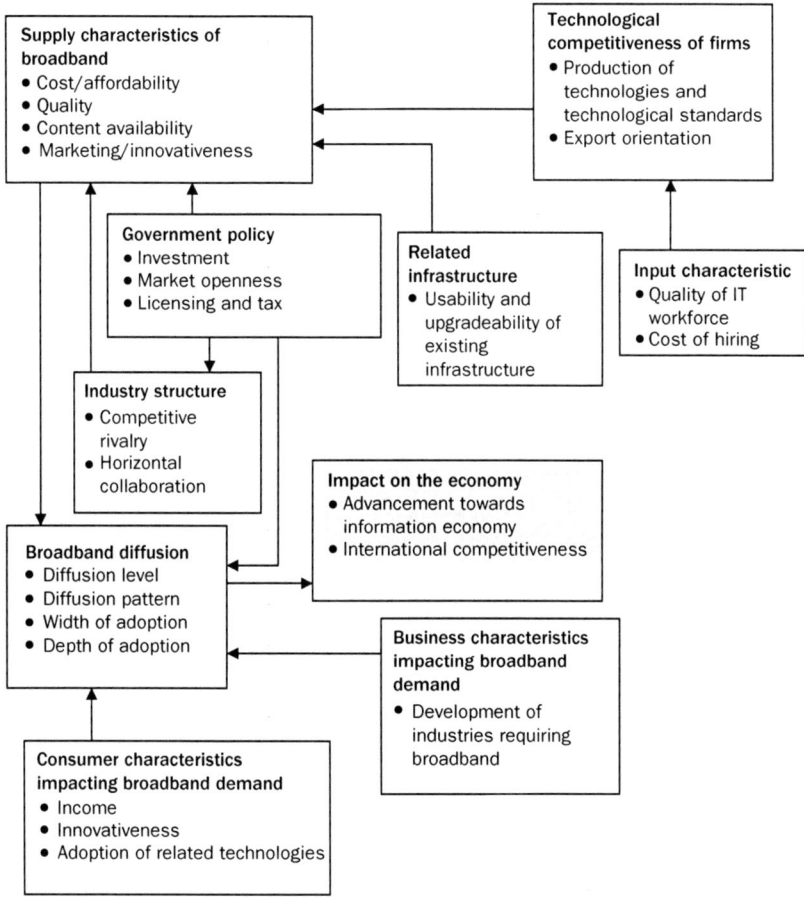

Business characteristics impacting broadband demand
- Development of industries requiring broadband

Consumer characteristics impacting broadband demand
- Income
- Innovativeness
- Adoption of related technologies

technology transfer, and other supply-push and demand-pull forces (King et al., 1994; Montealegre, 1999). Successful developing countries are those that use such means to attack barriers related to skills, information, market and infrastructures.

Government subsidies to technology-intensive industries (Newman, Hook and Moothart, 2006) and market innovation have played key roles in China. However, the state's deep entrenchment in the economy means that the government can play a more critical role in China than in many developing countries. For instance, whereas state-owned firms generate less than 7 per cent of GDP in India, according to UBS, the state accounts for at least 70 per cent of the Chinese economy (Pei, 2006).

In the 1990s, the Chinese government designed a series of programmes to accelerate development in the telecoms sector, including extensive re-engineering and intense competition. Telecom companies were forced to adapt to the rigorous disclosure requirements of the NYSE, NASDAQ and Hong Kong's Growth Enterprise Market (McDaniels and Waterman, 2000). The government initiative to launch its own national information infrastructure, known as the 'Golden Projects' (Tan, Meuller and Foster, 1997), established the key backbone infrastructure (Pyramid Research, 2001). The State Development and Planning Commission also ratified a high-speed internet project. The initial plan was to build a backbone network linking 15 major cities on the eastern seaboard including Beijing, Shanghai and Guangzhou (Lovelock, 2001).

In China, the government has also played a key role in promoting geographical equity. In line with its 'rural area informatization' policy (Business Wire, 2006a), the government set a goal of installing fibre-optic telecom lines to large villages by the end of 2000 (Lovelock, 2001). During the tenth five-year plan (2001–05), it targeted technological development as a priority task. Consequently, even western China has been transformed (Martinsons, 2005). The Sichuan province in western China provides a case in point, where the number of broadband internet users increased from 447 in 2000 to 780,000 in September 2005 (Economist Intelligence Unit, 2006).

China's state-owned companies have invested heavily in telecoms (*Economist*, 2006b). In 2000, this amounted to a $2.5 billion investment in broadband, with cumulative broadband-related investment estimated to be $24 billion by 2005 (Lovelock, 2001). A report in *Business China* describes the difference in the diffusion of broadband between China and India:

> So while India's bandwidth boom may be a triumph of its newly liberalised marketplace, it is serving a concentrated demand base rather than the population at large. China, on the other hand, is creating a broadband network that can reach all its regions and every community. (*Business China*, 2005)

Limited authorised content on the internet, however, may act as a roadblock to faster broadband diffusion in China (Clendenin, 2005b). For example, as discussed previously, there are Chinese laws banning politically and/or culturally objectionable content.

Industry structure

Telecom deregulation was among many preparatory reforms introduced by China for its World Trade Organization entry. By 2001, there were half a dozen carriers, dozens of foreign and domestic hardware providers, and many companies offering internet services and related products (Iritani, 2001). More recently, asymmetric DSL rollout from China Telecom and China Netcom has triggered broadband growth (Kirpalani, 2008).

Upgrading the cable television network to support broadband, however, will not be easy. Small entrepreneurs account for a significant share of the cable industry (Broadband Business Forecast, 2006). Internet access via cables requires substantial investment and technical expertise, which may be beyond the reach of small players. Moreover, if cable broadband is provided, the monthly charge is estimated to be about $20, which is unaffordable to poor households (ITU, 2005a). Thus, whereas low cable subscription fees (in the range of $1.28–3.20/month) and an active second-hand market for low-cost televisions (ITU, 2005a) have accelerated cable television diffusion, this industry is not easily upgradeable to broadband.

Input characteristics

An educated and skilled technology workforce is critical for the development of national technological capability. It is important to note that China has a huge pull of high-quality low-cost engineers. Moreover, most of the technologies related to broadband are developed domestically in China.

Consumer demand

Broadband subscription is beyond the reach of the majority of Chinese. Having said this, it is also apparent that, in terms of technological innovation or the propensity to adopt modern technologies such as broadband, Chinese consumers are much further ahead of those in other developing countries. For instance, in broadband applications, the Indian market is several years behind China's. For online games, India in 2006 was arguably comparable to China in 2001 (Business Wire, 2006b). Other indicators related to broadband demand, such as e-commerce and internet advertising, reveal similar patterns (Figure 9.2 and Table 9.1). Not surprisingly, width and depth of internet and

Table 9.1 A comparison of indicators related to broadband in China and India

Indicator	China	India
No. of broadband users, Q1, 2006 (million)[†]	34	1
Broadband users per 1,000 people, Q1, 2006	26	1
Personal computers (PCs) in use, 2005 (million)[*]	81	16
PC users per 1,000 people, 2005	62	15
No. of internet users, 2005 (million)[*]	240	93
Internet users per 1,000 people, 2005	182	87
Online households, 2005 (million)[*]	24	8
Annual growth rate of PC ownership, 1995–2003[††]	34%	24%
Annual growth rate of internet penetration, 1995–2003[††]	90%	54%
Average monthly broadband charges, 2004 (US$)[¶]	16	20
Capital investment in telecommunications, 2004 (US$ billion)[*]	26	5
Fixed telephone lines in use, 2005 (million)[*]	317	50
Mobile telephone users, 2005 (million)[*]	429	66
National telephone calls, 2004 (million minutes)[*]	68,015	35
International outgoing telephone calls, 2005 (million minutes)[*]	1,260	696
Adult literacy rate (% ages 15 and above)[***]	91	61
Researchers in R&D (per million people, 1990–2003)[***]	633	120
Total population, 2003 (millions)[***]	1,300	1,071
GDP per capita, 2003 (US$)[***]	1,100	564
GDP per capita, 2003 (PPP)[***]	5,003	2,892
High-tech exports (% of manufactured exports, 2003)[***]	27	5
No. of cable television subscribers (2005)	150 million[†]	30 million[†††]
E-commerce revenue (US$)	69 billion (2005)[§§]	500 million (2006)[††] 900 million (2006)[§]
Internet advertising revenue (US$)	812 million (2006) 1,657 million (2007)[**]	18 million (2004–5)[††]

Sources: [*]Euromonitor (2008); [†]Basu (2006); [‡]ZDNet Research (2006); [¶]Mishra (2004); [§]Content Sutra (2006); [**]China Economic Net (2008); [††]Sen (2005); [‡‡]Bhardwaj (2006); [§§]China Daily (2005); [***]UNDP (2005); [†††]ITU (2005a).

broadband adoptions are higher in China than in India (Kshetri, 2002). The growth of broadband applications such as e-commerce, blogging, instant messaging and internet-based phone, internet protocol television (IPTV) and video calls has triggered broadband demand in China (*Economist*, 2006b). For instance, the number of blogs in China was in the range of 1–2 million in 2005 (French, 2005). In early 2006, China was estimated have 30–38 million bloggers (Lim, 2006) compared about 9 million in India (Sengupta, 2006; see also United Press International, 2006). Another estimate puts the blogger population at 47 million at the end of 2007 (Chinaview, 2007). What is more, Chinese bloggers upload video and audio content in addition to comments (Madden, 2006).

China also leads the world in terms of the proportion of people watching music videos and television programmes on broadband. According to a worldwide study by Ofcom, in 2006, 76 per cent of Chinese broadband users watched downloadable or streaming music videos, and 70 per cent watched television on broadband (Ofcom, 2006). In terms of the depth of adoption, a 2005 study by Ipsos Insight indicated that Chinese broadband users spent 17.9 hours online per week, compared with 4.4 hours for Indians (Burns, 2006).

At this point, it is worth noting that although broadband prices are falling rapidly, few peasants and poor households can afford the services in China (*Economist*, 2006b). At the same time, shared broadband access is more common in China (*SinoCast China*, 2006c) than in industrialised countries, thus dampening demand.

Business demand

China's business users are rapidly deploying broadband applications. Much of that activity is centred on carrier-delivered ethernet services (heavyreading.com, 2008). In particular, demand from firms with bandwidth-intensive businesses such as online gaming is growing rapidly (Marcial, 2006). Indian broadband players, on the other hand, are focusing on a narrow range of business customers, such as those in the IT-enabled outsourcing market (Business China, 2005).

Government agencies in China are also among early adopters of ICT applications including broadband. Perhaps because of this, China outperforms industrialised countries such as Switzerland, UK, Singapore and Germany in e-governance (West, 2002). In most developing countries, the government is the single biggest user of ICT (Nidumolu et al., 1996).

Advanced e-government programmes in China have led to rapid growth of broadband demand from federal and local government agencies. To take one example, Motorola worked with the Chinese Ministry of Public Security in a project to deploy video surveillance systems connected via fibre in major cities (Lindstrom, 2006). Indeed, to support initiatives such as Golden Projects and e-government, China rapidly expanded its broadband communications network in the late 1990s (Lovelock, 2001). The Beijing 2008 Olympics is expected to be an important driver of broadband in China. In particular, the Beijing 2008 Olympic Games have been an important component of China Netcom's broadband strategy (*SinoCast China*, 2006d).

National technological capabilities

China has built national technological competence in a fairly short period of time. Chinese technological firms also have a high degree of export orientation. More to the point, since 2004, China has become the world's biggest ICT exporter and, compared with India, a greater proportion of its manufacturing exports are high-tech (Table 9.1). China and Taiwan produce nearly half of the world's broadband-related products (bharatbook.com, 2006). Chinese telecom equipment vendors Huawei and ZTE have sold fixed and mobile broadband equipment in a number of developing and developed countries across all five continents. Within China Huawei and ZTE have supplied most of the network equipment needed for deploying broadband and related applications such as IPTV (*SinoCast China*, 2006e, 2006f, 2006g).

Related infrastructure

Traditionally, a high proportion of the telecom investment in China went to the 'most modern' available infrastructure because the government wanted 'nothing but the best'. For instance, in the late 1980s and early 1990s, Shanghai Telecom undertook projects to lay under the city's streets one of the world's largest fibre-optic cable networks that contained much more bandwidth than needed for simple telephone services (McGill, 2001). Such investment in the 'most modern' infrastructure is a part of Chinese national initiatives to develop telecom infrastructure and high-speed data networks.

China's well-developed rail networks are also driving broadband growth. For example, the Ministry of Railways has over 35,000 kilometres of fibre (Lovelock, 2001), while its China Railcom division owns the network resources. As of 2003, China Railcom provided services to more than 500 cities. As of February 2005, China Railcom had over 1 million broadband subscribers (arabic.china.org.cn, 2005). In 2006, China Railcom accounted for a 2 per cent share in the Chinese telecom market (Chan, 2007).

Compared with broadband, China has a much higher cable television penetration (Table 9.1). The existing cable television infrastructure thus could be upgraded to broadband connectivity as has been seen in several OECD countries (ITU, 2005a).

Supply characteristics of broadband

In China, economies of scale are driving down costs of broadband services and related technologies to end-users. China's broadband prices are among the lowest in the world. An hour on a broadband connection in an internet café in a small Chinese town cost about 13 cents in early 2006 (*Economist*, 2006b).

Inferior mainstream media, which tend to be 'very boring and not very good at discovering new talent' (Madden, 2006) can be expected to fuel broadband diffusion. It can be suggested, therefore, that innovative marketing and content availability could drive broadband diffusion. In China, for instance, Shanghai Telecom's IPTV products offer many interactive features that are available with cable providers (Morris, 2006). Broadband growth is also being driven by the emergence of non-telecom broadband providers such as local municipal governments and real estate developers (Pan, 2006). Some analysts have, however, expressed concerns over the poor quality of broadband services (*SinoCast China*, 2006f). Insufficient broadband speed is also a concern for bandwidth-intensive applications such as IPTV (Morris, 2006).

Impacts of broadband diffusion

China is becoming an increasingly integrated information society. Thanks to broadband, some rural communities already have access to e-government, tele-education and telemedicine services. Students in rural Chinese villages are participating in distance-learning courses via very small aperture terminal satellite broadband (Lee, Bratton and Shi, 2005).

Thanks to broadband, low as well as high-skilled workers in China are actively participating in the design and development of innovative new products, thus breaking traditional geographical barriers (*Plant Engineering*, 2006). When internet and broadband connectivity are available in rural areas, the costs will fall even lower (Helm and Kripalani, 2006), thus further increasing the potential for outsourcing. Some analysts argue that when highly-skilled workers with broadband internet access in China and India begin to compete with industrialised countries, the theory of comparative advantage may need modification (*Business Week*, 2004b).

Discussion and conclusion

The foregoing discussion provided a framework for understanding broadband diffusion in the developing world. China's strong economy, growing household income and better penetration of base and related technologies such as PCs, availability of value-added services such as voice over internet protocol and IPTV (Table 9.1 and Figure 9.1) and consumer innovation are driving broadband growth. China has made an extensive use of related networks owned by railway and gas companies to provide broadband. Investment in cable broadband could contribute to the fuller exploitation of underutilised potential.

In China, the government has influenced demand and supply of broadband through its deep entrenchment in the economy. Meaningful government interventions have produced better results – in terms of penetration level as well as geographical equity.

10

The Chinese internet protocol television market

The market for internet protocol television (IPTV) in China is sizeable and clearly gaining speed. Chinese consumers seem to embrace IPTV enthusiastically. Like many sectors of the Chinese economy, the Chinese IPTV industry, however, has a number of idiosyncratic features. This chapter examines the facilitators and barriers of the Chinese IPTV industry. We analyse how the characteristics of Chinese consumers, IPTV suppliers, and businesses in the IPTV value chain as well as those related to substitute technologies combine with the complex political setting to shape the Chinese IPTV landscape. The insights developed in this chapter are important for understanding the rapidly developing trajectory of the Chinese IPTV industry as well as other high-technology industries.

Introduction

China has arguably grabbed the global spotlight in the market for internet protocol television (IPTV) (Wilson, 2007a). Although estimates vary across sources (not least because of the differing criteria used to define IPTV), different surveys concur that the Chinese IPTV market is sizeable and growing increasingly rapidly. As of November 2006, China Telecom had trial IPTV networks in 23 cities while China Netcom (CNC) had similar networks in 21 cities (*SinoCast China*, 2006f). By the end of 2006, 60 per cent of China's major cities were IPTV-ready (Chan, 2006). Similarly, China's IPTV subscribers increased by 240,000 in 2005 (In-Stat, 2006b) to 5 million in early 2007 (PRWeb, 2007); this is forecast to reach 10 million in 2008 (PRWeb, 2007) and 23 million by 2012 (Heiser, 2007; IHS, 2007). Ashling (2005) has suggested that China will be the world's largest IPTV market by 2010, while a study by Informa Telecoms and Media suggests that China will be the world's biggest IPTV market by 2012 (digitaltvnews.net, 2007). It is estimated that global IPTV spending will reach $1 billion in 2009, of which China will account for one quarter (*America's Network*, 2006). The market research firm In-Stat has estimated that China's IPTV will generate annual revenue of US$888 million by 2010 (Broadcast Engineering, 2007).

The Chinese IPTV industry, as is the case in other economic sectors, has a number of idiosyncratic features. In spite of its fast growth, the Chinese IPTV landscape still has plenty of wind in its sails. Factors such as the state's entrenchment in the economy; regulation on culture and communications, and restriction on foreign content; a deep thirst for a domestic IPTV standard; and the availability of pirated and illegally redistributed cable or IPTV content, are having a visible impact on China's IPTV industry.

In this chapter, we investigate factors facilitating and hindering the growth of the Chinese IPTV industry. In the remainder of the chapter, we

Table 10.1　IPTV in China: A timeline and future perspective

Time	IPTV milestone
May 2005	Shanghai Media Group (SMG) gets first IPTV licence and launches first IPTV service in Harbin (Liu, 2006)
December 2005	No. of IPTV subscribers: 240,000 (In-Stat, 2006b)
May 2006	CCTV gains second IPTV licence
October 2006	Four IPTV licences issued (China Economic Net, 2007)
December 2006	China Telecom's IPTV network in 23 cities and China Netcom's in 21 cities; of major cities IPTV-ready (Chan, 2006)
Early 2007	No. of IPTV subscribers: 5 million
February 2007	Beijing Television Station obtains the sixth IPTV licence (China Economic Net, 2007)
Mid-2007	China Netcom builds the IPTV network based on the home-grown AVS standard for 3,500 households in Dalian (*SinoCast China*, 2007c)
Mid-2007	In a landmark case filed by the MPA Asia-Pacific (MPAA), a Beijing court orders Sohu Internet Information Service Co. to pay US$140,000 in damages to Hollywood movie studios and apologise
2010	China is expected to be the world's largest IPTV market (Ashling, 2005)

draw upon the literature to examine factors driving the development of the technology industry. We then investigate factors driving the development of the Chinese IPTV industry. The final section provides discussion and implications.

Development of a technology industry: a theoretical framework

In examining the factors that can help IPTV thrive, one would do well to recall the comment by Adams (1996): 'like fire technology depends on its environment to flare or die'. A technology's ecosystem and environment are influenced by numerous factors. First, the diffusion of a technology is influenced by the nature of domestic demands and inputs (Linder, 1961; Vernon, 1966) such as consumer preferences, income, nature of input, infrastructures and government regulations and technological economies of scope (a function of prior national experience with previous generations of technology). Second, the importance of industry structure on the performance of an industry has been emphasised in the prior literature (Porter, 1990). Of special interest to this chapter is the development of related and supporting industries (Porter, 1990) such as broadband, PCs, cable, telecom equipment, etc. Industrial organisation theory holds that industry structure determines a firm's strategy and performance (Bain 1956; Porter, 1990). Competition level, size and distribution of IPTV suppliers, as well as the nature and structure of related industries fall under this category. Finally, transfer and export conditions such as trade policy, export orientation of firms, strategic regulation, and market size also affect an industry's growth (Tilton, 1971; Beise, 2001).

Up to this point, we have mainly concentrated on economic factors. Institutionalists have also recognised that the success of an innovation is tightly linked to the context provided by institutions in an economy (Storper and Walker, 1989; Sabel and Zeitlin, 1997). By institutions, we mean 'the macro-level rules of the game' (North, 1990: 27), which can be formal (e.g. laws and regulation) as well as informal (e.g. social norms and culture). Institutions provide the 'cognitive, normative, and regulative structures' (Scott, 1995: 33) that determine institutional preference for an innovation such as IPTV.

We argue that the above factors in the context of the IPTV industry can be captured in terms of the characteristics of consumers, the government, IPTV suppliers, businesses in the IPTV value chain and suppliers of substitute technologies. Our main focus is thus on these factors from the standpoint of the Chinese IPTV industry.

Factors driving the development of the Chinese IPTV industry

As a visual aid, Figure 10.1 schematically represents factors influencing the development of the Chinese IPTV industry. We discuss the building blocks of the framework in this section.

Consumer characteristics
Chinese consumers' propensity to adopt technology

Prior research has indicated that Chinese consumers have a higher propensity to adopt new technologies than their income level explains. For instance, in the mid-1980s, the penetration rates of consumer durables in China were about the same as South Korea, Japan and then USSR

Figure 10.1 A framework for understanding the development of the Chinese IPTV industry

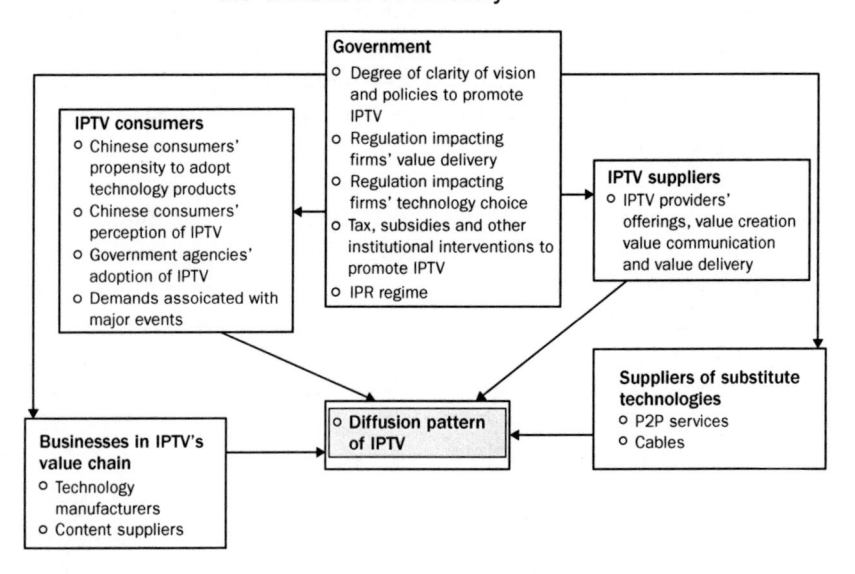

IPTV consumers
- Chinese consumers' propensity to adopt technology products
- Chinese consumers' perception of IPTV
- Government agencies' adoption of IPTV
- Demands assoicated with major events

Government
- Degree of clarity of vision and policies to promote IPTV
- Regulation impacting firms' value delivery
- Regulation impacting firms' technology choice
- Tax, subsidies and other institutional interventions to promote IPTV
- IPR regime

IPTV suppliers
- IPTV providers' offerings, value creation value communication and value delivery

Businesses in IPTV's value chain
- Technology manufacturers
- Content suppliers

Diffusion pattern of IPTV

Suppliers of substitute technologies
- P2P services
- Cables

(Sklair, 1994). In terms of the technology achievement index constructed by the UNDP (2001), China's rank of 45 (out of 72 economies considered) puts it in the group of 'dynamic adopters' of new technologies and ahead of other developing countries with higher per capita GDP such as Bolivia, Colombia, Peru, Paraguay, Jamaica and Tunisia.

The most relevant issue concerns the adoption of technology directly related to IPTV. One of the most straightforward explanations for the rapid IPTV diffusion in China is the country's broadband leadership. The proportion of Chinese internet users with broadband access increased from 6.6 per cent at the end of 2002 to over 50 per cent by early 2006 (*Economist*, 2006b) and to about 66 per cent by July 2006 (msnbc, 2006). Unlikely as it sounds, by early 2006, a Chinese internet user was more likely to have a broadband connection than his/her US counterpart (Koprowski, 2006). As of February 2008, China had about 122 million broadband users (Hui, 2008). Ownership of personal computers, however, is much lower in China. This means that IPTV can represent a means of access to internet-based information and social networking services (Wilson, 2007b).

Chinese consumers' perception of IPTV

There are reasons to believe that because of unique economic, cultural and political factors, Chinese consumers' value perception of IPTV may differ from consumers in industrialised countries. For instance, surveys have shown that the internet component rather than the video entertainment of IPTV is likely to be more popular in China (Wilson, 2007b). One such area is online gaming. In 2004, China had 20.3 million online game players, which generated US$298 million (Weitao, 2005). A study by CCID Consulting (2007) estimated China's online game market in 2007 to be worth 11.35 billion yuan (US$1.6 billion), which was 75 per cent larger than in 2006. The study also suggested that over 50 per cent of internet users played online games in 2007 (CCID, 2007). Another estimate has suggested that the number of online game players will reach 57 million by 2009, generating US$1.3 billion (Weitao, 2005). For this reason, China's online gaming operators, such as Shanda Entertainment Interactive, have made IPTV a top priority (Weitao, 2005).

Government agencies' adoption of IPTV

Household consumers are not the only IPTV consumers. Indeed, in most developing countries, the government is the single biggest user of information and communication technology (ICT) (Nidumolu et al., 1996). Government agencies in China are also among the early adopters of ICT applications. Indeed, in terms of e-governance, China outperforms industrialised countries such as Switzerland, UK, Singapore and Germany (West, 2002). As is the case in other ICT sectors, advanced e-government programmes of federal and local government agencies are likely to contribute to the growing demand for IPTV. CNC, for instance, uses IPTV to provide distance education to Communist Party members in rural Henan (*SinoCast China*, 2006f).

Demand associated with major events

Additionally, from the standpoint of IPTV demand, part of China's fascinating character stems from the fact that the country occupies an important global position. For instance, China is hosting crucial global events. Many industry observers expect 2007 and 2008 to be growth years for IPTV. Major events such as Beijing Olympic Games (Chan, 2006) and the 2010 World Expo in Shanghai (Heiser, 2007) are likely to contribute to the further growth of the IPTV network.

Government's characteristics

Vision and policies to promote IPTV

China's state strategies toward ICT have been to balance economic modernisation and political control (Kalathil, 2003). In this regard, it is important to note that the Chinese Communist Party's basis of legitimacy has shifted from Marxism-Leninism to economic development and nationalism (Zhao, 2000).

It is worth noting that the Chinese government wants to promote the development of IPTV as this technology is 'aligned to its long-term plan of unifying broadband, internet and television' (IHS, 2007). Consistent with the new form of legitimacy, while the Chinese government is showing its willingness to transform provincial television stations from 'conduits for propaganda' into 'cash-rich conglomerates' (Wehrfritz, Hewitt and Ansfield, 2005), confused policy has been a major roadblock to realising the full potential of IPTV (Stephen, 2007). Most high-technology sectors are plagued with a lack of clear policies and regulative uncertainty. Indeed, one business group has notably asked the Chinese government to issue clearer regulations (Kalathil, 2003).

China Telecom and China Netcom have both built trial IPTV networks in a number of cities, but are unable to sign up commercial users because of the lack of IPTV regulations (Weitao, 2007). There has been no clear policy on IPTV (*SinoCast China*, 2007c). A report published by the State Administration of Radio, Film and Television (SARFT) in 2007 was 'vague' on the government's position on IPTV (Sun, 2007).

There has reportedly been inter-ministerial fighting over whether IPTV should be promoted at all, due to the country's heavy spending on cable (Frater, 2006). Moreover, as it is in most countries, broadcasting and telephony have separate regulatory bodies in China. Television broadcasting regulators want television providers to retain their rights to provide IPTV services (Wang, 2006). Both broadcasting and telecoms operators want to enter into the IPTV market (*Asia Pulse*, 2006c). Most obviously, broadcasting authorities are concerned that IPTV could hurt cable television operators (Weitao, 2007). In December 2005, the broadcasting authorities in Quanzhou (Fujian) ordered the closure of an IPTV service jointly run by China Telecom and Shanghai Media Group (SMG), despite the fact that SMG had a licence to operate IPTV, issued by the SARFT (Weitao, 2007).

Regulation impacting firms' value creation and delivery

As noted above, deep government entrenchment is a common thread through all sectors of the Chinese economy. Companies delivering IPTV – China Telecom, China Netcom, and the IPTV licensees, led by the first national IPTV licence holder, Shanghai Media Group (SMG) – are all government-owned. Prior research indicates that state-owned firms tend to place a higher emphasis on political and social goals rather than on market share and profits.

There are reasons to believe that political pressure rather than business sense may be driving IPTV launch and product offerings (Yan, 2007). Both China Netcom and China Telecom are operating services with average revenues per user of US$4–10 a month (Wilson, 2007b, 2007c, 2007d). This low monthly IPTV subscription fee is then split with the content provider, Shanghai Media Group (Stephen, 2007). This raises the possibility that IPTV providers may not be motivated to provide high-quality services.

Another uncomfortable reality is that while much of China runs as a free-market economy, culture and communications are among the most regulated sectors (Frater, 2006; Wilson, 2007a). Because of concerns about the ideological dangers (Chan, Ellis and Styles, 2005), foreign content is highly restricted (Chan, 2006) and cannot be broadcast live (Dickie, 2005; Wilson, 2007d). It is hard to imagine the availability of foreign content in the near future.

Regulation impacting firms' technology choice

Another point to bear in mind is China's deep thirst for domestically developed ICT standards. In the Chinese policy landscape, there has been a strongly expressed desire for the representation of 'Chineseness' in ICT. A high level of advocacy for national self-reliance and domestic development of technology exists among Chinese policy-makers, researchers, scientists and military leaders (Simon, 2001). Foreign technology imports and the outflow of royalties have been a focus of concern among Chinese policy-makers (Einhorn, 2004a). Due to ICT standards such as those related to mobile telecoms and DVDs, China has in the past paid a considerable amount of royalties (*SinoCast China*, 2007d). For instance, China is the world's largest producer of DVD players. Estimates suggest that by adopting its own technology, the country could save $2 billion a year in royalties being paid to an 18-company consortium (Calbreath, 2004). There is also the matter of national pride in having domestically developed technological standards and setting standards for the world. The upshot of these tendencies is that Chinese telecom companies face pressure to adopt domestic standards irrespective of their potential for success.

China's past attempts to set standards for the world have been unsuccessful. Since the 1980s, China has made several attempts to develop a Chinese computer operating system, but has failed because of the rapid developments inherent in the global software industry (Goad and Holland, 2000). As regards devising standards, China has had more misses than hits; for example, its attempt in the mid-1990s to introduce its CD standard, Super Video CD to the world also faced foreign market resistance as well as a lack of strong consumer support within the country. Efforts are ongoing to create Chinese standards in computer operating systems, audio-video compression and third-generation (3G) data standards (ZDNet Asia, 2003). More recently, the Ministry of Information Industry has been encouraging domestic R&D institutions, telecom operators and manufacturers to contribute to the research into IPTV standardisation (*Asia Pulse*, 2007). More generally, Chinese firms are flexing their muscles on the standardisation front.

The Chinese internet protocol television market

There is a desire to avoid repeating past 'mistakes' in the emerging IPTV industry. In addition, the feeling in China is that patent charges on IPTV standards – MPEG-4 and H.264 – are huge. Thus, China's push to develop its own standards is motivated by its desire to avoid paying licence fees to MPEG LA and the ITU for using the H.264 standard (Sin, 2007). According to a vice director of Sinovac Biotech's SVA central research academy, the patent cost for each audio-video coding standard (AVS) decoding system is US$0.13; the cost for each H.264 is US$2.50 (Shenshen, 2007). The latter, however, provides better high-definition video reception (Shenshen, 2007).

It is thus easy to imagine why the Chinese want their own IPTV standard. China is determined to promote home-grown encoding technology, AVS and DMB-T digital broadcast technology (Sin, 2007). The video aspect of the digital audio and video encoding and decoding technical standard (AVS), a home-grown standard for IPTV, won approval to become the national standard in March 2006 (*SinoCast China*, 2007c). China owns full intellectual property rights (IPR) over AVS, which is considered as part of the ITU's global IPTV standard (*SinoCast China*, 2007d). China plans to make its AVS-supported IPTV network available to 4 million users by 2009 (*Asia Pulse*, 2007; *SinoCast China*, 2007d). Some analysts argue that among ICT standards originating from China, AVS is the one most likely to achieve widespread adoption after the home-grown 3G cellular standard Time Division – Synchronous Code Division Multiple Access (TD-SCDMA) (Asakawa, 2007). IPTV related companies from mainland China have also teamed up with those from Chinese Taipei to develop the AVS standard further (China IT & Telecom Report, 2007a).

In January 2006, China's Ministry of Information Industry approved AVS as China's national standard for digital video coding. Some experts have suggested that AVS is comparable with international standards (e.g. H.264, MPEG-4, Microsoft's VC-1), and is usable in satellite broadcasting, digital television, cable television, and IPTV as well as mobile television (China IT & Telecom Report, 2007b).

The IPTV networks of China Netcom and China Telecom are mainly based on foreign standards. Both companies, however, are building AVS-based pilot networks (*SinoCast China*, 2007d). China Netcom completed IPTV tests based on AVS in Dalian and reported results equivalent to H.264 (*SinoCast China*, 2007e). In 2007, China Netcom built an AVS-based IPTV network for 3,500 households in Dalian (*SinoCast China*, 2007c). In February 2008, China Netcom's Dalian Branch deployed the commercial use of the network (China Netcom, 2008). The company expects to have 6 million AVS-based IPTV users in 5–6 years (Yan, 2007).

Some observers, however, believe that compared with H.264, AVS is immature and may even entail higher costs (Yan, 2007). It is important to note that immature technologies may create risks for the reliability and quality of any IPTV network (David and Steinmueller, 1994: 230). China Telecom uses the foreign H.264 standard and said it would not use AVS in the short term (Clendenin, 2007b).

Tax, subsidies and other institutional interventions to promote IPTV

Government subsidies to technology-intensive industries (Newman, Hook and Moothart, 2006) have played a key role in China's technological development. Government subsidies in various forms are also flowing through the IPTV operators (Wilson, 2007b). For one thing, the government's policy of 'rural informatisation' is expected to promote geographic equity in IPTV (Business Wire, 2006a). The government also aims for all households to have access to digital technologies by 2015 (MediaNet Press Release Wire, 2007). Analysts have suggested that 'more resources will … be allocated to making IPTV a success' (IHS, 2007).

Intellectual property rights regime

The growth of the high-technology industry is tightly linked to the nature of the IPR regime. It is also essential to recognise that the high rate of cable or IPTV signals being illegally redistributed via peer-to-peer services may jeopardise IPTV suppliers (trackthat.wordpress.com, 2007). Estimates suggest that China has 70–80 IPTV companies operating in a grey area (Frater, 2006). The country's software piracy rate exceeded 90 per cent in 2004 (Evans, 2004). Even now, nearly all movies watched in Chinese internet cafés are illegally downloaded (Coonan, 2007). This means that many Chinese consumers have been benefiting from IPTV without having to pay for it, or by paying much lower than the market value.

At the same time, however, a constellation of factors linked to the Chinese economy and politics is pushing through a fundamental change in the country's IPR regime. Massey observes:

> The problem of IPR protection in China today is thus in one crucial respect very different from that which we faced in 1986: central government leaders and their policies no longer ignore or promote the infringement of intellectual property. (Massey, 2006)

As noted in Chapter 1, China's formal institutions related to IPR have changed drastically. The battle to promote IPR, however, is about more than just creating IPR-friendly political and economic institutions at the central level. In most cases, compared with formal institutions, de-institutionalisation and re-institutionalisation of social practices, cultural values and beliefs occur very slowly (North, 1990; Clark and Soulsby, 1999; Ibrahim and Galt, 2002: 109; Zweynert and Goldschmidt, 2006). North (1990: 6) notes that 'although formal rules may change overnight as the result of political and judicial decisions, informal constraints embodied in customs, traditions, and codes of conduct are much more impervious to deliberate policies'. To put things in context, a study by Kwong et al. (2003) found that the majority of Chinese do not consider buying pirated CDs as unethical. Similar findings are reported in studies conducted in other collectivist countries (Bagchi, Kirs and Cerveny, 2006) such as India and South Korea (Christie et al., 2003). Indeed, the attitudes of people involved in enforcement have also hampered IPR protection in China (Shen, 2005).

Some encouraging signs in the IPTV industry emerged in 2007 to suggest China's seriousness in fighting piracy. In a landmark case filed by the Motion Picture Association of Asia-Pacific (MPAA), a Beijing court ordered Sohu Internet Information Service Co. to pay US$140,000 in damages to Hollywood movie studios and apologise. These studios are now negotiating with Chinese IPTV providers such as BestTV to distribute movies through a video-on-demand model, which would give them multiple revenue streams (e.g. royalties, a share of the fees for every user request and advertising revenue) (Coonan, 2007).

IPTV suppliers

IPTV providers' offerings, value creation, value communication and value delivery

As noted earlier, the 'soft' concepts of management, such as marketing and consumer behaviour are not integrated into Chinese thinking (Borgonjon and Vanhonacker, 1992). As it is true for China's other economic sectors, Chinese IPTV firms thus may face problems in marketing their offerings. IPTV in China has yet to form mature business models. Currently, telecom operators, radio and television units and content providers have engaged in cooperative works but with unclear business models (PR Newswire, 2007). An analyst with iSuppli recently described that 'the [IPTV] industry as a whole [has] failed to develop a successful business model capable of being replicated in geographically dispersed urban centres' (Burns, 2008).

Yet, having said this, it is also apparent that Chinese IPTV providers have come up with some creative offerings. In Shanghai, for instance, customers get a free set-top box with a two-year contract. In Harbin too, China Netcom has waived set-top box fees (Wilson, 2007c, 2007d). SMG is also offering a service to consumers who want to upload their own video content (Wilson, 2007b). Similarly, Shanghai Telecom's service allows users to book medical appointments over IPTV (Morris, 2006). BestV, Shanghai Broadcasting Cooperation's digital media arm, is working with local banks to develop a bill-paying service. The company is also prototyping a service that lets investors track their stock portfolios. In addition, BestV is working on a service that would allow consumers to get healthcare advice and to make doctors' appointments (Wilson, 2007b).

Businesses in the IPTV value chain

Technology manufacturers

Companies such as ZTE, UTStarcom, Huawei and Alcatel Shanghai Bell have been developing IPTV technology (*SinoCast China*, 2006g). These companies have delivered IPTV technology based on low-cost set-top boxes, which is viable at a low average revenue per user (Wilson, 2007b). Most impressive of all, domestic manufacturers have been key suppliers of IPTV technologies. Shanghai Telecom uses ZTE's IPTV technology (Morris, 2006). ZTE has won contracts from operators to supply IPTV equipment and solutions in a number of cities and provinces including Beijing, Yunnan and Shaanxi (*SinoCast China*, 2006g). In August 2007, ZTE claimed that the company's IPTV market share by project size was over 50 per cent (*SinoCast China*, 2007f). China Telecom has also partnered with the local television manufacturer Sichuan Changhong to produce internet-ready television sets (Weitao, 2005).

Content suppliers

Observers argue that, in recent years, content providers have recognised the importance of content delivery on IPTV (Li, 2007). One of the biggest problems facing IPTV services in China, however, is the lack of content (PR Newswire, 2007; Wilson, 2007b;). Not only is there a lack of local content (Chan, 2006), but the government maintains tight controls over foreign content (Wilson, 2007a). While the government encourages internet use for education and business, it maintains a tight control over politically and culturally objectionable content such as pornography and items criticising communist rule (Jesdanun, 2008). In the words of an analyst from ABI Research, 'while the IPTV service is regarded by the industry as a potential revenue generator, lack of content may prove a short-term barrier to increasing uptake rapidly' (IHS, 2007).

Suppliers of substitute technologies

Peer-to-peer services

From the supplier perspective, there is another fear for the development of the Chinese IPTV industry – redistribution of IPTV content via peer-to-peer (P2P) services. P2P video-streaming has been successful in China; indeed, one estimate suggests that about half of its internet traffic is P2P streaming (Stephen, 2007). According to the research firm, IDC, P2P services illegally redistributing cable or IPTV content and are thus evading the original broadcasters' fees (trackthat.wordpress.com, 2007). As of early 2007, there were over 100 million users of P2P streaming and 5–10 million of them used IPTV services (Stephen, 2007). It is thus not clear whether users are willing to spend much on IPTV.

Cables

According to a State Administration of Radio, Film and Television regulation, foreign satellite television is restricted in China except for foreigner-oriented hotels and residences, hotels of or above a three-star rating and other authorised facilities (e.g. journalism schools) (Wilson, 2007e). China is among a very few countries where IPTV penetration has exceeded satellite (Cotriss, 2007). Most commonly viewed channels are delivered by cable (Wang, 2006). As of early 2007, there were 12 million digital cable subscribers (Stephen, 2007). For instance, most commonly viewed channels (e.g. HBO) are already delivered by cable. It has thus been a problem for IPTV providers to differentiate IPTV from cable providers (Wang, 2006). Notwithstanding the availability of relatively low-cost set-top boxes, IPTV is expensive compared with standard cable services that offer many channels for less than RMB15 (US$2) a month (Dickie, 2005).

Nonetheless, IPTV providers have more attractive offerings than those of cable television suppliers. For instance, as of early 2006,

China Netcom's IPTV business offered 41 satellite television channels, 14 customised channels and pay-on-demand programmes for RMB60 (US$7.50) a month (Lau, 2006a). The novelty value of IPTV has also helped drive subscriber growth. Moreover, compared with cable providers, China's telecom operators have major advantages – they have more money and have deployed a newer infrastructure (Wilson, 2007e). Cable television is provided by thousands of small operators and the cable industry is thus fragmented. Many are municipally owned (Wilson, 2007e). Moreover, cable services in China are much below the level that is taken for granted in industrialised countries (Wilson, 2007b).

Discussion and implications

The foregoing discussion provides a framework for understanding China's IPTV terrain. We analysed how characteristics of Chinese consumers, IPTV suppliers, businesses in the IPTV value chain and government policy and regulations are shaping the Chinese IPTV landscape. The insights developed in this chapter have important implications for management practices and public policy.

Horizontal collaboration

Horizontal coordination among companies in the IPTV supply chain is needed to enhance the delivery of value. Because of restrictions placed on foreign content, there is an obvious need for more local content. While some companies (e.g. UTStarcom) are working with the government and media authorities to generate more IPTV content (Chan, 2006), local content providers need a bigger push.

Efforts directed towards communicating values

The experience of some companies (e.g. Carrefour) (Child, 2006) indicates that considerable efforts need to be directed towards communicating values and educating Chinese consumers about the benefits of products that are new to China. The relatively low acceptance of IPTV (M2 Presswire, 2007) appears to be due to Chinese consumers failing to understand its values and benefits. IPTV companies must therefore work together to communicate the benefits of IPTV to potential consumers.

Influencing government policy by producing complementarity effects

As indicated above, confused policy has been a major roadblock to realising the full potential of IPTV (Stephen, 2007). Firms in the IPTV value chain should team up to influence government policy on this technology. To succeed in China and other emerging economies, it is important for foreign companies to understand the government's internal goals (Oliver, 1991: 165) and ways of producing complementarity effects (Potter, 2004: 479). Speaking of China, Microsoft's Greater China CEO Tim Chen noted that 'if a foreign company's strategy matches with the government's development agenda, the government will support [the company], even if they don't like [it]' (Kirkpatrick, 2007). For example, IPTV investment strategies that contribute to the government's ICT goals, such as the informatisation of rural areas and increasing household penetration of digital technologies, are more likely to win government approval.

Regional variation in IPTV diffusion

As is the case with other technology industries, the development pattern of the Chinese IPTV industry is likely to be characterised by a high degree of national heterogeneity. There are at least three critical reasons for such variation. First, and most importantly, China's coastal regions and big cities are more developed economically and technologically than the inland and rural areas (Kshetri and Cheung, 2002). For this reason, analysts believe that the Chinese IPTV market in large cities may resemble the markets of South Korea and Japan, while much of the network in the rest of China is underdeveloped and too unreliable to support IPTV (Clendenin, 2007b). Second, as noted above, the central government has provided a significant empowerment to regulatory agencies involved in IPR issues such as the State Administration of Industry and Commerce, the State Administration of Press and Publications, the Intellectual Property Rights Office and the State Pharmaceutical Administration (Yang, 2002). Not all provincial and local agencies are equally motivated to enforce IPR laws. For this reason, the piracy rate of IPTV is likely to vary across China. Third, studies have suggested that consumers in more developed coastal areas are more individualistic than those from inland (Koch and Koch, 2007). Such individualistic populations are less likely to engage in piracy than collectivist populations. It is also important to note that the existence of strong personal ties is a reason why the rule of law is 'ignored with impunity' (Bratton, 2007). This means that a high level of individualism in more developed Chinese cities is likely to strengthen formal and informal institutions related to IPR.

Changing IPR attitudes and the IPTV industry

While the Chinese government's focus is diverted to more 'urgent issues' associated with the country's significant transformation (Yong, 2006), a number of recent developments and trends indicate that pro-IPR actors are likely to become more powerful and the government is likely to further strengthen IPR related regulative institutions. The government has set a goal of capturing 50 positions in the Fortune Global 500 list, compared with 11 in 2004 (August, 2004). Most obviously, more substantive measures related to IPR may be needed to achieve this goal (see Chapter 1). China's changing attitudes to IPR are likely to benefit suppliers in the IPTV industry.

11

China's nanotechnology prowess

The Chinese economy is rapidly shifting into a higher technological gear. Among many visible examples to illustrate such a shift, one is particularly telling: the country's achievement on the nanotechnology front. The need to open up the black box of the Chinese nanotechnology industry is essential for ultimately improving the insights into the nature of the Chinese high-technology industry. It would also help IT professionals surmount or overcome the biases associated with their experiences in dealing with other sectors of the Chinese economy. Moreover, Western media have mainly focused on the human rights abuses in China and have failed to note recent progress in the Chinese high-technology industry. This chapter examines important idiosyncrasies associated with the Chinese nanotechnology industry and the country's sources of comparative advantage in this industry. The explanations offered shed some light on the nature and drivers of the Chinese nanotechnology industry.

Introduction

Although China lags behind the industrialised world in the overall development of its high-technology industries and markets, the country has developed sufficient prowess to compete globally in some areas. Among many visible examples to illustrate China's evolution as an innovation powerhouse, one is particularly telling: the country's achievement on the nanotechnology front. Implausible as it sounds, in some areas of nanotechnology, China has already overtaken the USA (Flemings, 2007).

Assessing China's prowess in nanotechnology is not a straightforward exercise. At a higher level of analysis, global technology analysts disagree as to the nature of the growth trajectory and competitiveness of Chinese high-technology industries. One view is that Chinese high-technology firms have exhibited the same development pattern as those from Japan and South Korea. The opposite argument is that it may be more useful to regard China as a 'normal' emerging industrial power like Brazil or India. There is merit to both these views. Nevertheless such views should be seen as context-dependent rather than set in stone.

Whether China will become the world's nanotechnology superpower; what comparative advantages the country possesses in this sector and how they are being exploited; and how the USA and other countries will respond to China's growing nanotechnology capacity ('Nanotechnology in China', 2007), have been important and longstanding questions with a pressing policy and theoretical issue. This is because nanotechnology is being touted as the 'next big thing' (Niosia and Reidb, 2007). Part of nanotechnology's fascinating character stems from its potential key role in a wide range of applications. The areas of current and projected product and process applications of nanotechnology include those in life sciences, medicine, electronics, optics, information technology, telecommunications, aerospace and energy (Niosia and Reidb, 2007). For instance, the breakthrough in nanotechnology offers the potential to enable tiny medical devices to enter human cells. Unsurprisingly, nanotechnology has been a top priority for many governments around the world.

The theory of comparative advantage, which provides a useful framework to explain international specialisation, typically frames developing countries as those possessing comparative advantages in low-skilled and low-technology products. Some developing countries' increasing prominence on the global ICT map has, however, made this assumption hopelessly obsolete.

The Chinese nanotechnology industry and its drivers: a brief survey

Analysts consider China as the most likely contender for catching up with advanced countries in the nanotechnology race. China has over 800 nanotechnology-related companies (Singer et al., 2006). Many nanotech startups from the USA and other countries have already made inroads into China. Accordingly, estimates suggest that the nanotechnology products and systems markets in China stood at US$5.4 billion in 2005, and that it is likely to increase to US$31.4 billion by 2010 and US$144.9 billion by 2015 (Amity Edumedia, 2006).

In 2005, China's nanotech spending ranked fourth in the world behind the USA, Japan and Germany. In purchasing power parity terms, however, China's US$1.11 billion investment in nanotech research was second only to that of the USA (Wolfe, 2007). Likewise, China's estimated corporate nanotech spending amounted US$165 million in 2006, which was 68 per cent higher than in 2005.

Nanotechnology was first introduced into China in the 1980s. Since then, nanoresearch has been gaining speed in the country. Perhaps the best gauge of achievement in this area is publications in peer-reviewed journals. In 2005, in terms of the number of peer-reviewed journal articles published on nanotechnology, China ranked second only to the USA. Moreover, as noted by Robert Cresanti, the undersecretary for technology at the US Department of Commerce who visited China in 2006, there has been 'a dramatic increase in the quality of nanotechnology papers published by Chinese scientists' (AFX News, 2006b). According to a report released by the Georgia Institute of Technology, China is now the world's foremost publisher of nanotechnology research (Toon, 2008).

China is also a nanotechnology patent leader (Singer et al., 2005). In terms of the number of nanotechnology patents, China ranked third in 2003, behind the USA and Japan. The number of nanotechnology patents filed annually by Chinese organisations increased from less than 1,000 in 2001, to more than 2,400 in 2003, which further increased to over 4,600 by March 2005. Some estimates suggest that 12 per cent of all nanotechnology patents worldwide are held in China (Gumbel, 2007). Despite this, it is also apparent that Chinese nanotechnology-related patent activities are very limited in the USA. Indeed, between 1995 and 2005, Chinese organisations filed only 21 patents with the US Patent Office, putting it 12th worldwide.

Nanotechnology's footprints are getting bigger in China. There are plenty of examples to illustrate this point (Table 11.1). Many Chinese companies have been able to increase profits by adding a 'nano' label to their products. Examples include nano-gas, nano-cups, nano-toothpaste and nano-beer (Wolfe, 2007). Haier, which manufactures over 15,000 varieties of household electrical appliances across 96 product lines in China and 30 overseas production factories, and which exports to more than 100 countries, is among some of the highly visible Chinese high-technology companies capitalising on the country's leadership in nanotechnology. The company has incorporated a series of nanotechnology-derived materials and features into refrigerators, televisions and computers (Singer et al., 2005). This and similar cases suggest that Chinese firms are embracing technology enthusiastically.

Table 11.1 The Chinese nanotechnology industry: some notable achievements

Nanotechnology applications	Chinese achievements
Nanomedicine	2003: Chinese doctors treated 18 bone disease patients by implanting nanobones 2005: Artificial nanobone applications in about 30 hospitals 2007: A new drug delivery method was developed by Chinese Academy of Sciences researchers uses 'nano-sized molecules to carry the chemotherapy drug doxorubicin to tumours, improving the effectiveness of the drug in mice and increasing their survival time' (Nanowerk, 2008)
Nanoelectronics	Haier incorporated nanotechnology-derived materials and features in refrigerators, television and computers (Singer et al., 2005)
Nanotubes	2004: Chinese scientists announced that they were capable of producing carbon nanotubes 60 times faster than their US counterparts
Nanoenergy	Chinese nanotechnology firms and H_2OIL developed 21 fuel additives using 'second generation' nanotechnology; the additives are expected to enhance cleanliness and efficiency of gasoline and diesel fuels The National Engineering Research Center for Nanotechnology in Shanghai is working on energy-efficient streetlights made from nanomaterials
Nanomaterials	Chinese researchers produced nano-coatings for textiles; such coatings are likely to make silk, woollen and cotton clothes water and oil-proof, prevent shrinking and protect from discoloration (Singer et al., 2005)
Nanosatellite	2004: A 25 kg nanosatellite was sent into space (Xinhua News Agency, 2004)

Nanotechnology as a national priority

President Hu Jintao and other Communist Party leaders would like to see the Chinese economy shift beyond low-margin contract manufacturing to high-end innovation. In a speech in April 2005, Chinese Premier Wen Jiabao said: 'Science and technology are the decisive factors in the competition of comprehensive national strength'. Chinese leaders have urged the country's scientists to seize the leading edge of nanotechnology and other emerging fields. Nanotechnology research is among the four science 'megaprojects', the main purpose of which is to catch up with the USA in terms of research by 2020 (Madrigal, 2008).

In 2005, China released a national plan for scientific development that calls for raising R&D spending to 2 per cent of GDP by 2010 – it is currently just above 1 per cent. Chinese leaders think that nanotechnology may enable China to 'leapfrog' wealthier nations (Steinbock, 2007). Nanotechnology has thus been China's priority since the late 1990s. Accordingly, China has a very strong national plan for nanoscience and nanotechnology, a National Steering Committee for Nanoscience and Nanotechnology, and a National Nanoscience Coordination Committee (Singer et al., 2005). The Chinese national nanotechnology plan has strong funding from programmes like the National 863 Hi-Tech Research and Development Plan and the Knowledge Innovation Program. It is suggested that compared with the supporting role of the Indian government, Beijing 'orchestrates' the development of industries and technologies (Steinbock, 2007).

To accelerate R&D efforts, China has established numerous national centres for R&D in nanotechnology (Singer et al., 2005) such as the National Center for NanoScience and Technology, the National Engineering Center for Nanotechnology and its Application, and various centres for nanotechnology commercialisation. The Center for Nanotechnologies at the Chinese Academy of Sciences (CAS) in Beijing opened in 2000, uniting over a dozen CAS institutes and several university laboratories. As of mid-2007, about 3,000 researchers from 50 universities, more than 20 CAS institutes and 300 enterprises in China were engaged in nanoscience and technology R&D. In addition, nanotechnology bases have been established in Beijing and Shanghai for promoting spin-offs and technology transfers.

Chinese nanotech enterprises are also engaged in a number of international collaborations. In 2002, CAS launched a joint project with the US company, Veeco Instruments. CAS and Veeco agreed to establish a nanometer technology centre. The centre provides Chinese researchers with access to Veeco's nanotech instruments, including atomic force and scanning-tunnelling microscopes. Similarly, in July 2007, the China–Korea Nanotechnology Research Center was established in Beijing. China's other notable collaborations in nanotechnology R&D include those with the Finnish Funding Agency for Technology and Innovation; the Minister of Alberta Innovation and Science; and the UK National Science and Innovation Forum.

Kaleidoscopic comparative advantage and the Chinese nanotechnology industry

The theory of comparative advantage provides a useful framework to explain international specialisation. According to this theory, developing countries are typically framed as possessing comparative advantage in low-technology products. However, it is increasingly recognised in the economics literature that it is more useful to regard comparative advantage as industry or occupation-specific, rather than in terms of high-skilled and low-skilled. To this end, certain developing economies have in recent years built up a highly-skilled workforce and, through specialisation, made themselves central to the production of a wide range of high-technology products. Jagadish Bhagwati's theory of kaleidoscopic comparative advantage stands out as one of the most developed efforts to conceptualise this centrality in certain technology sectors. He argues:

> What we are facing now is a new and steadily encroaching economic universe in which the nature of comparative advantage is becoming thin, volatile, and kaleidoscopic and is creating vulnerabilities for industries, firms and workers ... The margins of competitive advantage have, therefore, become thinner: a small shift in costs somewhere can now be deadly to your competitiveness. We used to call such industries 'footloose' ... In the old days, few considered such industries to be the norm. Today, they are the norm. (Bhagwati, 1998)

The basic idea behind kaleidoscopic comparative advantage is thus simple. As developing countries such as China shift into a higher technological gear, the industrialised world's comparative advantages become thin and vulnerable. At a higher level of analysis, as noted earlier, foreign R&D investment in China is growing rapidly.

More to the point, the Chinese nanotechnology industry has several comparative advantages (Table 11.2). We also noted earlier that a key trend in the Chinese technological landscape has been a rapid increase in the number of Chinese educated abroad who are returning home. In 2004, the National Engineering Research Center for Nanotechnology in Shanghai hired a director who had worked for IBM and NASA's Ames Center for Nanotechnology.

Table 11.2 China's advantages in the nanotechnology industry

China's advantage	Explanation	Remarks
First-mover advantage	Chinese researchers have been working in nanotechnology since the 1990s	Nanotechnology was introduced in China in the 1980s
Huger pool of low-cost and competent researchers/ engineers	A large proportion of Chinese scientists/ engineers are trained in the USA, Europe and Japan; some are world leaders in their fields	The annual cost of employing a chip design engineer in 2002 was US$28,000 in Shanghai and US$24,000 in Suzhou (US$300,000 in the USA)
Natural resources	China has important mineral and biological resources which are critical for developing nanomaterials	
Huge domestic market	No need to invest heavily in advanced technology to compete in international markets with world-class nanotech companies	Nanotechnology-related sales are largely restricted to the domestic market

The developing world market for Chinese nanotechnology products

As is the case with other high-technology sectors, Chinese firms are working on nanotechnology-related products that fulfil needs specific to developing countries. To take one example, researchers at China's Tsinghua University are testing a nanotech bone scaffold in patients. Experts say that this application of nanotechnology is especially relevant for developing countries, where the number of skeletal injuries resulting from road traffic accidents is high.

Barriers facing the Chinese nanotechnology industry

It is important to note, however, that the Chinese nanotechnology industry is facing a number of roadblocks (Table 11.3). More broadly,

Table 11.3 Some barriers facing the Chinese nanotechnology industry

Disadvantage	Explanation	Remarks
Underdeveloped basic science	Nanotechnology combines experts from many different fields to solve problems; China spends very little in basic research compared with end-stage product development	China's basic research budget is only 5 per cent of its total R&D budget; as technology in any highly advanced field evolves at an exponential rate, it can be hard for emerging players like China to catch up
Weak intellectual property regime	Problems of regulation and rigorous enforcement	
Lack of venture capital funding	China has a limited ability to attract venture capital and other forms of mature investments in nanotechnology	
Focus on low end of the nanotechnology spectrum	Most companies and scientists are concentrating on nanomaterials	Due to a lack of major breakthroughs, Chinese researchers lag behind those from developed countries in nanoelectronics and nanobiotechnology

China's ability to develop sophisticated high-technology products has been limited. For instance, most of the 300 companies and 7,000 scientists engaged in nanotechnology development in China are concentrating on nanomaterials (Niosia and Reidb, 2007). More sophisticated applications such as nanoelectronics and nanobiotech developments, on the other hand, have lagged behind the developments of other countries such as the USA (Niosia and Reidb, 2007). A lack of more mature investment financing has been a top concern for the development of more sophisticated nanotechnology applications.

As discussed previously, Chinese nanotechnology researchers have achieved much on the patenting front. But the number of patents tells only part of the story. From the Chinese perspective, what matters in its global competitiveness in nanotechnology is not simply the breadth of nanotech R&D activities, but the extent of their sophistication and how

R&D outcomes are used to increase the value addition. Put simply, the battle to overtake the industrialised world in nanotechnology is about more than just the number of nanotech patents.

Another point to bear in mind is the inherently complex nature of nanotechnology, which is built on many progenitor technologies with bases in a wide range of science areas such as molecular biology, electronics, materials science and physics (Niosia and Reidb, 2007). From the Chinese standpoint, the inescapable fact is that the country's basic science areas lag far behind those of industrialised world.

Finally, Beijing's failure to take measures to protect intellectual property rights (IPR) has raised the ire of Western technology companies. There are reports that foreign companies' research on nanotechnology has been stolen by Chinese scientists (Parloff, Chandler and Fung, 2006). Controlling IPR violations may thus prove to be a challenge of another magnitude.

Concluding remarks

This chapter examined important idiosyncrasies in the Chinese nanotechnology industry. The explanations offered shed some light on the nature and drivers of the Chinese nanotechnology industry. For one thing, there has been a far greater achievement in the Chinese nanotechnology industry than might at first appear. Thanks mainly to the Western media's tendency to focus on human rights abuses in China, they have failed to note recent progresses in the important sectors of the Chinese high-technology industry such as nanotechnology (Koo, 1998). Given the wide range of nanotechnology applications and China's increasing global influence, China's progress in this technology is likely to have significant global implications. China is already the world's biggest exporter of information and communication technology products.

A related point is that by not looking at China, Western IT firms active in nanotechnology risk missing out on key opportunities. More to the point, partnering with China in this industry is critical to develop products and technologies for the Chinese market. It is important to note that many industrialised world-based nanotechnology firms have neglected to pay enough attention to the progress in the Chinese nanotechnology industry. For instance, as of mid-2006, while the top 30 global nanoelectronics players had filed patents in the USA, Japan and Europe, only four – Samsung, Infineon, Philips and IBM – had filed such applications in China.

China's progress in nanotechnology is likely to increase the value addition of its export items. Nonetheless, for the near future, China's competitive advantage – both domestically and abroad – is likely to be on the low end of the nanotechnology spectrum (e.g. as a manufacturer of nanomaterials). Long-term success on this front hinges on having well-developed basic science research.

At the same time, it is important to note that China is overcoming some of the barriers discussed above. As noted previously, a weak intellectual property regime has been one of the underlying problems facing the Chinese nanotechnology industry. Numerous forces discussed earlier are completely reshaping the Chinese IPR landscape. The Chinese government has taken measures to strengthen regulative institutions. Of equal importance in the changing IPR landscape is the pressure of local entrepreneurs. The Chinese nanotechnology industry provides a visible example of how local entrepreneurs are creating institutions to protect IPR. For instance, the Nanometer Technology Center established in Beijing is actively involved in protecting IPR (Singer, Salamanca-Buentello and Daar, 2005).

Second, the Chinese venture capital landscape, which was one of the major weaknesses facing the Chinese nanotechnology industry, is also changing rapidly. Indeed, China is one of the fastest-growing markets for venture capital investment. One estimate has suggested that, in 2005 alone, venture capital investment in China rocketed to US$1.17 billion (Balfour, 2006b). This is a phenomenal growth compared with US$325 million in 2002 and US$88 million in 1999 (UNDP, 2001).

A final consideration with the Chinese nanotechnology industry is Chinese technology firms' ability to learn and catch up. For instance, it is suggested that in integrated circuits, China leapfrogged five generations in less than 10 years and became current with the state of the art. As China has already made such great leaps on the nanotechnology front, there are reasons to believe that the country may be able to seize and maintain that position.

12

Chinese technology enterprises in developing countries: sources of strategic fit and institutional legitimacy

While Chinese technology companies do operate in developed countries, their most impressive performances have been in developing markets. Experts argue that China's current technology, know-how and capabilities are more effective for competing in developing countries than in industrialised countries. In this chapter, we draw upon resource-based theory and institutional perspectives to investigate the relative performance of Chinese technology firms in the developing world compared with in the industrialised world. Chinese high-technology products tend to have low cost and be relevant in the context of the developing world. Moreover, Chinese high-technology companies seem to possess soft resources that help them to make decisions that are isomorphic with respect to institutions in developing countries and compatible with business parameters in these countries. Thus, the chapter argues that both the hard and soft resources of Chinese technology companies tend to have a higher degree of strategic fit in the context of the developing world compared with in the industrialised world.

Introduction

There are persuasive arguments for thinking that third-world multinational corporations (MNCs) tend to do better in developing countries than in the developed world (*Economist*, 2005b). The rapidly increasing level of trade between developing countries illustrates this point. One estimate suggests that over 40 per cent of developing countries' exports go to other developing countries and that trade among them is increasing at 11 per cent annually (Panitchpakdi, 2006). Third-world MNCs are familiar with the business terrains of other developing countries thanks mainly to economic, cultural and political proximity, and thus experience a lower degree of foreignness associated with dissimilarity or lack of fit (Kindleberger, 1969; Hymer, 1976) in the operating contexts of these countries.

More to the point, experts have suggested that China's current technology, know-how and capabilities are more effective for competing in developing countries than in industrialised countries.[1] While Chinese technology companies' businesses in developed markets are growing, their performances have been phenomenal in developing markets, generating the money needed for reinvestment (Dolven, 2004; Table 12.1). Referring to the Chinese PC manufacturer, Lenovo's operations in the global PC market, especially the developing world, a Gartner analyst recently noted that 'as the battle shifts from the US to the rest of the world, Lenovo has the advantage' (Hamm, Roberts and Lee, 2005).

Most published studies have focused on trade and foreign direct investment by Western and Japanese companies. These traditional theories have limited applicability in explaining trade and investment originating from developing countries in general, and the trade and investment between developing world countries in particular (Park, 2001). We thus know very little about the exact nature of the mechanisms involved in trade and investment within the developing world. This is particularly the case for technology-related trades originating in developing countries as the studies concentrate mostly on traditional low-value added products. For instance, the members of the Organization for Economic Cooperation and Development (OECD) account for 83 per cent of the world's expenditure on research and development (R&D) and 98 per cent of the global patent industry (OECD, 2004).

By analysing Chinese high-technology firms' internationalisation activities in the developing world, this chapter examines how third-world MNCs' resources fit in other developing countries and how they acquire legitimacy. As there are no generalisable definitions of 'high-technology' (Gardner et al., 2000) or 'high-technology firm' (Bowles, 2004), we offer some clarifying definitions for the purposes of this paper. Following Bowles (2004) and the National Academy of Engineering (1996), a high-technology firm is defined as a firm characterised by the utilisation of the latest technology in its production, involved in the creation of new technology, which has allocated a relatively high proportion of revenue on R&D or has a high proportion of scientists and engineers in its workforce. Such a firm is likely to be engaged in the design, development and manufacturing of information and communications technology (ICT) products such as computer hardware or software; electronic devices involving microelectronics, semiconductors, electronic equipment, optical devices, data and digital communications; medical devices (involving ICT); energy technologies and electric vehicles. Likewise, a developing country is defined as a low-income, a lower middle-income or an upper middle-income country according to the World Bank's classification.[2]

In the remainder of the chapter, we first briefly review Chinese technology companies' operations in the developing world. Next, we discuss theoretical foundation. Then, we examine the development of Chinese technology companies' operations in the developing world. The final section provides the conclusion.

Chinese high-technology products in the developing world: a brief survey

Since 2004, China has become the world's biggest exporter of ICT products. Chinese technology companies have operations all over the world (Table 12.1 lists some highly visible Chinese technology companies). High-technology products account for a higher proportion of China's exports to the developing world compared with the developed world. For instance, while 75 per cent of Africa's exports to China are primary commodities (Siddiqi, 2006), in 2005, machinery, electronic and other high-technology products accounted for over half of China's exports to Africa (*SinoCast China*, 2006h). What is more, high-technology products are growing faster than other product categories in China's exports to developing countries (*SinoCast China*, 2006h). Perhaps because of the low cost and appropriateness, a number of developing countries appear to embrace Chinese technology and know-how enthusiastically. For instance, African countries have adopted Chinese technologies in a range of industries from agriculture to advanced telecommunications and even satellites.

We contend that mechanisms associated with the attractiveness of Chinese technology products in the developing world vary with the degree of Chinese innovation or the proportion of Chinese intellectual property embedded in the products (Table 12.2). In the remainder of this section, we briefly review internationalisation by dividing the Chinese

Table 12.1 The operations of selected major Chinese high-technology players

Company	Size	Internationalisation activities	R&D
Huawei (network equipment)	Revenue: US$5.6 billion (2004) (Einhorn and Reinhardt, 2005) 24,000 employees (2005) (*Economist*, 2005f)	International revenue: US$2.3 billion (40% of total revenue) (Einhorn and Reinhardt, 2005) Operations in over 70 countries Over 3,000 employees are overseas nationals (*Economist*, 2005f).	Half of its employees do R&D Patent filings doubled each year during the 1990s No. of patents in 2005: 2,300 (Simons, 2006), expected to be 3,000 a year from 2006 (*Economist*, 2005a)
ZTE (network equipment)	Revenue: US$4.1 billion (Einhorn and Reinhardt, 2005) 21,000 employees (zte.com.cn, 2006)	International revenue: US$1.6 billion (2004) (40% of total revenue) (Einhorn and Reinhardt, 2005) Operations in more than 60 countries (zte.com.cn, 2006)	2,300 patents filed in China, Europe and the USA (Ortolani, 2005)
Lenovo (PC)	PC revenue: US$13 billion (2004) (Lenovo, 2005) 19,000 employees (2004) (Lenovo, 2005) 14 million units in annual volume (2004) (Lenovo, 2005) Market share: 28% in China in 2002 (CNET News, 2003b), over 25% in 2005 (four times that of Dell (Tucker, 2006)	Selling motherboards, other hardware products and accessories in Europe and Asia Aimed to increase out-of-China sales to 30% of total by 2006 (CNET News, 2003b) from was 5% in 2002 (Desker Shaw, 2003)	A joint venture R&D centre established with Intel in 2003 (UNCTAD, 2005)

Table 12.1 The operations of selected major Chinese high-technology players (*Cont'd*)

Company	Size	Internationalisation activities	R&D
Haier (white goods)	Revenue: US$12.8 billion (2005); overseas revenue of US$2.8 billion (*Appliance Magazine*, 2005) 51,000 full-time employees and 58,800 sales agents worldwide (Haier, 2007) Manufactures 15,100 varieties of items (household electrical appliances) over 96 product lines (Haier, 2007) Domestic market share: 21% (overall), 34% (white goods), 14% (small electric appliances)	Exports products to more than 100 countries (Haier, 2007) Ranked fourth in the revenue from the global sales of white goods 15 manufacturing complexes, 30 overseas production factories, eight design centres (Haier, 2007)	Turned out 1.3 new products per day in 2001 (*People's Daily* 2001)
Ningbo Bird (cellphone)	Revenue: US$1.3 billion (2003) Over 10,000 staff (2003) In 2004, IC Insights, a market research agency, ranked Bird as no. 8 in the world's top 10 brands Made 20 million handsets in 2003 (Johnsson, 2004)	Sells handsets in over 30 countries including Western European countries such as France and Italy (Johnsson, 2004) Exports for 2004 estimated to exceed 2 million units	Out of 7 million phones sold in 2002, 1 million were designed completely by the company (Einhorn, 2003b)
TCL (cellphone)	Revenue: US$3.4 billion (2003) Brand equity: US$3.3 billion (2003) 12% share in the domestic wiring devices market	Overseas handset sales 380,797 units (July 2005), up from 101,268 units in July 2004 (AFX News Limited, 2005)	In 2002, doubled R&D staff to 1,000 and cut the number of cellphone modules bought from foreign companies (Dean, 2003)

Table 12.2 Sources of value of Chinese technology products in other developing countries

Degree of Chinese innovation in the technology product	Low	Moderate	High
Chinese value added	Low	Moderate	High
Quality of technology manpower needed	Low	High	Very high
Origination of the technology	Industrialised world	Industrialised world/China	China
Source of value to the developing world	Low-cost	Relevance and appropriateness	Low-cost and relevance
Some examples	Most ICT products manufactured in China use technologies originated from the industrialised world	Agricultural technologies, radio-jamming equipment given to Zimbabwe, CDMA450, etc	TD-SCDMA, nano-technology, etc

high-technology industry into three categories according to the level of the intellectual property.

Chinese technology products with a low degree of Chinese intellectual property

A 2003 study by the Japan Electronic and Information Technology Industries Association predicted that China would have the highest market share in the export of eight out of 12 major high-technology items for that year. These items included cellular phones, colour television sets, laptop computers, desktop PCs, PDAs, DVD players, DVD drives and car stereos. Most of these products are based on foreign technologies and thus contain little or no Chinese intellectual property (*Economist*, 2003a).

Let us consider cellular handsets. China's mobile phone export was 90 million units in 2003 and 100 million units in 2004 (*SinoCast China*, 2004e). In 2003, 42 per cent of China's mobile phones were exported to the Asia-Pacific region, mainly to developing economies (*SinoCast China*, 2004e). Although foreign multinationals account for most handset exports from China, local firms' shares are increasing rapidly. By early 2004, the Chinese handset manufacturer, Ningbo Bird had sold more than 3 million handsets in over 30 nations, including those in Southeast Asia, India, Russia, France, Italy and Spain. The company has been expanding to South America and the Middle East. TCL, China's second biggest domestic brand (see Chapter 12) similarly exports handsets to Thailand, Bangladesh, Indonesia and Vietnam. In 2002, the company exported 196,000 handsets to Southeast Asia. These Chinese handset makers have traditionally concentrated on the 'look and feel' features of handsets by employing the core technologies developed by foreign suppliers (Ramstad, 2003). Indeed, this has been the case for most high-technology products from China.

Chinese technology products with a moderate degree of Chinese intellectual property

Most Chinese firms start off using foreign technology and then develop their own technology and products (Gao, Woetzel and Wu, 2003; see also Table 12.1 for R&D). A number of Chinese high-technology products thus have a significant and growing amount of Chinese intellectual property. In the handset sector, for instance, Bird started its own R&D for new products in 2002. Out of the 7 million phones sold by Bird in 2002, 1 million were designed completely by the company (Einhorn, 2003b). Similarly, in 2003, increased revenue enabled TCL to double its R&D staff to 1,000 and cut the number of modules bought from suppliers such as Wavecom (Dean, 2003). Thus, an increasing proportion of handsets exported by Chinese companies contain Chinese intellectual property. In addition to handsets, other TCL branded electronic products are also exported to the Middle East, Russia, South Africa, and Southeast Asia along with a number of developed countries (Gao et al., 2003). Likewise, the Chinese white goods manufacturer, Haier has significant operations in developing countries (Table 12.1).

In the telecom equipment sector, Huawei and ZTE have won contracts in a number of developing and developed countries across all five continents. In the three-month period from November 2003 to January

2004, Huawei and ZTE signed telecom contracts in Ethiopia, Libya, Pakistan, India, Indonesia, Hong Kong, Brazil, Russia, Romania, Mali, Afghanistan and Iraq (Dolven, 2004). With operations in over 70 countries, Huawei is probably the most globalised Chinese technology player. The company is a market leader in CDMA450, which is based on CDMA technology by Qualcomm, the US-based company, with significant value addition by Huawei. This technology is deployed by a number of operators in Eastern and Central Europe, Russia and Southeast Asia.

There are three global third-generation (3G) cellular standards and Huawei is ready to build networks based on each of them. The company's global 3G journey began with its winning a US$116 million contract in December 2003 with Hong Kong's Sunday Communications to supply a network based on the W-CDMA standard. The company also constructed the 3G mobile network launched in the United Arab Emirates in February 2004. Huawei's other 3G equipment contracts include a deal with Netherlands' Telfort BV, announced in December 2004, and a deal with the Latin America network of Argentina's CoTeCal, announced in June 2005. In these contracts, Huawei faced fierce competition from telecom equipment suppliers from all over the world. The company also signed 3G agreements with Brazil's Embratel and Malaysia's Telekom Malaysia.

Huawei also has over 30 branch offices in Africa, and has deployed products in 40 countries across the continent (China.org, 2008b). In November 2004, it won contracts worth US$400 million to build mobile phone networks in Kenya, Zimbabwe and Nigeria (*Economist*, 2004b). Huawei is providing telecom equipment of all levels of sophistication to African telecom companies. For instance, in 2005, Huawei and Nigeria's Ministry of Communications signed a US$200 million agreement for the nationwide deployment of CDMA450, a low-cost wireless standard. Similarly, the company constructed Africa's first commercial 3G cellular network, launched by Mauritius' EMTEL in October 2004. Africa accounted for about 22 per cent of Huwaei's international revenue in 2004. In 2006, the company's revenue in Africa reached US$2.08 billion (China.org, 2008b).

ZTE, the second biggest Chinese network supplier, also has a presence in over 40 countries in five continents and has become a strategic partner with dozens of major international telecommunication operators. In the first quarter of 2005, ZTE won 12 contracts in ten countries across five continents. The company has set up over 10 training centres in Africa (China.org, 2008b).

Chinese technology products with a high degree of Chinese intellectual property

Thanks to their accelerated pace of technological and scientific advances, Chinese companies have developed sufficient prowess to compete globally in some technology sectors (see Table 12.1 for R&D). Some visible examples of such technologies include nanotechnology, open source software, cloning technology and cellular telecommunications. While many of such technologies are also attractive to the industrialised world, they seem to have special appeal to consumers in the developing world. The Time Division – Synchronous Code Division Multiple Access (TD-SCDMA) is a highly visible example of such a technology.

We noted earlier that Lenovo, the world's third-largest manufacturer of computers behind Dell and Hewlett-Packard, is in a good position to compete with global PC makers in developing markets. Lenovo is already selling motherboards, other hardware products and accessories in Europe and Asia. The company is also active in South Africa. According to the Kenyan newspaper, *Kenya Press Club*, Lenovo Group is looking to penetrate the laptop market in East Africa. Hamm, Roberts and Lee note:

> Lenovo ... will boast a rare combination of management savvy, technical expertise, low costs, and proven track record in the developing world. As markets like China, India, and Russia become more important, Lenovo could end up presenting Dell with its most serious competition in years – by boxing it off where demand is growing fastest. (Hamm, Roberts and Lee, 2005)

A communication satellite project with the Nigerian government is probably China's highest-profile technology project in the developing world. The package, developed by the China Research Institute of Space Technology, includes the satellite, launch vehicle and launch service. Having passed the preliminary design review in Beijing on 1 July 2005 (Aerospace News, 2005), China successfully launched 'Nigcomsat-1', the Nigerian communication satellite, on a 'Long March 3B' carrier rocket on 14 May 2007 (*People's Daily*, 2007).

The theoretical foundation

At least two processes may explain why Chinese high-technology firms are likely to perform better in developing countries than in developed countries: the strategic fit of Chinese high-technology firms' resources in the context of the developing world and institutional legitimacy. We draw upon two major research streams to explain these processes.

The resource-based view

The resource-based view provides a conceptual framework to assess the strategic fit of resources originating from China in the context of the developing world. Originally proposed by Birger Wernerfelt (1984) and later developed and refined by Jay B. Barney (1991) and other scholars, the resource-based view of the firm has found considerable support in the business literature. A major premise of the resource-based theory is that competitive advantage is a function of the resources and capabilities of the firm (Wernerfelt, 1984; Conner, 1991; Peteraf, 1993). Barney (1991) has listed four attributes of resources that can give rise to a firm's competitive advantage: value, rarity, imperfect imitability, and lack of substitutability.

Valuable resources help a firm exploit opportunities and/or avoid threats in the environment (Barney, 1991) and enable it to develop and/or implement strategies to improve its efficiency and effectiveness (Capron and Hulland, 1999). When we discuss the value of a resource, it is important to ask, 'valuable to whom'? As explained in the next section, low cost and relevant products may act as critical sources of value related to Chinese high-technology products to consumers in the developing world. Chinese technology firms' resources are also characterised by rarity, imperfect imitability and lack of substitutability, which can be attributed to the organisational routines, systems and cultures of Chinese firms and the government (Nelson and Winter, 1982), among others.

An institutional perspective

An economic system can be considered as a 'coordinated set of formal and informal institutions' (Dallago, 2002) influencing economic agents' behaviour (Matutinović, 2005). Put differently, all economic phenomena arguably have institutional components and implications (Parto, 2005). Institutions can be better understood in the context of the tasks for which they were created (Holm, 1995). Studies of institutions constitute a notable stream in management literature. Different theoretical contributions and various empirical studies have led to the accepted view that institutions play a critical role in international business. Institutions define parameters for business operations (Farazmand, 1999) and provide macro-level rules of the game (North, 1990). Approaching international business from the standpoint of institutions, we can thus capture complex factors influencing international business.

Of greatest relevance here is the concept of institutional distance. Institutional distance can be conceptualised as the difference in institutional profiles between two national cultures (Kostova, 1999; Kostova and Zaheer, 1999), which are functions of the regulatory, cognitive, and normative institutions in the two countries (Scott, 1995, 2001). Institutional distance is generally considered as an important aspect of foreignness (Kostova, 1999) and is negatively related to the degree of strategic fit (Zaheer, 1995; Szulanski, 1996; Brannen, Liker and Fruin, 1999). A greater distance between the host country and a multinational company (MNC) means that the latter has to manage more regulatory, normative and cognitive differences. To do this, the MNC has to develop appropriate strategies, organisational forms and procedures to account for such differences (Ionaşcu, Meyer and Estrin, 2004). In short, key to the concept of institutional distance is that it provides a basis for the degree of adaptation required to achieve strategic fit with the local environment (Jensen and Szulanski, 2004).

An organisation with a low institutional distance to a host country is likely to have organisational structure and practices that are isomorphic (or consistent) with the host country's regulatory, cognitive and normative institutions (Meyer and Rowan, 1977; Powell and DiMaggio, 1991) and those of other actors (George et al., 2006: 353). Organisational isomorphism, in turn, is positively related with institutional legitimacy (Deephouse, 1996).

Finally, while the resource-based theory focuses primarily on 'hard' organisational resources (Winter, 1990), the institutional perspective deals with 'people-embodied' and 'soft' resources such as strategic organisational practices that are value and meaning-based (Kostova, 1999).

Chinese high-technology products in the developing world

As a visual aid, Figure 12.1 schematically represents relationships between the fit of Chinese technology companies' resources with the

Figure 12.1 Mechanisms associated with the performance of Chinese technology companies in the developing world

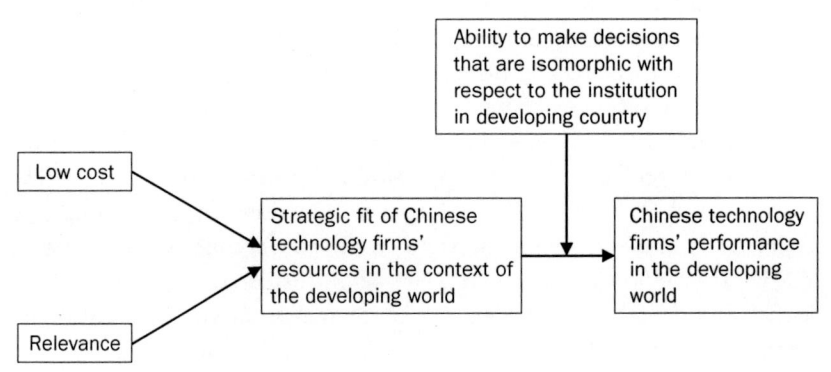

local environment and institutional legitimacy in the context of the developing world.

The resource-based view

Chinese technology products with a low degree of Chinese innovation

An important determinant of a multinational's performance in a developing country is its ability to cut costs. In this aspect, an important way in which Chinese technologies differ from technologies originating in the Western world is on the cost dimension. Low cost is a virtue of Chinese products, making them especially important for the developing world. China's giant pool of cheap labour and low-cost engineers allows Chinese high-technology firms to lower production costs. Most of China's technology companies compete internationally on price rather than technology, brands or after-sales support (*Economist*, 2005f). Chinese companies have won contracts in developing countries for a range of high-technology industries mainly by undercutting the prices of other multinationals and local rivals. For instance, Chinese manufacturers produce cellphone handsets at 15–30 per cent cheaper than the global benchmarks (Lynch, 2002). Similarly, Chinese telecoms vendors have won contracts in the developing world by undercutting competitors by at least 20–30 per cent (Dolven, 2004). To take one example, in an Indian tender in early 2005, the winning bid by Chinese equipment vendor ZTE was about half that Motorola's bid and one-fifth of Ericsson's bid. Likewise, Telecom Kenya reportedly reduced per fixed-line cost from US$200 to US$70 by importing digital switching equipment manufactured in China (China View, 2006).

Chinese technology products with a moderate degree of Chinese innovation

These are technology products that are either designed completely in China or Western technologies that have significant Chinese intellectual property. For these technologies, the source of value concerns appropriateness in the context of developing countries thanks mainly to China's economic, political and cultural similarity with these countries. Compared with Western technologies, which are often too advanced, Chinese technologies are arguably more suited to the needs of the developing world (Yu, 2006).

Huawei's CDMA450 is a good example of a technology relevant in the context of the developing world. The company is a market leader in CDMA450, which is a low-cost standard that requires fewer base stations than competing networks. A number of operators in Eastern and Central Europe, Russia and Southeast Asia have used this standard to provide wireless services with first-generation analogue equipment based on the Nordic Mobile Telephone standard. CDMA450 is the only technology commercially available to upgrade these networks. Huawei has sold CDMA450 equipment to operators in Portugal, Russia and Belarus (Einhorn and Burrows, 2004) as well as other operators in Asia, Africa, South America and Europe.

In the case of agriculture, for instance, Western technologies may be too capital-intensive for developing countries. For this reason, African countries have widely adopted Chinese technologies. For instance, an 11-member Chinese delegation visited Zambia in March 2006 to further cooperation in the agricultural sector. Zambia's Minister of Agriculture said that China can help the African economy build canals, dams and other irrigation projects with advanced Chinese technologies.[3] He also requested that the Chinese provide guidance in establishing farmer training centres.

In other cases, technologies borrowed from developed countries and modified significantly to meet Chinese needs may be attractive to other developing countries. China's export of censorship technologies to authoritarian regimes (Box 12.1) provides a compelling example of this type of trade.

Box 12.1 China's export of censorship technologies

The Chinese government learnt censorship techniques from Singapore. In June 1996, Beijing sent a senior information official to learn about the city state's state-of-the-art internet control and policing practices. In recent years, China has made significant strides in censorship know-how and technology and seems to be ready for exporting such techniques.

Beijing's provision of censorship technologies (such as phone-tapping, radio-jamming and internet-monitoring equipment) to Robert Mugabe's government in Zimbabwe provides a glimpse of this phenomenon. During the election campaign in 2004, a radio-jamming device located at a military base outside Zimbabwe's capital prevented independent radio stations from broadcasting. Similarly, the shortwave station, SW Radio Africa, an independent radio station based in Britain that employs exiled Zimbabwean journalists, experienced jamming problems in 2004. VOP, a shortwave station broadcasting from Madagascar, has also reported such problems. The BBC reported that Chinese intelligence officers visited Harare in 2005 to give further training in telecommunications and radio communications to Zimbabwean technicians.

South Africa-based Zim Online reported that in June 2005, Mugabe announced his government's plan to 'outlaw the dissemination through the internet of information and material it deems offensive' (Zim Online, 2005). Another Zim Online story in 2004 reported that Harare was seeking help and equipment from China to monitor e-mails and information exchanges. There have also been reports of arrests in an internet café for sending e-mails that criticised the government.

Experts say that China has tremendous potential to become a hotbed for advanced censorship technologies and is likely to capture a substantial share of the rapidly growing global market for such technologies, which is currently supplied by Western corporations. For instance, China's large market allows it to test

> **Box 12.1** China's export of censorship technologies (*Cont'd*)

> a number of experimental blocking features that cannot be done in most Western countries. In particular, as China is already a regional internet access provider for neighbours such as Vietnam, North Korea, Uzbekistan and Kyrgyzstan, these countries can easily be China's market for censorship know-how and technologies.

Chinese technology products with a high degree of Chinese innovation

Products with Chinese innovations tend to be cheap. While China's low-quality workforce lowers costs in the case of products with a low degree of Chinese innovation, China's huge pool of low-cost engineers contributes to cost reduction for products made with a high level of Chinese intellectual property. For instance, the annual cost of employing a chip design engineer (salary, benefits, equipment, office space and other infrastructure in 2002 was US$28,000 in Shanghai and US$24,000 in Suzhou, compared with US$30,000 in India and US$300,000 in the USA (Ernst, 2005). In addition to the fact that technology products originating from China are characterised by low cost, Chinese firms are also developing technologies that fulfil needs specific to developing countries. To take one example, researchers at China's Tsinghua University are testing a nanotech bone scaffold in patients. Experts say that this application of nanotechnology is especially relevant for developing countries, where the number of skeletal injuries resulting from road traffic accidents is high (Leahy, 2005). In this context, it is important to note that, in the number of nanotechnology patent applications, China currently ranks third in the world – behind only the USA and Japan.

Similarly, analysts believe that TD-SCDMA is likely to be adopted widely in developing countries thanks mainly to its potentially lower cost compared with competing standards (Reynolds, 2005). Despite its newness, TD-SCDMA is likely to perform better than its rivals on some dimensions and thus has a potential to attract some segments of global operators and customers, especially in the developing world. First, TD-SCDMA combines CDMA's spectrum efficiency and the asymmetric data-transfer capability of GSM. Compared with W-CDMA and CDMA 2000, TD-SCDMA uses the least frequency and is more suited for the dense cities of the developing world than the competing standards. Second, mobile phones and base stations installed with TD-SCDMA chips are likely to be cheaper than those based on competing standards.

Institutional legitimacy to Chinese high-technology products in the developing world

The performance of the technology firm is determined by its position with respect to the business parameters (Farazmand, 1999) and macroeconomic conditions (North, 1990) of the economy in which it is trading. The business parameters and macroeconomic conditions differ across developing economies. A Chinese firm is thus likely to face different business parameters in India and Zimbabwe. Moreover, it is likely that in a given economy, say Zimbabwe, the business parameters and macroeconomic conditions are likely to have different influences on a Chinese firm and a firm from an industrialised country.

Compared with firms from industrialised countries, Chinese technological firms have lower institutional distances and hence a higher degree of strategic fit in the context of developing countries (Zaheer, 1995; Szulanski, 1996; Brannen, Liker and Fruin, 1999). We can provide a number of examples to illustrate this point. For instance, while political, legal and corporate social responsibility pressure from consumers in their home countries have forced several Western companies to withdraw from some African economies, Chinese firms do not face such pressures. Put differently, while Western firms' actions and practices in developing countries are non-isomorphic with respect to institutions in their home countries, this may not be the case for Chinese firms. For example, in Sudan, Western companies, mainly those from the USA and Canada, were pressured to withdraw because of Sudan's civil war and charges of the use of slavery (Lyman, 2005).

To gather cognitive legitimacy and to reduce institutional distance, China is also strengthening cultural ties with developing countries. China International Radio, the voice of Beijing, launched a new FM station in Nairobi, Kenya, which broadcasts 19 hours a day in English, Chinese and Kiswahili. The voice of Beijing is thus competing with the BBC World Service and Voice of America. Similarly, in 2003, more than 6,000 African students received technical training in Chinese universities. It is also important to note that China's early relations with the continent included a large number of scholarships for African elites to study in China.

China has also created a sense of common membership in the developing world to win legitimacy and has emphasised the need for developing nations to unite together against the industrialised West. During a 2003 speech in Ethiopia, the Chinese prime minister Wen Jiabao said 'China is ready to coordinate its positions with African countries ... with a view to safeguarding the legitimate rights and interests of developing countries' (Legget, 2005). In a keynote address at a summit of Asian and African business leaders in Jakarta, the Chinese president Hu Jintao said that 'faced with both opportunities and challenges, we Asian and African countries must seize opportunities, strengthen cooperation to cope with challenges and seek common development' (IAfrica.com, 2005). Such connections have thus offered an alternative source of legitimacy and have reduced Chinese firms' institutional distance with the developing world, making the Chinese much more persuasive in their efforts to secure business deals in developing countries.

Farazmand notes:

> All organizations and institutions perform in a political or power environment through which the broad parameters are more or less defined, and any organizational action contradicting rather than enhancing, or conforming to, that environmental power structure is sanctioned by institutional means of the state, whether autonomous, dependent, mediating, or weak in dealing with powerful transnational corporations. (Farazmand, 1999)

In authoritarian regimes, the government plays a critical role in determining parameters for the business. It is widely recognised that when government policies play a role in international business, it is important to have the ability to influence government decisions (Osland and Cavusgil, 1996). Chinese firms have the ability to manipulate conditions to influence government decisions in developing countries, especially in authoritarian regimes.

High-technology has played a critical role in China's economic and foreign policy relationships with a number of authoritarian regimes. For instance, China's Huawei Technologies reportedly sold fibre-optic products to Saddam Hussein's government (Safire, 2002). It is also suggested that China offered Iran the satellite navigation technology needed to accurately aim its modified Taepo-dong missiles (Malmgren, 2005). Beijing has also offered to build an oil refinery for Venezuela (Malmgren, 2005). Landay (1997) quotes a US official: 'China has become the single most important source of technology that rogue countries cannot obtain from the West'.

We thus hypothesise that compared with Western multinationals, Chinese technology firms are likely to perform better in authoritarian regimes than in democratic ones. Why might this be the case? Institutional distance is a function of regulatory, cognitive and normative institutions in both types of economy (Scott, 1995, 2001). Compared with a Western country, China's institutional profile, at least related to the regulative component, is likely to be more similar to that of an authoritarian regime. Regulative institutions are related to regulatory bodies and existing laws and rules and consist of 'explicit regulative processes: rule setting, monitoring, and sanctioning activities' (Scott, 1995: 35). In short, a Chinese high-technology company is likely to make a decision that is isomorphic with respect to regulative institutions in the developing world.

In this context, it is important to note that the level of economic development is positively related to the level of democracy. Most authoritarian regimes are thus based in the developing world. According to Freedom House's 'Freedom in the World – 2006' report, only 11 out of 48 countries in Sub-Saharan Africa are in the 'free category', while 23 are 'partly free' and 14 are 'not free' (Freedom House, 2006). According to the Council on Foreign Relations, only 40 per cent of African countries are electoral democracies (Kempe, 2006). China has been able to secure political influence in countries avoided by Western nations because of poor governance and opaque political systems and a lack of civil liberty and political freedom (Thompson, 2005). For instance, Nigeria is fighting a continuous struggle with rebels who regularly disrupt oil production. The Nigerian government thus prefers to have a business partner that is indifferent to human rights. In October 2004, the governor of Nigeria's Kaduna Province, which was involved in sectarian killings and adopted a Shariah-based criminal law,[4] invited Chinese investors to set up businesses (Giry, 2004).

Beijing's official policy has been to respect other nations' sovereignty without being involved in their internal affairs. In direct contrast, the USA emphasises human rights (Giry, 2004). Beijing has thus been an attractive partner for leaders who do not want to implement economic or political reforms (Thompson, 2005). Many authoritarian regimes thus find China's modernisation model preferable to the free-market and democratic reforms advocated by the USA and the European Union (Brookes, 2006). Speaking of Africa, *Jane's Intelligence Review* noted that 'China is able to expand its influence in Africa partly because it is viewed with more credibility than Western states with imperialist legacies' (Hill, 2004).

At this point, it must be emphasised that private and public firms function side by side in many developing economies. In some African countries, for instance, the nationalisation process is still ongoing (*African Times*, 2004). For technology products, government purchases account for a significant proportion of imports in developing countries. In most developing countries, the government is the single biggest user of technology products (Nidumolu et al., 1996). Chinese technology companies have thus benefited from Beijing's influence on many authoritarian regimes, compounded by governments' greater control regarding technology-related decisions in such regimes.

Conclusion

To improve our understanding of the internationalisation activities of Chinese technology MNCs, this chapter examined Chinese high-technology enterprises' sources of strategic fit and institutional legitimacy in the context of the developing world. Chinese technology products and intellectual property seem to have tremendous appeal to consumers in developing countries compared with those in developed ones. As some technology sectors, such as cellular networks, are growing at much faster rates in developing countries than in developed countries, Chinese technology players can see new opportunities in the horizon. As to the source of institutional legitimacy, the approach adopted in this chapter captures China's strength stemming from its 'soft power'. Elements of this power include culture, political values, foreign policies and the economic attraction needed to persuade other nations to willingly adopt the same set of goals (Nye, 2004).

Notes

1. Comment of Kobus Van Der Wath, Founder and Managing Director of the Beijing Axis, quoted in Laschinger (2005).
2. The World Bank has divided its member economies according to their 2003 per capita gross national income (GNI). Low-income countries have per capita GNI of US$765 or less. The figures for lower middle-income, upper middle-income, and high income are in the range US$766–US$3,035, US$3,036–US$9,385 and US$9,386 or more respectively.

3. With regard to agricultural technology and know-how, China's African policy states: 'China will intensify cooperation in agricultural technology, organize training programs, carry out experimental and demonstrative agricultural technology projects in Africa and speed up the formulation of China-Africa Agricultural Cooperation Program' (Ministry of Foreign Affairs, 2006).
4. Sharia is traditional Islamic law, also known as Allah's Law.

13

Internationalisation of firms in the Chinese cellular industry

Handset manufacturers, mobile service providers and cellular technology producers based in China are rapidly internationalising their business activities. This chapter examines the globalisation pattern of the Chinese cellular industry from the angle of inward–outward connections in the internationalisation process. We propose a model that explains the factors, actors and moderators that have influenced the inward–outward connection in the internationalisation of the Chinese cellular industry.

Introduction

Chinese firms are rapidly emerging as powerful global players by emulating the successes of their Japanese and Korean counterparts and rapidly climbing the value chain. During 1985–2000, the proportion of primary products and resource-based manufactures in Chinese exports decreased from 49 to 12 per cent (UNCTAD, 2002). In recent years, China is becoming a more sophisticated developer and marketer of high-tech products (Kranhold, 2004; Lall and Albaladejo, 2004). China's high-technology exports as a proportion of manufacturing exports increased from 6 per cent in 1992 to 11 per cent in 1998, which further increased to about one-third in 2005 (Laudicina and White, 2005; Tong, 2005). A number of Chinese high-tech firms are already competing with established multinationals.

In no other industry sector is China's export performance more evident than in cellular technology. Chinese firms have demonstrated competence in all ingredients of the cellular industry. In handsets and cellular infrastructure equipment, Chinese firms have attacked market shares of Western multinational corporations (MNCs). In April 2008, China Mobile Communications launched a third-generation trial network based on the domestically developed Time Division-Synchronous Code Division Multiple Access (TD-SCDMA) standard in eight Chinese cities (Scent, 2008). TD-SCDMA is expected to capture a significant share of the 3G cellular markets both abroad and at home (Einhorn, 2003a).

China's handset exports increased from 22.8 million units in 2000 (GIS News, 2001) to 32 million in 2001, which further increased to 55 million units in 2002. During the first half of 2003, China exported 37 million handsets. Likewise, the total number of third-generation (3G) handsets exported from China was estimated at 4 million units in 2006, compared with 1.1 million in 2005 (ipr.gov.cn, 2006). The revenue generated by handset exports from China amounted US$3.7 billion in 2001 (Krikke, 2002). Not so long ago, handset exports from China were dominated by MNCs. For instance, only 500,000 of the 55 million (or 0.9 per cent of) handsets exported from China in 2002 were domestic brands. At a higher level of analysis, in 2005, about 90 per cent of high-tech products exported from China were manufactured by foreign multinationals (Tong, 2005).

Nonetheless, Chinese domestic companies are rapidly making inroads to international markets, mainly to cost-sensitive ones (Table 13.1). One estimate has suggested that the total export volume of domestic vendors was 20 million units in 2005, which was estimated to pass 32 million in 2006 (ipr.gov.cn, 2006). Most impressive of all, Chinese cellular players are also eyeing the developed-world market.

In 2002, the Chinese handset manufacturer, Ningbo Bird, sold 200,000 handsets, mainly in India, Pakistan, Malaysia, Hong Kong and Russia, and expected to sell 1 million units in 2003 (Weitao, 2003). Another player, TCL has already made inroads in Southeast Asian markets. Both Bird and TCL have engaged in joint ventures with cellular players in various countries. Other Chinese cellular players such as Kejian, Nanjing, Panda Electronics and Amoisonic also have various levels of presence in international markets.

The Chinese cellular industry also seems to be ready for tapping new value creation opportunities in the emerging 3G mobile technologies market. Analysts predict that following its launch, the TD-SCDMA technology will capture a significant market share globally.

Internationalisation pattern of the Chinese cellular industry: a brief survey

In 2005, exports of three Chinese companies in the cellular sector – Huawei, TCL and Bird – amounted to US$2 billion, US$620 million and

US$400 million respectively (Xinhua News Agency, 2006b). These figures were 86.0 per cent, 56.9 per cent and 67.4 per cent bigger than in 2004 for the three companies. Huawei and ZTE have snatched contracts from global telecom equipment manufacturers such as Cisco, Lucent, Nortel, Ericsson and Alcatel all over the world. These Chinese vendors have diversified businesses in a number of high-technology sectors including fixed and wireless phones, optical network equipment, intelligent network systems, and next-generation networks. Nonetheless, the majority of their international sales come from cellular infrastructures. They are developing products for developing as well as developed countries and according to all standards within the global cellular industry.

The international sales of Shenzhen-based Huawei grew by 117 per cent in 2004, and contributed to 41 per cent of its revenue of US$5.6 billion. Similarly, ZTE's foreign sales registered 170 per cent growth in 2004 to exceed US$4.1 billion. International sales accounted for over 40 per cent of ZTE's revenue. In 2005, Huawei's revenue from abroad was expected to surpass that from domestic customers.

In 2002, China had an annual handset production capacity of 200 million units where the local market's absorption capacity was 60 million units (Cellular News, 2003). This indicates an export requirement of 140 million units. In 2003, 37 domestic and foreign companies had licences to manufacture handsets (Cellular News, 2003). In 2007, China produced about 548 million handsets (Cellular News, 2008).

Although MNCs dominate handset exports, Chinese firms are rapidly making inroads into various forms of international markets (Table 13.1). Bird, TCL and Kejian have opened operations in Hong Kong (Online Reporter, 2003), and other domestic players, such as Panda and Amoisonic, have plans to expand overseas (Clark, 2003). As is the case in other sectors of the Chinese high-tech industry, China-based cellular players mainly focus on the middle and low-end segments of the global cellular sector and there are only a few well-known brands (Asia Pulse, 2005).

In August 2003, Bird entered India by launching four GSM[1] handset models in the US$109–261 price range (Reuters, 2003). The handsets were from entry-level to state-of-the-art models. Subsequently, the company introduced low-cost handsets priced below US$80 and phones for the CDMA[2] wireless local loop services (Reuters, 2003). Bird is also selling mobile phones in Pakistan (Weitao, 2003) and plans to enter Southeast Asian markets including Thailand, the Philippines, Malaysia, Indonesia and Singapore, and expects to sell 600,000–800,000 handsets

Table 13.1 International activities of selected players in the Chinese cellular industry

Players	Business	Internationalisation activities
TCL	Handsets	Focuses on mid and low-end GSM/GPRS phones; Asia-Pacific, Europe, the Middle East, Africa and Latin America are main export markets (ipr.gov.cn, 2006) 2002: Exported 196,000 handsets to Southeast Asia Announced plans to enter Russia Acquired Schneider, a bankrupt German television manufacturer Marketing blitz featuring a South Korean pop star Partnership with UK-based TTP Communications Capitalises on Alcatel's good brand awareness in Europe and Latin America (ipr.gov.cn 2006) 2006 Q1: Announced sales of 3.04 million handsets, about 90 per cent of which were contributed by Alcatel-branded handsets (China IT & Telecom Report, 2006)
Bird	Handsets	China's largest mobile phone manufacturer; focuses on mid and low-end GSM/GPRS phones 2002: Exported 200,000 handsets and planned to export 1 million units in 2003 2003: Launched three models of handset in India 2004: IC Insights ranked Bird eighth in the world's top 10 brands Handsets sold in India, Pakistan, Malaysia, Hong Kong and Russia Exported 3 million handsets, accounting for 20 per cent of the sales revenue in 2004 (O'Neill, 2005) Sold handsets in 30 nations and regions including France, Italy and Spain (dri.co.jp, 2004) Joint venture with French handset manufacturer Sagem SA Annual export volume of 6.11 million units in 2005 (ipr.gov.cn, 2006; Telecomasia.net, 2006) 2006 Q1: profit of US$3.34 million over revenues of US$256.375 million (China IT & Telecom Report, June 2, 2006)
China Unicom	Service provider	2000: Realised IPO globally and listed on NYSE and SEHK
Nanjing Panda Electronics	Handsets	Formed a joint venture with Beijing Ericsson Mobile Communication: Nanjing Ericsson Panda Communications Company Ltd is the largest Ericsson joint venture in China (pincn.com, 2005) Established a joint venture with Finland-based Microcell

Table 13.1 International activities of selected players in the Chinese cellular industry (*Cont'd*)

Players	Business	Internationalisation activities
Kejian	Handsets	2002: Paid US$3.2 million to sponsor the English Premier League soccer team Everton for two years 2004: Announced entry in India with a launch of four 'cutting-edge' handsets (convergenceplus.com, 2004) Has a manufacturing joint venture with Samsung Electronics which it is seeking to strengthen Entered into a long-term arrangement with India's Trust Telecom to market and support its handsets in South Asia (convergenceplus.com, 2004)
Huawei	Telecom equipment and handsets	2004: International sales grew by 117 per cent, contributing to 41 per cent of its revenue of US$5.6 billion 2005: Operations in over 70 countries Revenue from abroad was expected to surpass that from domestic customers Handset export reached 3 million units in 2005; about one-sixth of handsets were 3G ready (ipr.gov.cn, 2006)
ZTE	Telecom equipment and handsets	1997: Stocks listed on the Shenzhen Stock Exchange 2004: Foreign sales registered 170 per cent growth to pass US$4.1 billion (over 40 per cent of total revenue) 2005: Total overseas sales of handsets reaches 2.5 million units (ipr.gov.cn, 2006) As of 2007, has filed trademark applications in 155 countries and regions and registered in 108 countries (Ollier, 2007) Handset sales are bundled with equipment (ipr.gov.cn, 2006) Has relationships with operators in Brazil, Peru and Venezuela

annually in these countries (Einhorn, 2003b). The company has also expanded to European markets. In 2004, Bird established a strategic partnership with Siemens, which allowed it to utilise the latter's worldwide sales network in exchange for its 30,000-plus outlets across China (SinoCast China, 2004f). In 2002, TCL exported 196,000 handsets to Southeast Asian countries such as Vietnam, Thailand and the

Philippines (Dean, 2003). The company expected to sell over 20 million handsets annually. In early 2004, a TCL executive described the company's international position:

> Our mobile phones are on display in shops in Southeast Asia and South Asia, like Thailand and Bangladesh. And we are making steady progress in negotiations with agents in Russia, the UK, France, Germany, Greece and elsewhere. (nikkeibp.co.jp, 2004)

Chinese cellular players have also engaged in various types of joint venture, partnership or strategic alliance arrangements with foreign companies (Table 13.1). For instance, Bird has a 50-50 joint venture with Sagem SA, a French handset manufacturer (Dean, 2003). TCL was considering cooperation with Thailand's TA Orange and announced a partnership with UK-based design firm TTP Communications (Dean, 2003). Kejian has a cellular phone manufacturing joint venture with Samsung and is seeking a stronger partnership with the latter (Portelligent, 2002). Similarly, a Finnish firm, Microcell, has established a joint venture with Nanjing Panda Electronics. In 2002, Nanjing announced a plan to buy one-fifth of Beijing Ericsson Mobile Communication for US$11.3 million (Asian Mid Morning News, 2002).

In June 2000, China Unicom entered the international capital market through a global initial public offering (IPO), listing its property in the New York Stock Exchange (NYSE) and the Stock Exchange of Hong Kong (SEHK). The company completed its IPO of 2.83 billion shares, raising proceeds of US$5.7 billion (Wei, 2002).

In the 3G arena, Germany's Siemens AG began cooperating with China's Datang Telecom Technology & Industry in 1998 for the development of TD-SCDMA. Datang's proposal for the TD-SCDMA standard was accepted by the International Telecommunications Union (ITU) in May 2000 and by the 3GPP in March 2001 (3ginsight.com, 2001). In November 2001, Datang and Siemens signed a new agreement to cooperate for an additional three years on the further development of TD-SCDMA technology (3Gnewsroom.com, 2001). In 2003, Datang announced plans to invest more than US$120 million in support of the TD-SCDMA (John, 2002). About 150 Siemens engineers were working in the TD-SCDMA development and production centres in Shanghai and Beijing to develop the technology further (Siemens, 2003).

Managers in the Chinese cellular industry have also realised that products associated with Chinese companies are not perceived to be of high quality in either the domestic or the international markets. Some of the internationalisation activities of the Chinese cellular players are mainly targeted to the domestic market. For instance, to enhance its brand name, TCL launched advertisements featuring a South Korean pop star (Dean, 2003).

In August 2002, Kejian paid US$3.2 million in a two-year shirt-sponsorship deal with Everton, the English Premier League soccer team. In May 2003, Everton played two exhibition games in Shenzhen (Khermouch, Einhorn and Roberts, 2003). As Premier League games can attract audiences of 100–360 million, Kejian capitalised on its Everton ties to promote its brands in the domestic market (Hennock, 2003).

The inward–outward connection in the internationalisation process: relevant theories and past research

Turning to the main focus of our analysis, the ideas of inward and outward internationalisation can be very helpful in understanding how players in the Chinese cellular industry have developed their international operations. Before proceeding, it is important to note that the existing literature lacks a universally accepted definition of 'internationalisation' (McAuley, 1999). Johanson and Vahlne (1977) view internationalisation as an evolutionary process whereby a firm's activities increase internationally parallel to its commitment and knowledge. Similarly, Calof and Beamish (1995: 116) define internationalisation as the 'process of adapting firms' operations (strategy, structure, resources, etc) to international environments'. The internationalisation process can involve both inward and outward modes of activity (Welch and Luostarinen, 1993). According to the above definitions, Chinese cellular players' internationalisation activities include listing on the NYSE; forming joint ventures with foreign firms; sponsoring foreign sports teams to promote their brands; acquiring foreign firms; and exporting handsets.

Empirical evidence from advanced, newly industrialised and developing economies suggests that inward modes of internationalisation contribute to outward activities in various ways (e.g. Welch and Luostarinen, 1993; Luostarinen and Hellmann, 1994). Arguably, managers can gain experience through 'inward' modes, such as outsourcing, technology transfer and countertrade, which are essential for preparing them for outward internationalisation (Cavusgil, 1994).

Dunning (1981) proposed the investment development cycle (IDC) concept. IDC is based on the premise that the levels of nations' inward and outward investments are functions of their stages of development. In the case of firms from advanced nations, Cantwell (1989) extended Pavitt's (1987) theory of technological accumulation, which deals with the sectoral patterns of innovations. Tolentino (1993) extends technological accumulation to present a model that argues that the outward foreign direct investment (FDI) of developing countries begins with simple forms of manufacturing. Subsequently, the FDI trajectory evolves towards technologically advanced manufacturing investments, thereby allowing for independent technological development trajectories. Wells' (1983) product lifecycle model predicts that innovation in a developed country leads to technology transfer and diffusion to firms in developing countries. It is followed by local innovation and then by exports and FDI in some cases. Another model pertaining to firms from developing countries, suggested by Lall (1983), argues that firms in developing countries tend to have an ability to innovate in different products compared with firms in developed countries.

The evidence of licensing and franchising in Australian firms suggests that inward steps can play an important role in the outward process (Welch and Luostarinen, 1993). The work of Hwang (1994) on South Korea also found that inward activities contribute to outward internationalisation, although the study also indicated that the Korean government played a major role. The nature and extent of inward investment and technology transfer, however, differ significantly across time periods, sectors and firm types. In labour-intensive industries (e.g. textiles), the mechanisms for technological enhancement were assistance to Korean subcontractors from foreign buyers, as well as multinational corporations' (MNCs) inward investment in offshore processing. Foreign buyers helped Korean subcontractors to innovate and improve product quality, processes and design. The subsequent outward internationalisation in terms of FDI by Korean firms was found to be compatible with the IDC model. By contrast, the technological accumulation model is arguably appropriate for Korean outward internationalisation in developed countries. The acquisition of technology, access to market information, and responses to FDI decisions by rival companies were the key variables in the outward FDI decisions.

Major barriers to the inward transfer of advanced technologies include the preference of MNCs for retaining proprietary know-how and forming wholly-owned subsidiaries, as well as the lack of research and development (R&D) capabilities in the host country to generate new technologies (Young, Huang and McDermott, 1996). In the case of inward FDI in China, Lan and Young (1996) found the technology transfer fairly restricted because of limited technological capacities in the host country, poor partner firm assimilation, and policy distortions.

The study of Young, Huang and McDermott (1996) also considers the importance of inward internationalisation as a mechanism of facilitating competitive catch-up in Chinese MNCs. For instance, from 1979 to the early 1990s, Shougang, a steel company, imported 600 items (444 of them had advanced patented technologies). In most cases, the imports did not involve complete sets of equipment, and thus required the know-how of adoption and absorption. The study found three factors that were critical to the achievement of technological capabilities: the accumulation of development funds including foreign exchange; exploitation of a range of skills existing within the firm; and upgrading human resources through professional training and the encouragement of entrepreneurship.

Internationalisation of the Chinese cellular industry

Chinese equipment vendors such as ZTE and Huawei have already demonstrated success in foreign markets (Clark, 2003). As suggested by the technological accumulation model, outward internationalisation of Chinese MNCs is evolving towards technologically advanced investments. It is apparent from our review that the cellular industry is one such sector.

Low labour cost, open-door policy, huge market and location in Asia have provided China with synergistic advantages in the inward internationalisation process. Factors, actors and moderators related to the inward–outward connections in the internationalisation of the Chinese cellular industry are presented in Figure 13.1.

Chinese cellular companies' competence: cost, quality and leadership

Thanks to low labour cost and economies of scale in the domestic market, Chinese companies have competitive advantage related to low cost. For instance, in 2003, Ningbo Bird had the biggest share in the

Figure 13.1 Internationalisation of the Chinese cellular industry: actors, factors and moderators

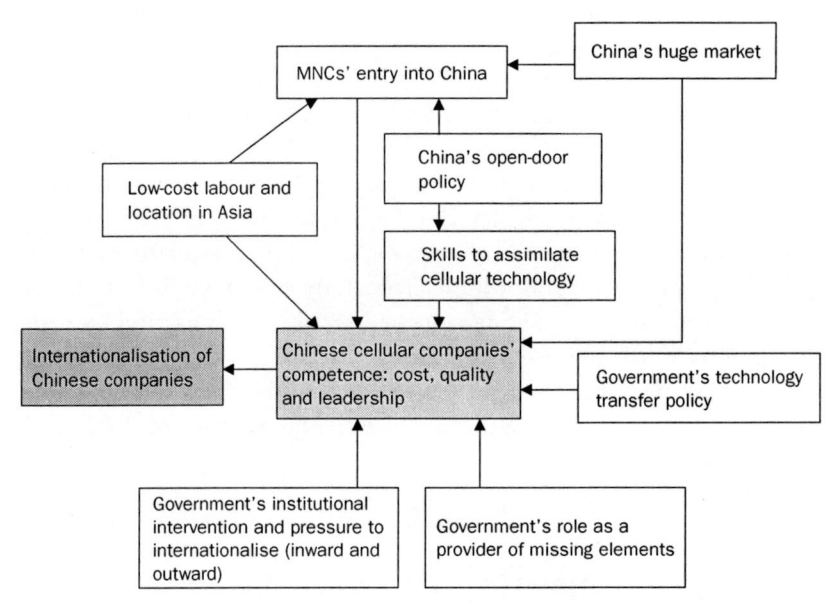

domestic market (15 per cent in the first half of 2003) and TCL displaced Ericsson as the third largest handset manufacturer (Choong, 2003). These players benefited from the economies of scale. In 2002, Chinese manufacturers were able to produce handsets 15–30 per cent cheaper than the global benchmarks (Lynch, 2002).

Cost advantage alone, however, may not be sufficient to compete in the global market. In terms of GSM mobile technology quality, Chinese handset manufacturers have closed the gap with MNCs (Wuzhou, 2003). A Portelligent study has indicated that the 2G and 2.5G^3 phones manufactured in China were technically comparable with those from MNCs (Merritt, 2003).

The government's technology transfer policy in the cellular sector: foreign and domestic sources

Although Chinese companies are typically weak in creating technologies, they are active in acquiring foreign technologies. Technology transfer from foreign companies is supplemented with domestic creation and transfer of technologies.

Although studies dealing with technology transfer in the Chinese cellular sector are limited, the experiences of firms in computing and telecom sectors indicate that, before China's entry to the World Trade Organization (WTO), the transfer of foreign technologies to domestic firms was a prerequisite to accessing China's cheap workforce or enormous market. Three of the most critical technology areas are computers, telecommunications and aerospace (Hollings and McMillion, 2000). These sectors have experienced the highest level of US commercial technology transfers to China (Bureau of Industry and Security, 1998a). In the mid-1990s, of the top US investors in China, half were involved in joint ventures producing electronics, telecommunications and computing (Schoenberger, 1996).

Product superiority is important but not sufficient to succeed in the Chinese market. In computing industry, for instance, Microsoft was able to establish itself as the 'standard' operating system in China only in exchange for helping Chinese programmers to create a Chinese language version of the Windows software – a significant transfer of technological know-how (Barnathan, Crock and Einhorn, 1996). Similarly, in the mid-1990s, Intel had over an 80 per cent share in the Chinese CPU market. The successes, however, were accompanied by numerous cooperative developments and commercial offset agreements (Bureau of Industry and Security, 1998b).

Technology transfers in the telecom sector seem to be still more interesting mainly because of the jurisdiction of the Chinese military and Ministry of Information Industry (MII) over a range of radio frequencies (Bureau of Industry and Security, 1998a). For foreign telecom companies to gain access to the military-controlled bandwidth, 'the PLA [People's Liberation Army] and its [Chinese joint venture] partners expect a significant infusion of capital and technology' (Mulvenon, 1997). The Chinese partner for a GTE joint venture to build a national paging network, for instance, was the Guangzhou Guangtong Resources Co., a PLA-affiliated company (Bureau of Industry and Security, 1998a). This partnership was necessary for GTE to gain access to the required radio frequencies and distribution system (Clifford, Roberts and Engardio, 1997).

Domestic creation and transfer of technology has supplemented transfers from foreign firms. China has been active in creating technologies domestically. Modernisation of the science and technology sector became one of the four dominant goals of the Chinese Communist Party after Deng Xiaoping came to power in the late 1970s (Wang, 1993). To achieve this goal, Deng endorsed the concept of 'technology as a commodity' and formed a large number of research institutes. By 1987, there were over 5,000 state-operated research units in China, employing over 230,000 scientists (Baark, 1988). The transfer of technology from state-operated research units to industrial enterprises has been fuelling rapid technology diffusion in China.

MNCs' entry into the Chinese cellular sector: motivation and attraction

Activities related to inward internationalisation in the Chinese cellular sector date back to the early 1980s, following China's open-door policy. The Swedish company, Ericsson, for instance, began cooperation with Beijing in 1983 (Ericsson, 2003b). In 1985, Nokia opened an office in Beijing to market its NMT450 analogue cellphone system (Chinanex.com, 2003b). In 1987, Motorola also entered the Chinese market, setting up an office in Beijing (Chinanex.com, 2003a) and selling analogue phones.

Intense competition and shrinking profit margins have forced MNCs to enhance efficiencies and lower costs. China's low cost has been a major attraction to MNCs. The Bureau of Industry and Security observes:

> So, why are US and other foreign high-tech firms in China? The answer heard most often in our interviews and survey of press reports is that one cannot be in China, lest a competitor get a foothold first. (Bureau of Industry and Security, 1998a)

Of equal importance in the discussion of China's technological developments is China's location in Asia. Quoting a World Technology Evaluation Center report, the Bureau of Industry and Security writes:

> What is driving the rush to China? The motivation does not appear to be profits. Even the American giant, Motorola, appears not to be making much return on its huge investments in China, and is reinvesting in China whatever revenues are realised from its joint ventures. The primary motivation is also not necessarily the availability of labor at low cost, although this is a big factor. Rather, it is to be nearer to the fastest growing electronics markets, which are now in Asia, and where the market demand and government support for electronics is significant. According to a recent study on China's electronics sector, 'In fact, all US electronics companies are increasing their Asian investments in R&D to take advantage of

favorable industrial-government partnerships and engineering workforces that are highly motivated and well trained (frequently in the United States).' (Bureau of Industry and Security, 1998a)

Skills required for the assimilation of cellular technology

China's open-door policy also contributed to the manpower required to assimilate cellular technology. As noted above, an increasing number of Chinese are returning back home in recent years (Bureau of Industry and Security, 1998b).

Chinese handset manufacturers initially concentrated on less technology-intensive components such as casings and screens. The handset modules were sourced from Western suppliers such as the French company Wavecom. More broadly, China depended on electronic components (e.g. computer chips) imported from the USA and the European Union to manufacture laptops and mobile phones (Sparshott, 2005).

Thanks to a wide availability of technical manpower, Chinese handset manufacturers shifted their focus from the 'look and feel' features of handsets (Ramstad, 2003) to sophisticated R&D activities. To further consolidate their market positions, Chinese cellular players are intensifying R&D activities. The growing strength of Chinese firms on the R&D front is powerfully illustrated in the sharp drop in its share of imports from the USA and EU in advanced electronic components such as computer chips, from 27 per cent in 2004 to 12 per cent in 2000 (Sparshott, 2005).

Huawei and ZTE invest over 10 per cent of their revenue in R&D. They have established R&D centres in a number of foreign locations. Around half of Huwai's 34,000 employees do R&D work. The company's patent filings almost doubled each year during the 1990s and reached 2,400 in 2005. Similarly, ZTE has filed 2,300 patents in China, Europe and the USA. Likewise, in 2002, TCL doubled its R&D staff to 1,000 and cut the number of modules bought from foreign companies. In the same vein, out of 7 million phones sold by Bird in 2002, 1 million were completely designed by the company.

Institutional intervention and pressure to internationalise

The government's active and meaningful intervention has thus played a pivotal role in the emergence of Chinese cellular players. Starting in the mid-1990s, the Chinese government designed a series of programmes to accelerate the development of the telecom sector. These interventions led to extensive re-engineering of and intense competition in the mobile telecom sector. China Unicom was formed in 1994 to compete with the then-monopoly China Telecom. China Unicom was licensed for mobile, paging, data, internet and long-distance communication (James, 2001). China Mobile, China Unicom and a number of high-tech firms were forced to follow the rigorous disclosure requirements of the NYSE, NASDAQ and Hong Kong's Growth Enterprise Market and to dismiss redundant workers (McDaniels and Waterman, 2000).

In the early 2000s, the Chinese government began to focus on the mobile phone industry's balance of trade and also started tracking the competition in the industry to make sure that it was under control (Dean, 2003). In the summer of 2003, the Ministry of Commerce and MII organised a meeting of Chinese handset manufacturers in Fujian province to encourage handset manufacturers to increase their exports (Dean, 2003). In response to this policy, many original design manufacturers (ODMs) are moving to neutralise the balance-of-trade argument by shifting production to China. In June 2003, South Korean ODM Pantech finalised a joint venture in Liaoning. In August, Telson, another South Korean ODM, announced it was investing in a production joint venture in Shandong (Dean, 2003). The MII is also considering restricting ODM imports so as to push out weaker players relying on the resale of imported handsets (Dean, 2003).

Government as a provider of other missing elements needed to internationalise

The Chinese government has traditionally offered cheap credit to state-backed enterprises. Since the late 1990s, the government has narrowed its priority focus to the cellular industry. In August 2003, the government gave US$85 million to the TD-SCDMA Industry Group, which is formed by eight domestic companies developing products and services based on TD-SCDMA (China.org, 2003). In 2003, the Chinese government also set a target to increase the domestic handset manufacturers' share of the market to 85 per cent by 2006 (Merritt, 2003). This goal, however, failed to materialise. Indeed, Gartner, the US-based research firm, estimated that in the second quarter of 2007, Nokia, Motorola and Samsung, the top three foreign brands, together held more than a 50 per cent share of the mainland market (Chung, 2007).

State-owned players have benefited tremendously from government's cheap credit and soft loans. For instance, in March 2005, TCL received US$723 million in credit for exports including handsets. Beijing is also providing soft loans to network providers from developing countries to purchase equipment from Chinese suppliers. In February 2005, Huawei and ZTE announced that they would receive US$600 million and US$500 million respectively in financing from the Export-Import Bank of China to help overseas buyers purchase their equipment. Likewise, in April 2005, the China Development Bank provided Nigeria with a loan of US$200 million for the nationwide deployment of CDMA450 wireless access technology to be built by Huawei.

Discussion and conclusion

Our analysis clearly establishes the significance of inward internationalisation activities in the subsequent outward internationalisation of Chinese firms in the cellular market. Of course, factors such as cheap labour and economies of scale associated with the domestic market have contributed to the competence of Chinese players in the global market. The exports of the Chinese cellular industry are mainly going to countries that are economically closer. Although Chinese players have developed a reasonable level of product quality and leadership in the cellular arena, their products are still perceived as inferior to those of players from advanced countries. On the bright side, low cost is a major competitive advantage for Chinese firms. They are thus better able to capitalise on the high price elasticity of consumers in low-income countries. In advanced economies, on the other hand, winning the trust of telecom operators that purchase handsets and installing after-sales service networks may prove to be a challenge of another magnitude (Dean, 2003).

The above sources of Chinese competitive advantage – cheap labour and economies of scale associated with the domestic market – are less relevant in the case of high-technology exports. In the 3G arena, for instance, technology superiority is required to succeed in the global market mainly because the demand for such technology is likely to originate from more advanced countries.

The Chinese government has supplied various elements required for the outward internationalisation process, which has helped these players sustain an enviable momentum in their internationalisation activities. For instance, the government has provided money; helped in the transfer of domestic as well as foreign technologies; removed weak players; and provided pressures to internationalise.

Chinese cellular players are also facing a number of roadblocks in domestic as well as international markets. Poor branding is among the Chinese players' most glaring shortcomings. The inescapable fact is that Chinese consumers associate foreign brands with superior quality and functionality, which enables foreign brands to command a price premium. Even in developing countries such as India, consumers have a deep-rooted perception of China as a technologically backward country. In developed markets such as the USA, analysts suggest that the news about faulty products originating from China is likely to pose a threat to Chinese brands, including those in the cellular sector such as TCL (Madden, 2007).

Second, although many Chinese firms have started setting up offices in various foreign locations, they lack operations (e.g. after-sale services) outside China. Moreover, Chinese cellular companies lack familiarity with global practices (Chai, 2003).

Third, Chinese cellular players are much smaller in size compared with their foreign counterparts. For instance, Bird's revenues and profits in 2003 were US$1.3 billion and US$29.5 million compared with Nokia's US$36.1 billion and US$4.4 billion respectively.

Fourth, foreign multinationals are fighting back by strengthening their weak areas. Nokia, for example, has set up its own grassroots network in some 300 Chinese cities and has developed phones specifically-targeted at the Chinese market, with models that make it easier for Chinese consumers to write text messages in Chinese characters.

Notes

1. The global system for mobile phone communications (GSM) was the first commercially available digital standard mainly used in Europe and Asia.
2. CDMA (Code Division Multiple Access) technology allows many conversations to be carried over one frequency. It does this by separating each communication into groups of bits, tagging each group with a different code, and sending the communications in groups of bits mixed together. CDMA is mainly used in North America (Bernatchez, 2008).
3. Second-generation (2G) refers to digital wireless standards that concentrated on improving voice quality, coverage and capacity. It was originally designed to support voice. 2.5G builds upon the 2G standards by providing increased bit-rates and bringing limited data capability.

14

Chinese high-technology firms' outward-oriented mergers and acquisitions: a case study

International mergers and acquisitions (M&As) involving Chinese businesses are a relatively new phenomenon. In recent years, however, there have been some high-profile international M&As involving Chinese technology firms such as Lenovo and TCL. In particular, TCL's M&A activities in Europe did not work out as planned due to a number of unforeseen difficulties. This chapter investigates the enabling factors and motivations associated with Chinese businesses' outward M&A activities and examines the determinants of M&A performances. Specifically, we provide three M&A case studies of TCL in Europe – the Schneider acquisition in Germany, the TCL-Thomson deal, and the TCL-Alcatel deal.

Introduction

According to the Boston Consulting Group, a fourth wave of Chinese companies' overseas mergers and acquisitions (M&As) is rising.[1] While Chinese firms' outward-oriented M&A activities have been relatively slow in the past (Lange, 2005), there have been a number of high-profile M&A activities in recent years. Among Chinese firms engaged in outward-oriented M&A activities, Today China Lion (TCL) is a highly visible example. TCL acquired the television assembly business of the French company Thomson SA, the handset manufacturing business of Alcatel and the bankrupt German company Schneider.

Among Chinese companies involved in M&As, TCL is far more entrepreneurial and is among the few Chinese companies engaged in high-profile international M&A activities (Einhorn, Roberts and Matlack, 2003). For instance, a survey of 800 companies involved in domestic M&As found that 86 per cent of them invested in firms within their own city and 91 per cent invested in firms within their own province (Milberg, 2004). Likewise, during the mid-1980s until 2000, most international M&A deals involving Chinese businesses were inward-oriented (Hemerling, 2006).

The Chinese context from the standpoint of M&A activities

The Chinese government has been pressuring as well as encouraging Chinese firms to expand overseas (Powell and Roston, 2005; Hemerling, 2006). To facilitate the internationalisation process, Chinese regulators have been rapidly liberalising laws related to foreign investments by Chinese companies (Lange, 2005).

There has been significant progress in property rights (Pei, 1998), which has facilitated M&As (Gu, 1997; Sun, 1997; Perotti, Sun and Zou, 1999). The Chinese government is also gradually relaxing its acquisition rules (Sun, 2006). Consequently, foreign direct investment in the form of cross-border M&As is gaining popularity in China (Peng, 2006).

At the 15th Party Congress in 1997, then President Jiang Zemin announced an initiative to boost M&A activities (Burdekin, 2000). In 1998, the authorities targeted 2,000–3,000 enterprises for bankruptcy or M&A (World Bank, 1999: 30). Especially in the light industry sector, the objective of these M&As was to form large conglomerates similar to Korean *chaebols* (Dana, 1999).

Notwithstanding the reforms discussed above, some analysts argue that the Chinese government has exercised its power over these firms in a 'chaotic way', which has hindered their success overseas as well as on the home front (Gilboy, 2004). This has been true for some state-owned Chinese firms' M&A activities. Nolan and Yeung (2001) examined M&A activities of two large state-owned steel firms. Both firms in their study faced political and social pressure to diversify their businesses and engaged to 'administrate mediated' M&A activities, which were top-down in nature. Factors such as limited business skills in the new non-core businesses, a lack of economies of scale, distraction from their core competences and lack of human and financial capital led to competitive disadvantages. In both cases, profits made by the firms in their core businesses were offset by losses in the newly acquired non-core businesses (Nolan and Yeung, 2001).

There is an interesting contrast here between inward and outward-oriented M&A activities involving Chinese businesses. Chinese companies are gravitated towards foreign firms to overcome their weaknesses in technology and marketing functions such as branding, sales and distribution channels. Analysts argue that many Chinese companies are analysing their companies and competitors for possible acquisitions (Ramos, 2005a, 2005b).

TCL, for instance, capitalised on the marketing-related resources and technologies of Schneider, Thomson and Alcatel to enter into Europe and to some extent the USA. Here, TCL's acquisition of Thomson-RCA's television line and Alcatel's cellphone business allowed it to have access to Western brand names, distribution networks in Europe and Western technologies (Einhorn et al., 2004). Thus in China, the company continues to sell its TCL brand products, while in North America it sells RCA brand televisions and in Europe its Schneider or Thomson brand products (Seno, 2004). Foreign acquisitions have thus provided TCL with a unique opportunity to sell overseas. With such a business model, the company can over time co-brand its products to build awareness of the core brand and, after achieving a sufficient degree of association and awareness, phase out the acquired brand (Gao, Woetzel and Wu, 2003).

A note on TCL

TCL deserves special attention among Chinese technology firms. The company is the world's biggest television manufacturer and hit a world sales record with 2.3 million units in 2005. TCL also manufactures other consumer electronics products such as PCs, DVD players and household appliances (Einhorn, Roberts and Matlack, 2003). The company produced its first television set in 1992 (Chandler, 2004) and built its first cellphone in 1999. In 2007, the multimedia sector led by televisions accounted for about 55 per cent and cellphones for 13 per cent of the company's revenue (chinadaily.com.cn, 2008). When looking at the changes over time, we see that multimedia has always been TCL's main business. As early as in 2002, multimedia accounted for almost 50 per cent of the company's market share. The cellphone market was at that time its second biggest business line with an almost 40 per cent share.

TCL was founded in 1981 in Guangdong. Originally, the company was controlled by the Huizhou city government (*Economist*, 2003b). Indeed, until 1996, TCL was 80 per cent owned by the city of Huizhou (Chandler, 2004). In spite of its being owned by the city, however, the company was aggressively commercialised from the beginning (*Economist*, 2003b). Subsequently, the city government diluted its control over TCL. Prior to its overseas M&A activities, TCL was one of the first major manufacturers in which the Chinese government had cut its ownership level to a minority status (Ramstad, 2004). Without majority government equity, TCL can be considered to be a private company and hence subject to relatively little direct government influence. By the end of 2003, the Huizhou city government held a 41 per cent ownership share in TCL, which dropped to 25 per cent by the end of 2004 and 12.84 per cent by 2007 (China IT & Telecom Report, 2007c).

TCL's strategic investors include Toshiba and Philips, and its technology partners include Microsoft and Intel (Einhorn, Roberts and Matlack, 2003). By the end of 2007, Philips Electronics China BV became TCL's third largest shareholder with a 6.3 per cent share, while Toshiba China Co. Ltd became the ninth biggest shareholder with a 1.04 per cent share (China IT & Telecom Report, 2007c).

As noted above, while most Chinese firms prefer to invest within their own city or province (Milberg, 2004), TCL is engaged in a number of high-profile M&A activities at the global level. In 2007, TCL's revenue was 39 billion yuan (US$5.6 billion) with a net profit of 396 million yuan (US$55.7 million) (chinadaily.com.cn, 2008). The net profit, however, went into the red in 2005 and 2006. In 2005, TCL made a loss of US$40 million. As of 2004, TCL employed over 40,000 workers in its factories in China, Vietnam, the Philippines and Germany, making televisions, cellular handsets, notebook PCs, refrigerators and air conditioners (Chandler, 2004). By early 2008, TCL sold its products with brands such as Thomson, RCA and Alcatel in over 40 countries (Hoover's Company Records, 2008).

In 2002, TCL displaced Ericsson as the third largest manufacturer of cellular handsets (Choong, 2003). With their widespread grassroots distribution networks that comprised of thousands of employees and dedicated shops throughout China, TCL's consumer electronics businesses provided the company with a unique competitive advantage. The company took advantage of such networks to push products to end-consumers. TCL also introduced additional tactical elements to compete in Chinese cities. Handset retailers have traditionally favoured foreign brands over Chinese brands such as TCL, as they have much larger turnover and thus generate more revenue for distributors and sales agents. Thus, to compensate for these disadvantages, TCL paid higher commissions to distributors and experimented with a wide range of new channels and promotions. The company produced about 10 million handsets in 2003 (Ramstad and Pringle, 2004). TCL handset models are considered to have sophisticated design features to give them lower prices than their competitors (*Media Asia*, 2005). The company's diamond-studded phone casings were among their phone models with notable success.

After conquering the domestic market, TCL attacked the international arena. TCL brand electronics products are exported to the Middle East, Russia, South Africa and Southeast Asia along with a number of developed countries (Gao, Woetzel and Wu, 2003). TCL exports handsets to Thailand, Bangladesh, Indonesia, Vietnam and other countries. The company's overseas handset sales increased to 380,797 units in July 2005 compared with 101,268 units in July 2004 (AFX News, 2005). In 2004, TCL products had a 14 per cent share in Vietnam and 8 per cent in the Philippines (Chandler, 2004).

TCL is also becoming increasingly sophisticated in technologies. In the early years, the company bought handset modules from foreign companies. In 2002, TCL doubled its research and development (R&D) staff to 1,000 and cut the number of cellphone modules bought from foreign companies such as Wavecom (Dean, 2003).

TCL's outward M&A activities: enabling factors

Resources

Some Chinese companies have plenty of available capital (Hemerling, 2006). For instance, in 2002, TCL claimed a 19 per cent market share in China and was the only Chinese television producer to consistently gain market share and show a profit each year during 1998–2003 (Chandler, 2004). In the first three quarters of 2003, the company earned US$169 million on sales of US$4.2 billion (Chandler, 2004). In the first half of 2004, strong television sales in China and abroad led to a 44 per cent increase in TCL's earnings (Einhorn et al., 2004). TCL thus leveraged its increasing profits and market-leader position to buy Western brands and form partnerships (Seno, 2004).

Chinese players have also benefited tremendously from the government's cheap credit and soft loans. Government subsidies and credits have also been an important trigger for TCL's M&A activities. For instance, in March 2005, TCL received US$723 million in credit from the government.

Dilution of government control

It is suggested that 'M&A activity in China is being driven by continued restructuring and privatisation of former state-owned enterprises' (PricewaterhouseCoopers, 2005). In this regard, it should be noted that unlike the steel firms in Nolan and Yeung's (2001) study, being a private firm, TCL has been facing relatively less political and social pressure.

Deal-making expertise

Some experts argue that Chinese players face a serious challenge in overcoming the lack of modern management thinking. They maintain that Chinese firms lack entrepreneurship and management talent (Mourdoukoutas, 2004) and face difficulties working in international partnerships (*Economist*, 2005f). Others have argued that Chinese companies 'often arrive unprepared, overpay for acquisitions, fail to do their due diligence and aren't sure how their new Western holdings fit into their global strategies' (Theil, 2006). In sum, it is tempting to describe the business and management skills of Chinese players such as TCL as in their infancy.

With regard to Chinese companies' deal-making capability, however, it is important to note two trends. First, the world's top investment banks have provided Chinese companies such as TCL with the necessary expertise for M&A deals (Ramos, 2005a). In the case of the TCL-Thompson deal, for example, the two companies had to negotiate specific and detailed contracts for access to those parts of the business, including sales and existing intellectual property not being transferred to their joint venture. New intellectual property rights generated by the joint venture had to be licensed to one of the shareholders for use, with royalties being paid back to the joint venture (*International Financial Law Review*, 2005). The world's top banks helped TCL to understand these issues.

Second, Chinese enterprises such as TCL are building up their management, marketing and R&D teams by attracting staff from multinationals like Microsoft, Procter & Gamble, and Ogilvy & Mather (Chen 2004; Balfour and Roberts, 2005). To take one example, in 2005, a senior manager from Motorola with experience in the USA and Singapore joined TCL as a vice president as the latter offered him 'a more challenging job' (Balfour and Roberts, 2005). Indeed, a highly-professional management team is helping TCL to understand the local market and sales networks (Gao, Woetzel and Wu, 2003). The managing director of Asian M&As at investment bank Morgan Stanley in Hong Kong described that 'at this stage in China's development, large local companies are completely capable of using their competitive advantage to really go international and continue doing these deals' (Ramos, 2005b).

Cost advantage

Perhaps the most notable features of Chinese firms are that they are leaner, more flexible and have lower costs (Williamson and Zeng, 2004). Many Western companies are under tremendous pressure to cut costs, and Chinese buyers such as TCL can address this by moving a large proportion of the acquired company's production to China (Gao, Woetzel and Wu, 2003).

TCL's outward M&A activities: motivators

China's technology companies are under tremendous pressure to expand overseas (Gao, Woetzel and Wu, 2003) but are facing a number of problems. Nolan (2002) summarises views of experts who underestimate Chinese firms' potential to become global players:

> China's large firms will generally be unable to compete on the global level playing field with the World's leading system integrators. China should, therefore, focus on improving the position of its indigenous firms with global value chain of the world's leading companies. (Nolan, 2002: 120)

The basic ideas behind Chinese firms' acquisitions are thus simple. Through M&As, Chinese firms seek to build on their strengths and overcome weaknesses in the areas of branding, sales, marketing and technology (Ramos, 2005a).[2]

From the sellers' standpoint, the brand, technology and distribution network create a unique opportunity to work with the buyer to sell their loss-making businesses successfully. The sellers' revenues come in various forms including trademark royalties, technology royalties and distribution fees (Lange, 2005).

Branding and other marketing-related resources

Chinese firms arguably lack experience in developing and managing international operations (Lange, 2005). As noted earlier, China's import of Western or modern management techniques traditionally concentrated on the tangible and quantitative approach. Note that Mao viewed the market as 'embodiment of capitalism' and suppressed it with a central plan-based economic system (Dittmer and Gore, 2001). Indeed, the Chinese government is providing conflicting forces to local firms with respect to market orientation. Beijing wants its firms to succeed but the inescapable fact is that it also wants them to fulfil political and social expectations.

It is well established in the marketing literature that most developing world-based firms severely lack branding expertise and power. Chinese technology firms, for instance, are largely regarded as providers of basic value or as 'unknown names on a box' (Hirt and Orr, 2006). These companies thus compete on price (Mooney, 2005) and believe that they are forced to set prices lower than what their products deserve (Hirt and Orr, 2006).

Most Chinese companies are structured internally to favour sales functions and do not have a marketing department, and hence lack an understanding of branding (Madden, 2004). Chinese firms also lack familiarity with global practices (Chai, 2003), established international distribution networks (Lange, 2005), and after-sales services and systems to collect firsthand feedback from end-users in foreign countries (Zhou and Belk, 1993). For instance, in advanced economies, many Chinese handset manufacturers lack international after-sales service networks, which has been a major barrier to winning the trust of telecom operators (Dean, 2003).

Other factors contributing to Chinese brands' weaknesses in Western countries include the negative perception of the Chinese government regarding environmental issues and human rights, as well as concerns about outsourcing and domestic job losses (Tucker, 2006). It is also important to note that when brand names are not well known, the country of origin becomes relatively more important in the consumer's assessment of a brand (Ofir and Lehman, 1986). This means that one would not expect Chinese brands such as TCL to be successful in the developed world.

Moreover, Chinese brands like TCL face problems even in the domestic market. Chinese consumers pride themselves on driving foreign-brand cars and using mobile phones, computers and other high-tech items manufactured in the industrialised world (Euromonitor, 2004; Gilboy, 2004). Faced with such trends, TCL has introduced foreign elements in its marketing activities within China. As noted earlier, for instance, TCL's advertisements feature a South Korean pop star (Dean, 2003).

In the words of TCL's chief financial officer, the company wants 'to be the next Sony ... to be the next Samsung' (Crowell, 2005). It could, however, take several years and billions of dollars to build TCL's own brand (Lange, 2005). Acquiring international brands and distribution networks could put TCL on the fast track to achieve its global ambitions. Like other Chinese companies, TCL has sought to buy existing well-known brands (Hirt and Orr, 2006). Most obviously, TCL's appropriate M&A targets can be characterised by valuable brands, customer bases and distribution channels, but also by low prices.

Technology

To be sure, most Chinese firms are technologically weaker compared with their Western counterparts. Therefore, non-Chinese companies with more advanced technology are appropriate M&A targets for Chinese firms (Gao, Woetzel and Wu, 2003). It should also be noted that technology licence arrangements are very complex, sometimes entailing a large number of patents in a single product. In most cases, manufacturers with strong patent portfolios often share their patents among themselves through bilateral cross-licence agreements or multilateral patent pools (Lange, 2005). Chinese companies with relatively weak patent profiles such as TCL tend to be in a disadvantage in terms of their access to such patent pools. As Chinese companies expand overseas, there is a higher degree of risk associated with patent infringement claims (Lange, 2005). Foreign acquisition of technologically stronger companies is thus likely to reduce such risks. While some companies have been investing heavily to build their brands, TCL has chosen a different strategy. In the short run, the company wants to control several international brands by buying them rather than developing its own.

Cases related to TCL's M&A activities in Europe

Schneider acquisition in Germany

In an attempt to break into the European market, TCL International Holdings purchased Schneider Electric Appliance Co., based in Durkheim,

Germany, for US$8.2 million in September 2002. This low price was possible as the 113-year-old firm had gone bankrupt in early 2002. The acquisition price included Schneider's plants, its distribution network of chain stores, hypermarkets and mail-order business, as well as the trademark rights to a series of brands including Grundig, Schneider and Dual (Gao, Woetzel and Wu, 2003). Schneider had three colour television production lines, with an annual production capacity of 1 million units. In 2001, the company's revenue surpassed €200 million (US$192 million) (China IT & Telecom Report, 2002). The company's major markets were Germany, the UK and Spain.

Branding and other resources related to marketing

The acquisition was partly motivated to capitalise on the Schneider brand and access Schneider's worldwide distribution network (*People's Daily*, 2002). TCL is also using its Schneider brand to position its products in the high-end segment of the Chinese market (Gao, Woetzel and Wu, 2003). According to TCL officials, the company hopes to introduce its own brand to the European market in the future (Gao, Woetzel and Wu, 2003).

Performance of the Schneider deal

It is suggested that TCL did not do enough research regarding the local production costs (Jun, 2005). Critics have also questioned Schneider's brand reputation. Some have also argued that Schneider severely lacked brand recognition in Europe (Theil, 2006). Moreover, local managers are likely to know their home market better than outsiders, and for that reason, it would have been in TCL's interest to recruit the best German managers and let them make strategic decisions (Bacani, 2004). TCL, however, failed to do so.

TCL-Thomson deal

In November 2003, TCL and Thomson Electronics formed a US$560 million television manufacturing joint venture known as TCL-Thomson Electronics (TTE). The signing of the agreement took place in the presence of the French prime minister and the Chinese president (*International Financial Law Review*, 2005). The deal was considered as a marriage between TCL's cost advantage and Thomson's strengths in brands, distribution and research networks in Europe and the USA (*Economist*, 2003b). Analysts saw TCL's control over the reputed Thomson television brand as a signal of the intention of Chinese companies such as TCL to move 'beyond low-margin contract manufacturing' (Shih, 2006a, 2006b).

TCL contributed its television and DVD manufacturing plants in China, Vietnam and Germany as well as its R&D centres and sales network to the joint venture. Thomson contributed its television manufacturing plants in France, Mexico, Poland and Thailand, all DVD sales business, and all television and DVD R&D centres (*Medialine*, 2003).

TTE was expected to have sales of US$3.5 billion and a production of over 18 million television sets in 2004 (Chandler, 2004). Indeed, TTE became the world's largest producer of traditional cathode ray tube televisions (Escobar, 2005). In 2003, it manufactured at least 6 million more sets than any other company in the world (Chandler, 2004).

In July 2004, TCL acquired majority control (67 per cent) of TTE (Crowell, 2005) and finally, in early 2005, TCL acquired full ownership of the venture (Ramos, 2005a). TCL thus became the first Chinese company to acquire majority control of a well-established global business from a Western multinational (*International Financial Law Review*, 2005). The TCL-Thomson deal has been one of the biggest in which a Chinese company has taken control of a Western enterprise. TCL controlled Thomson's television plants and its subsidiary based in Indiana, USA (Chandler, 2004). Thomson also provided a receivables purchase facility to TTE to help to meet working capital needs (Lange, 2005). TTE employed about 200 former Thomson workers, including product development engineers in the new company (Maurer, 2003; *Indianapolis Business Journal*, 2004).

Technology

A major motivation behind TCL's acquisition of Thomson's television unit was to accelerate its development of television technology (Hirt and Orr, 2006). Importantly, however, both Thomson and TCL were relying on cathode ray tube technology for their production lines – a technology that analysts predict will be replaced over the next decade by newer display technologies (Tagliabue, 2003). TCL now depends on Thomson's rear-projection technology to make thin television sets (Escobar, 2005).

Branding and other resources related to marketing

Thomson licensed its trademarks and Thomson and RCA brands (including the image of Nipper, RCA's famous little dog) to the new venture (Crowell, 2005) for 20 years (Lange, 2005). The combined company would use the RCA brand in North America, the TCL brand in Asia, and Thomson in Europe (Tagliabue, 2003).

In 2003, the Thomson brand had about 8 per cent market share in Europe (Malester and Tarr, 2003) and the company's RCA brand had about 18 per cent of the market for television sets in the USA (Tagliabue, 2003). Meanwhile, the TCL brand had a 17 per cent share in China (Malester and Tarr, 2003). The Chinese managers thus planned for the deal to give TCL footholds in Europe and the USA. The trademark and brand licences were to give TCL the flexibility to manage its brand portfolio and take advantage of the value of the acquired brands (Lange, 2005). It was expected that, over time, the company could build its own brands.

Performance of the deal

TCL's chairman said TTE could make a profit within 18 months, but the company failed to achieve this goal (*SinoCast China*, 2006i). TCL experienced difficulty in integrating the Thomson business with its operations (Einhorn, Wassener and Reinhardt, 2005; Lindblad, 2005). It took a year for TCL and Thomson executives to overcome differences in language, salary and other issues and to develop a 'coherent production strategy' together (Ramstad and Li, 2006).

TCL expected that TTE would be profitable by the end of 2004 (Chandler, 2004). But in 2005, TCL Multimedia reported a second-quarter loss of US$6 million (Lindblad, 2005). In October 2006, TTE announced that most European joint venture operations with Thomson would be 'shrunk, sold, closed, or returned' including a factory in Poland and distribution networks (*Economist*, 2006c).

TCL began formulating a restructuring plan for its European business in May 2006 (Lau and Mitchell, 2006). In November 2006, TCL announced restructuring and adjustment in TTE's European employees and sales offices in Europe, including Germany, Spain, Italy, Sweden and Hungary (*SinoCast China*, 2006i). According to the TCL chairman, the reshuffle gained support from local governments and Thomson. In a research note, Goldman Sachs viewed 'the restructuring plan as positive' and stated that the pullout from Europe would allow TCL to focus on the cash-generative Chinese market (Pang, 2006). As a 'sign of fracturing ties', TCL announced it would release Thomson from the obligation to hold a 30 per cent stake in TCL Multimedia for five years (Lau, 2006b).

While Thomson's trademark could be used for 20 years in North America, according to the modified agreement, TTE remained free to use Thomson's trademark in Europe only until the end of 2008 (AFX – Asia, 2006a). Subject to paying the agreed licence fee, however, TTE was allowed to use Thomson's trademarks in some regions, including Russia and Ukraine, until 2013 (*SinoCast China*, 2006j). However, TCL could not repay its debt and the French law required it to file for insolvency (*China Daily*, 2007b). In May 2007, TCL declared its European operation 'insolvent' (Economist Intelligence Unit, 2007). The TCL-Thomson deal is cited as an example of a Chinese company that 'failed miserably in overseas expansion' (seekingalpha.com, 2007). Factors contributing to the failure include TCL's lack of international M&A experience, a lack of capable managers in the company to help improve communications with European employees, and the inability to transfer its cost advantage due to labour practice-related differences in the two economies (Rugman and Li, 2007).

TCL-Alcatel deal

In October 2004, TCL Communications and Alcatel formed a US$128 million joint venture partnership in the cellphone business known as TCL and Alcatel Mobile Phones Limited (TAMP). Alcatel's phone business was losing money and market share and only accounted for a small portion of the company's overall business (Dano, 2004). The Alcatel-TCL deal thus combined a struggling phone company with a Chinese technology firm looking to expand its business globally.

TCL held a 55 per cent equity stake and Alcatel owned the remaining 45 per cent (*People's Daily*, 2004). This deal also was signed in the presence of the presidents of China and France (Seno, 2004). In May 2005, TCL announced that it would purchase Alcatel's share in their joint venture to enhance the integration of its businesses (Chen, 2005; Einhorn, Wassener and Reinhardt, 2005). As of 2006, Alcatel owned less than 5 per cent of TCL Communications (Einhorn, 2006c).

Access to technology and marketing-related resources

A major motivation behind TCL's acquisition of Alcatel's handset business was the need to gain cellphone patents (Hirt and Orr, 2006). TCL also hoped that the deal would give the company the branding power needed to compete with multinationals such as Nokia and Motorola (Einhorn, 2006c). Through TAMP, TCL expected to expand its businesses to Western and Central Europe, Africa and the Middle East (Dano, 2004).

Performance of the venture

For the first nine months of 2005, TCL Communications lost US$166 million (Einhorn, Wassener and Reinhardt, 2005). There were also reports of clashes in managerial values, personality and culture. Furthermore, Alcatel staff and TCL managers disagreed on strategic and tactical issues related to managing TAMP (Jie, 2006).

Discussion and conclusion

To an optimist, TCL's European acquisitions are an indication of Chinese companies' global ambition. The unsuccessful deals, mainly the French acquisitions, however, put a heavy financial burden on TCL. The accumulated loss from TCL's European operation amounted to €203 million as of September 2006 (Pang, 2006). The company attributed losses partly in regard to the acquisitions in France – Thomson's television unit and Alcatel's cellphone unit (Hirt and Orr, 2006). In 2005, the company announced a net loss of US$40 million, and for mid-2006 a net loss of almost US$100 million. In January 2007, TCL stated that it expected to have made a net loss in 2006 too.

Across the above examples we see that the sellers were, in the French cases, large multinationals with a number of product lines seeking to sell an unprofitable business, or in the case of Schneider, to sell a bankrupt company to Chinese investors. The sellers, however, had fairly strong international brand, good technology and established international distribution channels (Lange, 2005). TCL bought certain business lines from them (e.g. cellphone branding from Alcatel and television business from Thompson and Schneider) with the aim of making them into successful ventures.

Chinese technology companies' steep learning curve

Chinese companies are arguably on 'a steep M&A learning curve' and are learning their lessons 'the hard way' (*The Banker*, 2006). The TCL chairman said that the company's European M&A activities had taught a valuable lesson (Jie, 2006). Because of a lack of in-house management and marketing skills, some analysts argued that the biggest obstacle for Chinese companies such as TCL would be to find competent managers to manage the newly acquired business (Gao, Woetzel and Wu, 2003). Chinese buyers such as TCL can receive some support related to operational, financial, marketing, distribution and other management aspects from the seller (Lange, 2005). Nonetheless it is unrealistic to rely completely on the seller.

It is suggested that Chinese companies such as TCL often pay more attention to the brand, market and low purchasing price in M&A activities, but lack understanding of local laws and regulations, high labour cost and strict labour regulations (Jun, 2005). TCL also lacked experience of operating in Europe's fragmented and highly competitive markets (Theil, 2006). For instance, TCL's market approach failed to understand the necessity for more efficient R&D and inventory management to deal with rapid price fluctuations (Jie, 2006).

TCL's chairman admitted that the Thomson business was harder to manage than he expected (Lau, 2006b). In Europe, the company also failed to predict consumer responses and lacked an understanding of workforce laws (Lau, 2006b). Indeed, the TCL chairman admitted that 'the costs of firing the workforce and restructuring are legislated in Europe and are much more complicated than we imagined' (AFX International Focus, 2006). As the TCL chairman also confessed, 'our product, technology and management lagged behind the changing European market' (Jie, 2006).

Chinese technology firms' lack of resources to buy profit-making companies or established brand names

In terms of revenues and profits, Chinese companies are relatively small compared with their Western counterparts. For instance, Nokia is about 52 times as large as TCL in terms of revenue and 34 times when we compare revenue generated by cellphone business. Likewise, in 2005, Nokia manufactured about 24 times as many cellphones as did TCL.

Clash of managerial values, personality and culture

The nature of the integration process required for the success of M&A is a function of cultural differences between the merging firms (Lodorfos and

Boateng, 2006). TCL's painful adjustments with the newly acquired businesses are also a result of clashing managerial values, personality and culture. There are reports that while Alcatel staff preferred formulating detailed plans, TCL managers tended to take immediate actions without much analysis (Jie, 2006). With regard to selecting staff, the French side preferred educated and professional managers, while the Chinese emphasised entrepreneurship and pragmatism (Jie 2006). For the Chinese management team, the concerns about managing European counterparts who received salaries higher than their own also grew equally complex. For instance, the annual salaries of vice presidents for most Chinese companies typically vary from US$20,000 to US$30,000 (Economist Intelligence Unit, 2007). Speaking of these income disparities from the perspective of Chinese managers, an executive of the Hina Group reportedly said 'they think, "how can we have someone reporting to us who makes US$100,000 a year?"' (Economist Intelligence Unit, 2007). Most obviously, TCL's management also encountered such problems in its European operations. In Schneider's case, TCL failed to allow German managers to make strategic and tactical decisions (Bacani, 2004). Likewise, because of language issues and disagreements regarding salary and other issues as noted above, it took a long time for TCL and Thomson executives to develop a common production strategy (Ramstad and Li 2006). Faced with such realities, some argue that Chinese firms are more likely to succeed by focusing on M&A activities in developing Asian countries, which are more likely to welcome Chinese companies (Economist Intelligence Unit, 2007).

European perspective: the weaknesses of Chinese firms

For the European firms, in addition to the opportunity to sell their brand, technology and distribution network (Lange, 2005) and save jobs in loss-making enterprises, gaining access to resources that are considered to be unique to Chinese companies is a tempting proposition. These strengths, however, have turned out not to be unique after all. For instance, many Taiwanese contract-manufacturers have been utilising China's huge pool of cheap labour more effectively than Chinese firms such as TCL (*Economist*, 2003b). Likewise, Chinese companies such as TCL fall far behind their Japanese counterparts in terms of process engineering (*Economist*, 2003b). Indeed, Chinese firms such as TCL tend to increase labour and capital to solve problems rather than making improvements in processes (*Economist*, 2005f).

It is also important to note that in most cases, the companies being sold were operating in a highly-competitive market and were under enormous pressure from low-cost Asian manufacturers. Indeed, the sellers had started outsourcing their manufacturing activities to cut costs. In each case, however, the business line that was sold remained unprofitable (Lange, 2005).

European policy-makers viewed the M&As between TCL and their loss-making enterprises as a formula for saving jobs. TCL's restructuring, which entailed management changes and job cuts (Lau and Mitchell, 2006), was a severe blow to the employees in the acquired ventures.

Structural factors

TCL's European operation also faced some structural problems. European customers have stopped buying old-fashioned cathode ray tube televisions and started switching to new LCD screens (Yam, 2006). The TCL chairman acknowledged that the company focused on traditional cathode ray tube models and that the company's response to consumer demand for flat panel screens had been 'lethargic' (AFX News, 2006c). As noted earlier, TCL depends on Thomson's rear-projection technology to make thin television sets (Escobar, 2005). This may not be a technologically smart move as consumers are switching to digital television (e.g. Japan is planning to switch to digital television in 2007).

The cases discussed above shed some light on patterns associated with Chinese firms' outward M&A activities. TCL's M&A targets have been long-established European companies facing problems. The French companies wanted to sell some lines of their non-profitable businesses and the German manufacturer was already bankrupt before being acquired.

Notes

1. It is argued that the first wave (1986–96) focused on offshore investment. The second wave (1996–99) was associated with Hong Kong's return to China and enterprises started paying attention to overseas expansion. The third wave (starting in 2000) entailed domestic expansion and Chinese companies investing in the stocks of their foreign partners' joint ventures (China-embassy.org, 2006).
2. There are other advantages as well. TCL, for instance, wanted to avoid European quotas on the importation of Chinese TV sets and other legal barriers set up by the EU (Gao et al., 2003).

15

Concluding remarks

The rapidly transforming Chinese high-technology industry and market

In this book, we examined a number of idiosyncratic and unusual features associated with institutions and other ingredients shaping the Chinese high-technology industry and market. Of equal importance in the discussion of recent developments in the Chinese high-technology industry are major structural shifts taking place in terms of institutions, market, technology, product, actors and focus (Chapters 1 and 7). In this regard, we analysed contexts, mechanisms and processes associated with such shifts.

As we have demonstrated, the Chinese economy is clearly shifting into a higher technological gear. The country's achievements in cellular technology (Chapter 4), open source software (Chapter 8) and nanotechnology (Chapter 11) are among many visible examples that illustrate such a shift. Indeed, China has grabbed the global spotlight in these sectors.

As noted earlier, the most profitable niches in international IT markets are heavily skill-dependent (Campbell, 2001). In this regard, the current five-year plan's emphasis on human resource development and the increasing number of foreign-educated Chinese joining the country's technology workforce are likely to drive the growth of the Chinese high-technology industry further (Chapter 6). At the same time, the spillover effects created by foreign multinationals' research and development (R&D) activities have given a major boost to the growth of the country's high-technology talent. Note too that some multinationals are performing cutting-edge R&D in China.

Ten years ago, Koo (1998) noted that the 'progress in China has been scarcely noted in the Western media and overshadowed by the focus on the human rights abuses as perceived by the West'. This observation remains generally true today. The Western media have neglected to pay enough attention to the transformations underway in the Chinese high-technology industry.

The growing size, sophistication and impacts of the Chinese technology industry are being felt at the global level. For instance, some analysts argue that the rise of China is one of the three major challenges facing the USA (de Blij, 2005). Indeed, in the USA, the 2004 report of the President's Council of Advisers on Science and Technology said: 'The entry of China into the high-technology arena has created a new level of nervousness on the part of many industry and academic professionals' (Stokes, 2005). Likewise, respondents in a 2003 survey carried out by Nikkei among R&D managers of a large number of Japanese companies indicated that by 2013, China will have technological prowess comparable with that of Japan and Germany in almost every field (Sigurdson, 2004). For one thing, this technological advancement will allow the Chinese to exercise more control over the policy and direction of the increasingly multi-polar world.

The growth of the Chinese technology industry: current debate

Global technology analysts disagree as to the nature of the growth of the Chinese high-technology industry. For instance, some consider today's Chinese high-tech firms as comparable to those of Japan in the 1950s and 1960s and South Korea in the 1970s (Greenfeld, 2003; Zainulbhai, 2005). Zeng and Williamson (2003) have warned that many Western managers have held the 'dangerous misconception' that Western high-tech businesses are 'immune to competition from Chinese companies'. Vogelstein (2004) has more forcefully argued that 'if any country has the intellectual talent, financial resources, and will to one day build microprocessors as good as Intel's, it is China'.

Others maintain that it may be more useful to regard China as a normal emerging economy like Brazil or India instead of considering it like Japan, South Korea or Taiwan. Gilboy (2004) suggests that the Chinese government exercises its power over its firms in a 'chaotic way', thus hindering the success of Chinese firms. The second commonplace observation is that Chinese technology firms mainly engage in low-value assembly and the country competes on price rather than on high-technology or brand name at the global level (Mooney, 2005). Nolan summarises views of experts who play down China's potential to become a global economic superpower:

> China's large firms will generally be unable to compete on the global level playing field with the world's leading system integrators. China should, therefore, focus on improving the position of its indigenous firms with global value chain of the world's leading companies. Nolan (2002: 120)

The research presented in this book suggests that both views are right but incomplete. Our findings demonstrate how China's unique combination of institutional and market conditions has harboured a paradoxical nature. The framework developed in this book exposes the two significant one-sided perspectives on the current debate regarding the Chinese high-technology industry. Lampton (2005) has noted that 'China can be weak and strong simultaneously'. The same can be said of its high-technology industry.

Global economic and political implications of China's technological prowess

At the global level, China's emergence as a high-tech superpower expands opportunities for international specialisation. It can thus be considered as a positive sum game. Nonetheless, like all other international economic activities, there will be winners and losers associated with China's determination to climb the technology ladder.

Competitors in the international export market for ICT products

China competes with Japan, Korea and economies of the Association of Southeast Asian Nations (ASEAN) in the international ICT export market. From 1989 to 2000, the proportion of US imports coming from China more than tripled to 8.4 per cent. At the same time, Japan's share fell by half to 11 per cent and that of the four Asian 'tigers' – Hong Kong, South Korea, Singapore and Taiwan – declined by one-third to 8 per cent (Chandler, 2001).

There is a notable contrast between the trade structures of China and Japan. Broadly speaking, China specialises in labour-intensive products, while Japan possesses design expertise and focuses on capital-intensive goods (Brooke, 2003). One estimate suggests that 'the areas of head-to-head industrial competition' between these two countries is 16 per cent (Hart-Landsberg and Burkett, 2004). The strengthening positions of Chinese firms may affect these dynamics. In general, however, China's imports consist of processed goods, parts and components, while its exports are mainly completed goods (Okamoto, 2005). With China's high-tech success, the range of products in which Japan has a competitive advantage over China is likely to narrow (Hart-Landsberg and Burkett, 2004).

Next, consider ASEAN economies. As is the case with China, ASEAN economies import high-technology and capital-intensive products from industrialised countries. There seems to be a significant overlap between exports of China and ASEAN countries. Estimates suggest that 15–25 per cent of exports to the USA and Japan from Thailand and Indonesia have faced intense competition from China (Ianchovichina, Suthiwart-Narueput and Zhao, 2003: 67). Similarly, another estimate has suggested that China's export overlaps with Indonesia and Singapore are 83 per cent and 38 per cent respectively (Yue, 2003). As China also imports from ASEAN economies, China's technological rise is likely to stimulate exports from ASEAN economies to China. Note, however, that most ASEAN economies' exports to China are concentrated in only a few product lines (Herschede, 1991; Lardy, 2003; Voon and Yue 2003).

Newly industrialised economies (NIEs) in Asia, such as South Korea and Taiwan, on the other hand, have trade structures that lie between those of Japan and China. The impact of China's technological rise on these economies is thus likely to be between those of Japan and developing ASEAN economies.

India, on the other hand, competes with China in software products. Although Indian software firms export to over 95 countries, more than 90 per cent of these exports go to the USA, the EU and Japan. India thus has a limited presence in developing countries. China's electronics exports to developing countries and increased bundling of software with hardware will help boost China's commercial software growth, especially to developing countries (Chapter 8).

Another mechanism by which China's technological advancement will affect its competitors is through the relocation of technology-intensive production to China. In the past, rising wages in Taiwan, South Korea, Hong Kong and Japan forced textile manufacturers in these countries to relocate their plants to China (Chandler, 2001; Wu and Chen, 2004). Consistent with history and theory, multinationals are finding it attractive to relocate R&D activities to China from developed economies (Chapter 5) (see also Rasiah, 2005).

Exporting ICT products to China

China heavily imports microchips and circuit-board components for its hardware manufacturing industry. China's trade deficit in components was US$62 billion in 2004 (*Los Angeles Times*, 2005). The EU and the USA have a trade surplus with China in microchips and circuit-board components (*Economist*, 2005g). During 1990–98, high-technology industry exports from the USA to China increased more than 500 per cent (Bairstow, 2000). The rapid advancement in the Chinese high-tech industry has led to a decline in China's imports relative to exports. Consequently, exports of these items from Europe and the USA to China are declining. Indeed, the USA enjoys only a 'very small surplus' in components (*Los Angeles Times*, 2005). In 2002, China overtook both Japan and Mexico to become the biggest supplier of high-technology products to the USA (Harbert, 2003).

In particular, Japan exports capital goods to China. Capital spending is growing rapidly in the country (Brooke, 2003). Chinese businesses' orders have driven the growth of capital equipment related firms in Japan (Sender, 2003). In the near future, Japanese firms in this sector may thus benefit (Hart-Landsberg and Burkett, 2004).

Moreover, Chinese consumers pride themselves on driving foreign-brand cars and use mobile phones, computers and other high-tech items manufactured in advanced countries (Gilboy, 2004). In fact, excluding the USA, China had a trade deficit of about US$12 billion in 2005 (Barboza, 2006). China imported more from countries such as South Korea, Japan, Malaysia, Saudi Arabia and Angola than it exported to these countries. In particular, Japan produces better consumer electronics that than does China (Sender, 2003). Nonetheless, as China climbs the technological value chain, Japanese exports of these items to China may decline.

In particular, Korea has had a trade surplus with China since the early 1990s. Some argue that the surplus will grow as Korean investment in China encourages new exports through intra-firm trade channels. It should, however, be noted that export losses resulting from China's production may exceed such gains (Hart-Landsberg and Burkett, 2004).

In the same vein, Russia exports high-technology military products to China. China's goal is to build domestic high-technology arms industries so as to reduce its dependence on Russia. It has already established a high-tech army unit which accounts for 15 per cent of the PLA's size (Marquand, 2005).

Importing ICT products from China

Many developing countries import ICT products from China. For instance, China's exports to Africa exceeded US$14 billion in 2005 and imports were about US$16 billion (africa-business.com, 2005). Chinese firms' exports to Africa include machinery, electronics and other high-tech products and finished goods. Imports from Africa, on the other hand, involve crude oil, iron ore, cotton, diamond and other natural resources (africa-business.com, 2005).

Because of China's seemingly bottomless source of labour, however, the country is less likely to follow the pattern of other Asian NIEs and relocate less technology-intensive industries to other lesser-developed economies, such as those in Africa. Nonetheless, a number of Chinese firms have factories in Africa, which mainly stitch garments (*Economist*, 2004b).

For developing countries with a capability to produce technology products, however, Chinese high-technology firms have been a threat to the development of domestic industries. In South Africa, for example, trade unions have complained about competition from China facing their domestic technology industry (*Economist*, 2004b).

Countries that consider China as a national security threat

Some countries, most notably India and the USA, are concerned about national security issues associated with China's technological advancement. Mearsheimer (2005) suggests: 'China cannot rise peacefully, and if it continues its dramatic economic growth over the next few decades, the United States and China are likely to engage in an intense security competition with considerable potential for war'.

Due to the advancement in the Chinese high-technology industry, the attempts by Western powers to restrict dual-use (civil and military) technology exports to China have been ineffective. For instance, the Chinese military has been able to access domestically developed supercomputers (Chapter 8). In August 2002, Legend developed China's first supercomputer capable of making over 1 trillion calculations per second, giving it the same operation speed as the 24th fastest computer in the world at that time (*People's Daily*, 2002). Indeed, in the past decades, China has undergone two military high-tech reforms (Marquand, 2005).

The role of high-technology has been critical in China's economic and foreign policy relationships with economies that the USA considers as enemies. In turn, this has made US efforts to restrict technology exports to hostile nations ineffective. China's growing ties with certain countries in the Middle East, Africa and Latin America have been described as a threat to US security (Klare, 2005). Landay (1997) quotes a US official: 'China has become the single most important source of technology that rogue countries cannot obtain from the West'. For instance, Beijing offered to build an oil refinery for Venezuela (Malmgren, 2005). Similarly, China's Huawei Technologies arguably violated the UN embargo regarding Saddam Hussein's government in Iraq by selling it fibre-optic products which were reportedly used in its air defence system (Safire, 2002). Furthermore, it has been suggested that China offered Iran satellite navigation technology which has enhanced Tehran's ability to aim missiles accurately (Malmgren, 2005).

Chinese technology companies' controversial operations abroad have further intensified the fear. For instance, some analysts think that some of Huawei's employees in foreign countries 'spy for China' (*Economist*, 2005g). In 2001, India's Research and Analysis Wing (RAW) suggested that Huawei's Bangalore-based engineers developed telephone surveillance equipment for Afghanistan's Taliban regime (Iype and Prema, 2002). RAW has also suggested that software developed at Huawei's Bangalore R&D centre was supplied to the Pakistani government through indirect channels (Liu, 2006). India's Intelligence Bureau has suspected that Huawei had ties with China's intelligence bodies and the PLA (Simons, 2006). Because of this, the Indian Telecom Ministry blocked the company from bidding as an equipment supplier for Indian telecommunication projects.

How should the world respond?

The real question, then, is how countries with different combinations of economic and institutional characteristics should respond to the rapid transformation underway in the Chinese high-technology industry. From the US standpoint, the New York City Mayor Michael Bloomberg (2008: 58) recently put the issue this way: 'The challenge that we face is not preventing China from catching up with where we are today, but preventing ourselves from slowing down'. Bloomberg's observation broadly provides a helpful perspective to all economies, especially industrialised ones, for responding to the development in the Chinese high-technology industry. Developing economies, meanwhile, can borrow a page from the lesson book of the Chinese high-technology development locus.

Finally, China's technological development has profound implications for businesses all over the world. Undoubtedly, China's technological advancement has increased opportunities for foreign multinational companies to enter the Chinese market or intensify their operations in the country. The risks associated with China's unique institutions, however, remain significant (Chapters 1–3). Success in the Chinese market thus hinges on having a creative way to manage these risks. However, there remains another point that is perhaps even more important: the internationalisation of Chinese technology companies (Chapters 12–14) is set to transform the competitive landscape in the global high-technology arena. And firms in the developing world have much to learn from the distinctive modus operandi of multinational technology firms based in China.

Bibliography

3g.co.uk (2003) 'Sharp supply GSM wireless phones to China', available at: *http://www.3g.co.uk/PR/April2003/5285.htm* (accessed 30 April 2008).

3g.co.uk (2006) 'China's first round investment on 3G equipments will reach RMB 10 billion to 20 billion', available at: *http://www.3g.co.uk/PR/July2006/3297.htm* (accessed 2 May 2006).

3gamericas (2008) '3G standardization process', available at: *http://www.3gamericas.org/English/Technology_Center/standardization_process.cfm* (accessed 30 April 2008).

3ginsight.com (2001) 'Third 3G technology arrives', available at: *http://www.3ginsight.com/3G_insight_23.htm* (accessed 30 March 2002).

Abramson, D. B. (2006) 'Urban planning in China: Continuity and change', *Journal of the American Planning Association* 72(2): 197–215.

Ackerman, E. (2006) 'eBay's China lesson: go local: stung by failures, US firms see value in finding partners', *San Jose Mercury News*, 24 December, available at: *http://www.mercurynews.com/mld/mercurynews/business/16311645.htm* (accessed 26 December 2006).

Adams, J. (2001) 'Virtual defense', *Foreign Affairs* 80(3): 98–112.

Adams, R. M. (1996) *Paths of Fire: An Anthropologist's Inquiry into Western Technology*, Princeton, NJ: Princeton University Press.

Adar, E. and Huberman, B. A. (2000) 'The economics of surfing', *Quarterly Journal of Electronic Commerce* 1(2): 203–14.

Adcock, A. (2005) 'Solutions for exporting R&D from China', *Managing Intellectual Property* 151: 35–7.

Aerospace News (2005) 'China's first satellite for a foreign buyer going on smoothly as scheduled', available at: *http://www.cnsa.gov.cn/English/news_release/show.asp?id=140* (accessed 30 April 2008).

Africa-business.com (2005) 'Chinese goods gain popularity in Africa', available at: *http://www.africa-business.com/features/chinese_goods.html* (accessed 30 April 2007).

African Times (2004) 'Zimbabwe', *African Times*, 15 September, p. 1.

AFX – Asia (2006a) 'China's TCL Corp to restructure European ops; cost seen at 45 mln eur', available at: *http://www.accessmylibrary.com/coms2/summary_0286-24390740_ITM* (accessed 29 April 2008).

AFX International Focus (2006) 'China's TCL to post another loss this year in European television ops – report', available at: *http://www.forbes .com/technology/feeds/afx/2006/12/12/afx3249339.html* (accessed 29 April 2008).

AFX International Focus (2007) 'China's internet market at 135 mln users, worth 83.5 bln yuan-researcher', available at: *http://www.abcmoney .co.uk/news/09200752685.htm* (accessed 26 December 2007).

AFX News (2005) 'China's TCL, Alcatel JV restructuring positive move – analysts', available at: *http://www.forbes.com/technology/feeds/afx/2005/ 09/02/afx2203302.html* (accessed 21 January 2008).

AFX News (2006a) 'Louis Vuitton sues Carrefour in China – report', available at: *http://www.forbes.com/business/feeds/afx/2006/04/19/ afx2682374.html* (accessed 30 April 2007).

AFX News (2006b) 'China closing gap with US on nanotechnology – report', available at: *http://www.forbes.com/markets/feeds/afx/2006/ 09/26/afx3047486.html* (accessed 30 April 2008).

AFX News (2006c) 'China's TCL struggling to bring Thomson color television ops to profit – report', available at: *http://www.forbes.com/ technology/feeds/afx/2006/08/31/afx2987337.html* (accessed 29 April 2008).

Ahlstrorn, D. and Bruton, G. D. (2001) 'Learning from successful local private firms in China: establishing legitimacy', *Academy of Management Executive* 15(4): 72–83.

Ahlstrom, D., Foley, S., Young, M. N. and Chan, E. S. (2005) 'Human resource strategies in post-WTO China', *Thunderbird International Business Review* 47(3): 263–85.

Alarcon, R. (1999) 'Recruitment processes among foreign-born engineers and scientist in Silicon Valley', *American Behavioral Scientist* 42(9): 1381–97.

Aldrich, H. (1999) *Organizations Evolving*, Newbury Park, CA: Sage.

Aldrich, M. (1999) *E-com Legal Guide: China*, Hong Kong: Baker and McKenzie.

Alford, W. P. (1995) *To Steal a Book Is an Elegant Offense*, Stanford, CA: Stanford University Press.

America's Network (2006) 'The digital revolution ARRIVES March', *America's Network* 110(2): 18–23.

Amity Edumedia (2006) 'Nanotechnology trends', available at: *http://www.amityedumedia.com/industry/nanoTrends.htm* (accessed 30 April 2008).

Amsden, A. H. (2001) *The Rise of the Rest: Challenges to the West from Late-Industrializing Economies*, New York: Oxford University Press.

Andrews, P. (2003) 'Courting China to persuade the middle kingdom to pick its software over Linux, Microsoft plays nice', *US News & World Report*, 24 November, p. 44.

Applewhite, A. (2003) 'Should governments go open source?', *IEEE Software* 20(4): 88–91.

Appliance Magazine (2005) 'Haier pushes for global brand', available at: http://www.appliancemagazine.com/news.php?article=9590&zone=0&first=1 (accessed 30 April 2008).

arabic.china.org.cn (2005) 'China Railcom broadband subscribers exceed 1 mln', available at: http://arabic.china.org.cn/english/scitech/120954.htm (accessed 30 April 2006).

Archibugi, D. and Michie, J. (1997) 'Technological globalization or national systems of innovation?' *Futures* 29(2): 121–37.

Arrow, K. (1962) 'Economic welfare and the allocation of resources for inventions', in R. R. Nelson (ed.) *The Rate and Direction of Inventive Activity*, Princeton, NJ: Princeton University Press, pp. 609–25.

Arthur, B. W. (1994) *Increasing Returns and Path Dependency in the Economy*, Ann Arbor, MI: University of Michigan Press.

Arthur, M. (2003) 'Share price reactions to work-family initiatives: An institutional perspective', *Academy of Management Journal* 46: 497–505.

Asakawa, N. (2007) 'TD-SCDMA: more standards to come', available at: http://techon.nikkeibp.co.jp/article/HONSHI/20070725/136763/ (accessed 1 May 2008).

Ashling, J. (2005) 'Internet companies invest in China', *Information Today* 22(9): 22–3.

Asia Africa Intelligence Wire (2006) 'R&D cooperation helps build capacity', available at: http://www.lexisnexis.com/us/lnacademic/search/homesubmitForm.do (accessed 30 April 2008).

Asia Pulse (2005) 'China's high-tech trade expected to top US$400 bln this year', available at: http://goliath.ecnext.com/coms2/gi_0199-5076712/CHINA-S-HIGH-TECH-TRADE.html (accessed 30 April 2008).

Asia Pulse (2006a) 'China's online transactions seen to reach US$125 bln in 2006', available at: http://www.bushwatch.com/archive2006/e-mail-20060719.htm (accessed 30 April 2007).

Asia Pulse (2006b) 'Transnational pharmaceutical giants investing in China R&D', *Asia Pulse*, 5 June, p. 13.

Asia Pulse (2006c) 'IPTV likely to get head start in China next year', available at: http://marketinfo.hktdc.com/content.aspx?data=china_content_en&contentid=744986&src=CN_BuNeTrSt&w_sid=194&w_ pid=630&w_nid=9929&w_cid=744986&w_idt=1900-01-01&w_oid=156&w_ji d (accessed 1 May 2008).

Asia Pulse (2007) 'China's CCSA to accelerate IPTV standard drafting', available at: *http://goliath.ecnext.com/coms2/gi_0199-6363918/CHINA-S-CCSA-TO-ACCELERATE.html#abstract* (accessed 1 May 2008).

Asia Times (2000) 'Analysis: the China-Taiwan military balance: part two, China's failure to close the technological gap', available at: *http://www.atimes.com/china/BA27Ad02.html* (accessed 3 May 2008).

Asiainfo Daily China News (1999a) 'China: Henan lures overseas graduates' *Asiainfo Daily China News*, 30 December, p. 1.

Asiainfo Daily China News (1999b) 'China: Hongqi Linux showing on the market', *Asiainfo Daily China News*, 24 August, p. 1.

Asiainfo Daily China News (1999c) 'China: Motorola's Linux strategy published', *Asiainfo Daily China News*, 8 September, p. 1.

Asian Mid Morning News (2002) 'E-Trade', available at: *https://institutional.etrade.com/permFiles/ETRADE%20Asian%20News%204Dec02.pdf* (accessed 30 April 2008).

Audretsch, D. B. and Feldman, M. P. (1996) 'R&D spillovers and the geography of innovation and production' *American Economic Review* 86(3): 630–40.

August, O. (2004) 'Huawei sets up in UK as China pushes globally', available at: *http://business.timesonline.co.uk/tol/business/article 1051765.ece* (accessed 1 May 2008).

Avermaete, T., Viaene, J., Morgan, E. J. and Crawford, N. (2003) 'Determinants of innovation in small food firms', *European Journal of Innovation Management* 6(1): 8–17.

Axelrod, R. (1997) *The Complexity of Cooperation*, Princeton, NJ: Princeton University Press.

Baark, E. (1988) 'The value of technology: a survey of the Chinese theoretical debate and its policy implications', *Research Policy* 17(5): 269–82.

Bacani, C. (2004) 'Walking with giants', available at: *http://www.cfoasia.com/archives/200404-05.htm* (accessed 22 April 2006).

Backoff, J. F and Martin, C. L., Jr. (1991) 'Historical perspectives: development of the codes of ethics in the legal, medical and accounting professions', *Journal of Business Ethics* 10: 99–110.

Bagchi, K., Kirs, P. and Cerveny, R. (2006) 'Global software piracy; can economic factors alone explain the trend?', *Communications of the ACM* 49(6): 70–5.

Baijia, L. (2004) 'Firms keen on software outsourcing', available at: *http://www.chinadaily.com.cn/english/doc/2004-05/12/content_ 329899.htm* (accessed 30 April 2005).

Bain, J. S. (1956) *Barriers to New Competition, Their Character and Consequences in Manufacturing Industries*. Cambridge, MA: Harvard University Press.

Bairstow, J. (2000) 'Let's open the door for China', *Laser Focus World* 36(6): 160.

Balabanis, G., Diamantopoulos A., Mueller, R. D. and Melewar, T. C. (2001) 'The impact of nationalism, patriotism and internationalism on consumer ethnocentric tendencies', *Journal of International Business Studies* 32(1): 157–75.

Balbus, I. D. (1977) 'Commodity form and legal form: an essay on the 'relative autonomy' of the law', *Law and Society Review* 11: 571–88.

Balfour, F. (2006a) 'China Mobile is growing rural', available at: *http://www.businessweek.com/globalbiz/content/dec2006/gb2006122 0_600529.htm* (accessed 30 April 2008).

Balfour, F. (2006b) 'Venture capital's new promised land', *Business Week*, 16 January, p. 44.

Balfour, F. and Roberts, D. (2005) 'Stealing managers from the big boys', *Business Week*, 26 September, pp. 54–5.

Balmond, S. (2003) 'Mobile industry to clamp down on youth marketing', *Precision Marketing* 16(9): 3.

Bar, F. and Borrus, M. (1987) 'From public access to private connections: network policy and national advantage', Working Paper No. 28. Berkeley, CA: Berkeley Roundtable on the International Economy.

Barboza, D. (2006) 'Trade surplus tripled in '05, China says', *New York Times*, 12 January, p. C.1.

Barboza, D. and Lohr, S. (2007) 'FBI and Chinese seize $500 million of counterfeit software', *New York Times*, 25 July, p. C.1.

Barme, G. (1999) *In the Red: On Contemporary Chinese Culture*, New York: Columbia University Press.

Barnathan, J., Crock, S. and Einhorn, B. (1996) 'Rethinking China', available at: *http://www.businessweek.com/1996/10/b34651.htm* (accessed 30 April 2008).

Barnett, R. W. (1972) 'China and Taiwan: The economic issues', *Foreign Affairs* 50(3): 444–58.

Barney, J. B. (1991) 'Firm resources and sustained competitive advantage', *Journal of Management* 17(1): 99–120.

Barnwal, R. (2006) '41% Online Indians prefer Indian language website', available at: *http://alootechie.com/News/1192.asp* (accessed 11 December 2006).

Barrett, C. R. (2005) 'Educational complacency will make US feel the pain', *USA Today*, 24 February, p. A.13.

Barshefsky, C. and Gresser, E. (2005) 'Revolutionary China, complacent America', *Wall Street Journal*, 15 September, p. A.20.

Basu, I. (2006) 'China excited for IPTV', available at: *http://www.physorg.com/news9569.html* (accessed 1 May 2008).

Basu, O., Dirsmith, M. and Gupta, P. (1999) 'The coupling of the symbolic and the technical in an institutionalized context: the negotiated order of the GAO's audit reporting process', *American Sociological Review* 64: 506–26.

Baum, J. and Oliver, C. (1991) 'Institutional linkages and organizational mortality', *Administrative Science Quarterly* 36: 187–218.

BBC News (2002a) 'China blocking Google', available at: *http://news.bbc.co.uk/1/hi/technology/2231101.stm* (accessed 28 January 2008).

BBC News (2002b) 'Google fights Chinese ban', available at: *http://news.bbc.co.uk/1/hi/technology/2233229.stm* (accessed 28 January 2008).

Beijing Review (1992) 'Overseas students welcomed home', *Beijing Review*, 31 August, p. 10.

Beise, M. (2001) *Lead Markets: Country-Specific Success Factors of the Global Diffusion of Innovations*, Heidelberg: Physica-Verlag.

Benford, R. D. (1993) 'You could be the hundredth monkey: collective action frames and vocabularies of motives within the nuclear disarmament movement', *Sociological Quarterly* 34: 195–216.

Benkler, Y. (2006) *Wealth of Networks*, New Haven, CT: Yale University Press.

Benson, M. and Simpson, G. R. (2001) 'Privacy measure for US is backed by trade group', *Wall Street Journal*, 18 January, p. B.6.

Berger, P. L. and Luckmann, T. (1967) *The Social Construction of Reality: A Treatise in the Sociology of Knowledge*, New York: Doubleday.

Bernatchez, E. (2008) 'What does CDMA mean?', available at: *http://cellphones.about.com/library/glossary/bldef_cdma.htm* (accessed 30 April 2008).

Bernstein, A. (2004) 'Shaking up trade theory', *Business Week*, 6 December, 3911: 116.

Besen, S. M. and Farrell, J. (1994) 'Choosing how to compete: strategies and tactics in standardization', *Journal of Economic Perspectives* 8(2): 117–31.

Betts, M. (2005) 'Linux goes global: Serious Linux deployments are popping up all over, from German insurers to Chinese banks', available at: *http://www.computerworld.com/action/article.do?command=viewArticleBasic&articleId=103202* (accessed 30 April 2008).

Bezlova, A. (1999) 'China's last chance to catch the high-tech train', available at: *http://www.atimes.com/media/AL04Ce01.html* (accessed 3 May 2003).

Bhagwati, J. (1998) 'A new epoch', in *A Stream of Windows*, Cambridge, MA: MIT Press, pp. 20–1.

bharatbook.com (2006) 'China sourcing report broadband communication,' available at: *http://www.bharatbook.com/detail.asp?id=9001* (accessed 30 April 2008).

Bhardwaj, P. (2006) 'Powering Indian e-commerce' available at: *http://www.atimes.com/atimes/South_Asia/HJ26Df02.html* (accessed 2 May 2008).

Biggs, S. (2006) 'Anti-piracy pledge fails to boot; Beijing's mandate requiring China-made PCs to carry legitimate operating systems will not significantly curtail illegal software', *South China Morning Post*, 25 April, p. 1.

Bishop, T. (2006) '$1.2 billion Windows deal with Chinese PC maker: China's president set to arrive at Microsoft headquarters today', available at: *http://seattlepi.nwsource.com/business/267036_msftchina18.html* (accessed 1 May 2008).

Blanc, H. and Sierra, C. (1999) 'The internationalisation of R&D by multinationals: a trade-off between external and internal proximity', *Cambridge Journal of Economics* 23(2): 187–206.

Blomstrom, M. and Kokko, A. (2001) 'Foreign direct investment and spillovers of technology', *International Journal of Technology Management* 22(5/6): 435–54.

Bloomberg, M. R. (2008) 'A race we can all win; The American system still has inherent advantages, but we can't slow down', *Newsweek*, 7 January, p. 58.

Bohr, N. H. D. (1963) *Essays, 1958–1962, on Atomic Physics and Human Knowledge*, New York: Interscience Publishers.

Borgonjon, J. and Vanhonacker, W. R. (1992) 'Modernizing China's managers', *China Business Review* 19(5): 12–17.

Boswell, C. and Baker, J. (2006) 'Securing value in China', *ICIS Chemical Business*, 1 October, pp. 31–2.

Bowles, R. (2004) 'Employment and wage outcomes for North Carolina's high-tech workers: North Carolina's employment surge in "high-tech" industries has captured many economic planners' focus; these industries may provide high-paying employment to workers in declining industries, such as textiles, furniture, and apparel', *Monthly Labor Review* 127(5): 31–9.

Branine, M. (1996) 'Observations on training and management development in the People's Republic of China', *Personnel Review* 25(1): 25–39.

Brannen, M. Y., Liker, J. K. and Fruin, W. M. (1999) 'Recontextualization and factory-to-factory knowledge transfer from Japan to the US: the case of NSK', in J. K. Liker, W. M. Fruin and P. Adler (eds) *Remade in America: Transplanting and transforming Japanese manufacturing systems*, New York: Oxford University Press, pp. 117–54.

Bratton, M. (2007) 'Formal versus informal institutions in Africa', *Journal of Democracy* 18(3): 96–110.

Bremmer, I. (2006) 'The world is J-curved', *Washington Post*, 1 October, p. B.3.

Bremner, B. (2006) 'ITU Telecom World 2006: a robust outlook', available at: *http://www.businessweek.com/globalbiz/content/dec 2006/gb20061204_046060.htm?chan=tc&campaign_id=rss_tech* (accessed 30 April 2008).

Bresser, R. K. F. and Millonig, K. (2003) 'Institutional capital: competitive advantage in light of the new institutionalism in organization theory', *Schmalenbach Business Review* 55(3): 220–41.

Brint, S. and Karabel, J. (1991) 'Institutional origins and transformations: the case of American community colleges', in W. Powell and P. DiMaggio (eds) *The New Institutionalism in Organizational Analysis*, Chicago, IL: University of Chicago Press, Cambridge Reports/Research International, pp. 337–60.

Brizendine, T. (2002) 'Software integration in China', *China Business Review* 29(2): 26–31.

Broadband Business Forecast (2006) 'China soon to top US in broadband penetration', available at: *http://www.accessmylibrary.com/coms2/browse_ JJ_B024_151_225* (accessed 30 April 2008).

Broadcast Engineering (2007) 'The IPTV picture', *Broadcast Engineering* 49(8): 22–4.

Brooke, J. (2003) 'Japan's recovering economy is relying heavily on China', available at: *http://query.nytimes.com/gst/fullpage.html?res= 9F02EFDA153BF932A15752C1A9659C8B63* (accessed 4 May 2008).

Brookes, P. (2006) 'Back to the Maoist future', *Weekly Standard*, 17 April, pp. 11–12.

Browning, L. D. and Folger, R. (1994) 'Communication under conditions of litigation risk: A grounded theory of plausible deniability', in S. B. Sitkin and R. J. Bies (eds.) *The Legalistic Organization*, Newbury Park, CA: Sage, pp. 251–80.

Buckley, C. (2004) 'Let a thousand ideas flower: China is a new hotbed of research', *New York Times*, 13 September, p. C.1.

Burdekin, R. C. K. (2000) 'Ending inflation in China: From Mao to the 21st century', *Cato Journal* 20(2): 223–35.

Bureau of Industry and Security, US Department of Commerce (1998a) 'US perspectives on technology transfers to China' available at: *http://www.bxa.doc.gov/DefenseIndustrialBasePrograms/OSIES/DefMarketResearchRpts/ChinaGuides/China2.pdf* (accessed 30 April 2001).

Bureau of Industry and Security, US Department of Commerce (1998b) 'US commercial technology transfers to the People's Republic of China', available at: *http://www.bxa.doc.gov/DefenseIndustrialBasePrograms/OSIES/DefMarketResearchRpts/ChinaGuides/China1.pdf* (accessed 3 May 2008).

Burns, E. (2006) 'Global internet adoption slows while involvement deepens', available at: *http://www.clickz.com/stats/sectors/geographics/article.php/3596131* (accessed 30 April 2008).

Burns, S. (2008) 'China's IPTV market disappoints: Licensing roadblocks deter investment', available at: *http://www.vnunet.com/vnunet/news/2207323/china-iptv-market-disappoints* (accessed 26 January 2008).

Burt, R. S. (1983) *Corporate Profits and Cooptation: Networks of Market Constraints and Directorate Ties in the American Economy*; New York: Academic Press.

Bush, P. (1987) 'The theory of institutional change', *Journal of Economic Issues* 21: 1075–16.

Business China (2001a) 'Lining up for Linux?', *Business China*, 12 February, p. 3.

Business China (2001b) 'About face', *Business China*, 5 November, p. 4.

Business China (2005) 'India vs China', *Business China*, 25 April, pp. 4–5.

Business CustomWire (2003) 'Telecom manufacturers to invest more in 3G R&D', available at: *http://www.interactivedata-rts.com/data_coverage.shtml* (accessed 30 June 2004).

Business Week (2004a) 'China.Net; China will soon be no. 1 in web users. That will unleash a world of opportunity', *Business Week*, 15 March, p. 22.

Business Week (2004b) 'How America can meet 'the China price', 6 December, p. 164.

Business Week (2005) 'China and India: how should the US respond?', *Business Week*, 19 September, p. 18.

Business Week (2006) 'Microsoft's China card', available at: *http://www.businessweek.com/technology/content/apr2006/tc20060426_405461.htm?campaign_id=bier_tca* (accessed 30 April 2007).

Business Wire (2006a) 'China Netcom 2006 interim results; strategic transformation to "broadband communications and multi-media services provider" yields remarkable results; high-growth businesses record robust growth', available at: http://goliath.ecnext.com/coms2/summary_0199-5689479_ITM (accessed 30 April 2008).

Business Wire (2006b) 'The online games market in India is starting to heat up due to an increase in internet users and broadband penetration', available at: http://findarticles.com/p/articles/mi_m0EIN/is_2006_April_21/ai_n16129705 (accessed 30 April 2008).

Business Wire (2007) 'By September 12th, 2006, online registered users in China had surpassed 25 million and the quantity of online commodities over 5 million – China e-commerce profit model report, 2006–2007', available at: http://findarticles.com/p/articles/mi_m0EIN/is_2007_March_28/ai_n18768785 (accessed 2 May 2008).

Business Wire (2008) 'Lenovo reports third quarter 2007/08 results', available at: http://findarticles.com/p/articles/mi_m0EIN/is_2008_Jan_31/ai_n24241135 (accessed 30 April 2008).

Businessworld (2006) 'Lagging behind', available at: http://www.businessworld.in/content/view/2398/2476/1/1/ (accessed 30 April 2008).

Calbreath, D. (2004) 'China invents ways to avoid paying royalties to tech companies', *Knight Ridder Tribune Business News*, 23 February, p. 1.

Calof, J. L. and Beamish, P. W. (1995) 'Adapting to foreign markets: explaining internationalization', *International Business Review* 4(2): 115–31.

Camenisch, P. E. (1983) *Grounding Professional Ethics in a Pluralistic Society*, New York: Haven Publications.

Campbell, D. (2001) 'Can the digital divide be contained?', *International Labor Review* 140(2): 119–41.

Campbell, J. L. (2004) *Institutional Change and Globalization*, Princeton, NJ: Princeton University Press.

Cantwell, J. A. (1989) *Technological Innovation and Multinational Corporations*, Oxford: Basil Blackwell.

Capron, L. and Hulland, J. (1999) 'Redeployment of brands, sales forces, and general marketing management expertise following horizontal acquisitions: A resource-based view', *Journal of Marketing* 63(2): 41–54.

Carbone, J. (2004) 'Is India the next China?', available at: http://www.allbusiness.com/electronics/computer-electronics-manufacturing/6358809-1.html (accessed 30 April 2008).

Cardinal, L. Sitkin, S. and Long, C. (2004) 'Balancing and rebalancing in the creation and evolution of organizational control', *Organization Science* 15: 411–31.

Carney, C. P. (1999) 'The (not so) long march? China's pace of change', *Asian Affairs, an American Review* 25(4): 236–48.

Cavusgil, S. T. (1994) 'From the editor in chief', available at: *http://ciber .bus.msu.edu/jim/abstract/abs-v2n3.htm* (accessed 30 April 2002).

CCID (2007) 'CCID Consulting analyzes China's internet industry', available at: *http://www.prnewswire.com/cgi-bin/stories.pl?ACCT=104& STORY=/www/story/12-28-2007/0004728407&EDATE* (accessed 26 January 2008).

Cellular News (2003) 'Report on China's top 5 handset manufacturers', available at: *http://www.cellular-news.com/story/9442.shtml* (accessed 30 March 2004).

Cellular News (2008) 'Chinese handset production growth slows', available at: *http://www.cellular-news.com/story/30889.php* (accessed 1 May 2008).

Chae, B. and McHaney, R. (2006) 'Asian trio's adoption of Linux-based open source development', *Communications of the ACM* 49(9): 95–9.

Chai, W. (2003) 'Chinese IT players gain ground', available at: *http://news.com.com/2100-1011-994508.html* (accessed 22 April 2006).

Chan, A. (2006) 'Human resource management in China: past, current and future HR practices in the industrial sector', *The China Journal*, No. 55: 164–5.

Chan, C. S. (2006) 'China IPTV hungry for content', available at: *http://www.businessweek.com/globalbiz/content/dec2006/gb20061207_ 674387.htm*.

Chan, J. K., Ellis, M. and Styles, A. (2005) 'The TV content challenge', *China Business Review* 32(4): 46–9.

Chan, W. (2007) 'The Chinese telecom/IT outsourcing market and Infosys' experience', paper presented at ChinaTel Summit and VC Forum 2007, Honolulu, HI, 13–14 January.

Chandler, C. (2001) 'A factory to the world; China's vast labor pool, low wages lure manufacturers', *The Washington Post*, 25 November, p. A.01.

Chandler, C. (2004) 'TV's Mr Big', *Fortune*, 9 February, pp. 84–6.

Chandler, D. (1996) *Facing the Cambodian Past*, Chiang Mai: Silkworm Books.

Chang, P. and Deng, Z. (1992) 'The Chinese brain drain and policy options', *Studies in Comparative International Development* 27(1): 44–60.

Charny, B. (2003) 'In China, Motorola gets boxed in', available at: *http://news.com.com/2100-1037_3-997167.html* (accessed 30 April 2004).

Chau, F. (2007) 'TD-SCDMA declared ready for action', *Telecom Asia*, 18(1): 6.

Chen, C. and Zweig, D. (1993) 'The impact of the open policy on higher education in China', paper presented to the Annual Meeting of the Association for Asian Studies, Los Angeles, CA, 27 March.

Chen, K. and Dean, J. (2006) 'Low costs, plentiful talent make China a global magnet for R&D', *Wall Street Journal*, 13 March, p. A.1.

Chen, Y.-C. (2004) 'Restructuring the Shanghai innovation system: the role of multinational corporations' R&D centres in Shanghai', paper presented at the ASIANLICS International conference on 'Innovation Systems and Clusters in Asia – Challenges and Regional Integration', Bangkok, 1–2 April.

Chen, Z. (2005) 'TCL to buy-out partner Alcatel', available at: *http://www.chinadaily.com.cn/english/doc/2005-05/18/content_443549.htm* (accessed 22 April 2006).

Cheung, K. and Lin, P. (2004) 'Spillover effects of FDI on innovation in China: evidence from the provincial data', *China Economic Review* 15: 25–44.

Child, P. N. (2006) 'Lessons from a global retailer: An interview with the president of Carrefour China', *McKinsey Quarterly*, Special Edition: Serving the new Chinese consumer, pp. 70–81.

China Business Information Network (1997) 'Fujian welcomes returned students to apply for leading posts', *China Business Information Network*, 29 August, p. 1.

China Business Information Network (1999) 'China: most government-sponsored students return to China', *China Business Information Network*, 24 December, p. 1.

China Daily (2005) 'E-commerce chalking up strong growth: Survey', *People's Daily Online,* available at: *http://english.people.com.cn/200602/20/eng20060220_244208.html* (accessed 1 May 2008).

China Daily (2006) 'Chinese 3G standard finds support', available at: *http://english.people.com.cn/200612/05/eng20061205_328450.html* (accessed 30 April 2008).

China Daily (2007a) 'China patent applications total four million', available at: *http://www.chinadaily.com.cn/bizchina/2007-12/25/content_6347571.htm* (accessed 4 May 2008).

China Daily (2007b) 'TCL declares European unit insolvent', available at: *http://www.china.org.cn/english/BAT/212043.htm* accessed 21 January 2008).

China Daily (2008) 'Firms' attention to patents is still not enough', available at: *http://english.ipr.gov.cn/ipr/en/info/Article.jsp?a_no=176873&col_no=928&dir=200802* (accessed 2 May 2008).

Bibliography

China Economic Net (2007) 'Beijing television obtains China's 6th IPTV license', available at: *http://en.ce.cn/Industries/Consumen-Industries/200702/07/t20070207_10351047.shtml*.

China Economic Net (2008) 'Olympics to fuel ad-spending in China', available at: *http://en.ce.cn/Industries/MI/200802/03/t20080203_14455428.shtml* (accessed 2 May 2008).

China Economic Times (2000) 'High technology affects national security', available at: *http://www.china.org.cn/english/GS-e/668.htm* (accessed 30 April 2008).

China Internet Network Information Centre (CNNIC) (2001) 'Semi-annual survey report on the development of China's internet', available at: *http://www.cnnic.gov.cn/develst/e-cnnic200101.shtml* (accessed 26 May 2001).

China IT & Telecom Report (2006) 'Major Chinese handset makers report profit in Q1 despite declining market share', available at: *http://www.lexisnexis.com/us/lnacademic/results/docview/docview.do?docLinkInd=true&risb=21_T3644726229&format=GNBFI&sort=RELEVANCE&startDocNo=1&resultsUrlKey=29_T3644726242&cisb=22_T3644726241&treeMax=true&treeWidth=0&csi=167603&docNo=1* (accessed 30 April 2008).

China IT & Telecom Report (2007a) 'Mainland China and Taiwan to cooperate over TD-SCDMA, IPTV and AVS development', available at: *http://goliath.ecnext.com/coms2/gi_0199-6775695/Mainland-China-and-Taiwan-to.html* (accessed 1 January 2008).

China IT & Telecom Report (2007b) 'AVS may become an international standard for IPTV at year's end', available at: *http://internetprotocoltelevision.com/2007/08/07/voice-and-video-over-the-net-iptv-daily-home-front-page/* (accessed 1 January 2008).

China IT & Telecom Report (2007c) 'TCL Corporation', available at: *http://www.alacrastore.com/storecontent/markintel/INTERFAX_WEEKLY_REPORTS-50103647* (accessed 29 April 2008).

China Staff (2003) 'Student returnee policy pays off', *China Staff* 9(10): 41.

China View (2006) 'With China's technology, rural Africa may soon get digitized', available at: *http://news.xinhuanet.com/english/2006-01/24/content_4094516.htm* (accessed 9 April 2007).

China.org (2000) 'High technology affects national security', available at: *http://www.china.org.cn/english/GS-e/668.htm* (accessed 30 April 2008).

China.org (2003) 'Government sets to fund 3G standard', available at: *http://www.china.org.cn/english/scitech/79882.htm* (accessed 30 March 2004).

China.org (2008a) 'Microsoft launches Chinese-language map search engine', available at: http://www.china.org.cn/english/business/240334.htm (accessed 30 April 2008).

China.org (2008b) 'Chinese companies in Africa zealously repay local societies', available at: http://www.china.org.cn/english/international/238213.htm (accessed 29 April 2008).

China-embassy.org (2006) 'China sees fourth wave of overseas M&A', available at: http://www.china-embassy.org/eng/gyzg/t251469.htm (accessed 21 January 2008).

China Netcom (2008) 'China Netcom first commercialized AVS-IPTV in Dalian', available at: http://www.china-netcom.com/eng/news/news080318.htm (accessed 1 May 2008).

Chinanex.com (2003a) 'Motorola in China', available at: http://www.chinanex.com/company/forCompany/motorola.htm (accessed 30 March 2004).

Chinanex.com (2003b) 'Nokia in China', available at: http://www.chinanex.com/company/forCompany/nokia.htm (accessed 30 March 2004).

Chinaview (2007) 'China's blogger population reaches 47 mln', available at: http://news.xinhuanet.com/english/2007-12/29/content_7336026.htm (accessed 30 April 2008).

Chinese Education and Research Network (2001) 'Zhejiang to regulate internet banners', available at: http://www.edu.cn/20010101/22583.shtml (accessed 11 December 2006).

Choong, A. (2003) 'Is India the new China?', available at: http://asia.cnet.com/reviews/handphones/wirelesswatch/0,39020107,39148302,00.htm (accessed 21 January 2004).

Christensen, C. M., Raynor, M. E. and Anthony, S. D. (2003): 'Six keys to building new markets by unleashing disruptive innovation', available at: http://hbswk.hbs.edu/item.jhtml?id=3374&t=innovation &noseek=one (accessed 30 April 2008).

Christensen, T. (1996) 'Chinese realpolitik', *Foreign Affairs* 75(5): 37–52.

Christie, P. M. J., Kwon, I. G., Stoeberl, P. A. and Baumhart, R. (2003) 'A cross-cultural comparison of ethical attitudes of business managers: India, Korea and the United States', *Journal of Business Ethics* 46(3): 263–87.

Chung, O. (2007) 'China's phone makers in speed dial mode', available at: http://www.atimes.com/atimes/China_Business/IK03Cb01.html (accessed 30 April 2008).

Clark, D. (2003) 'Going global: China's handset manufacturers take on the world', available at: http://www.sinomedia.net/eurobiz/v200308/story0308.html (accessed 30 March 2004).

Clark, D. and Harwit, E. (2004) 'WiFi in China', paper presented at the Pacific Telecommunications Council's Annual Meeting, Honolulu, HI, 11–14 January.

Clark, E. and Soulsby, A. (1999) *Organisational Change in Post-Communist Europe*, London: Routledge.

Clark, N. (2002) 'Innovation systems, institutional change and the new knowledge market: implications for third world agricultural development', *Economics of Innovation & New Technology* 11(4/5): 353–48.

Clark, R. (2004) 'Double edge to Asia's trade barriers'. *America's Network*, 1 June, p. 14.

Claypool, G. A., Fetyko, D. F. and Pearson, M. A. (1990) 'Reactions to ethical dilemmas: a study pertaining to certified public accountants', *Journal of Business Ethics* 9: 699–706.

Clemens, E. S. and Cook, J. M. (1999) 'Politics and institutionalism: explaining durability and change', *Annual Review of Sociology* 25: 441–66.

Clendenin, M. (2002) 'China's homegrown 3G spec alive but delayed', *Electronic Engineering Times*, 2 September, pp. 30–1.

Clendenin, M. (2003) 'China's 3G gambit spurs global moves', *Electronic Engineering Times*, 21 April, pp. 1–2.

Clendenin, M. (2005a) 'Strong R&D fuels handset design firm (China Techfaith Wireless Communication Technology)', *Electronic Engineering Times*, 23 May, p. 12.

Clendenin, M. (2005b) 'Internet television gets running start in large Asia markets – Hong Kong service thrives; China, India face roadblocks', *Electronic Engineering Times*, 19 December, p. 1.

Clendenin, M. (2007a) '3G is still trying to hurdle a great wall in China', available at: *http://www.eetimes.com/showArticle.jhtml?articleID= 197000868* (accessed 30 April 2008).

Clendenin, M. (2007b) 'IPTV drives network upgrade in China, but demand still uncertain', *Electronic Engineering Times*, 16 July, p. 18.

Clendenin, M. and Mannion, P. (2002) 'China puts 3G spec on speed dial', *Electronic Engineering Times*, 11 November, pp. 1–2.

Clifford, M. L., Roberts, D. and Engardio, P. (1997) 'How you can win in China: the obstacles are huge but surmountable', *Business Week*, 26 May, pp. 66–8.

CNET News (2003a) 'Shanghai: School's out for Microsoft Office', available at: *http://www.news.com/2100-1012_3-5068050.html* (accessed 12 May 2008).

CNET News (2003b) 'Can China's Legend go global?', available at: *http://news.com.com/Can+China's+Legend+go+global%3F/2009-1069_3-995529.html* (accessed 2 April 2004).

CNETAsia (2003a) 'Can China's homegrown 3G measure up?', available at: *http://www.zdnetasia.com/news/business/0,39044229,39150203,00.htm* (accessed 30 April 2007).

CNETAsia (2003b) 'Nokia aims to overtake Motorola in China', available at: *http://asia.cnet.com/newstech/communications/0,39001141,39142511,00.htm* (accessed 30 April 2008).

CNN (2003) 'Intel: China PC market overtaking US', available at: *http://edition.cnn.com/2003/TECH/ptech/10/16/china.intel.reut/* (accessed 30 April 2004).

CNN Money (2006) 'China, India attracting offshore R&D', available at: *http://money.cnn.com/2006/02/16/smbusiness/offshore_rd/index.htm* (accessed 9 May 2008).

Cohen, J. R. and Pant, L. W. (1991) 'Beyond bean counting: establishing high ethical standards in the accounting profession', *Journal of Business Ethics* 10: 45–6.

Colvin, G. (2005) 'Can Americans compete? Is America the world's 97-lb. weakling?', available at: *http://www.fortune.com/fortune/articles/0,15114,1081269,00.html* (accessed 13 May 2006).

COMMWEB (2006) 'World's digital divide is narrowing', available at: *http://wcn.cnt.org/news/2006/04/27/world%E2%80%99s-digital-divide-narrowing/* (accessed 30 April 2008).

ComputerWeekly (2001) 'AOL in internet project with China's Legend', available at: *http://www.computerweekly.com/Articles/2001/06/11/180787/aol-in-internet-project-with-chinas-legend.htm* (accessed 11 December 2006).

Computing Canada (2000) 'Windows close on Microsoft in China', *Computing Canada*, 21 January, p. 34.

Comtex News Network (2006) 'Novartis creates new strategic biomedical R&D center in Shanghai', available at: *http://www.webwire.com/ViewPressRel.asp?aId=23179* (accessed 9 May 2008).

Conner, K. R. (1991) 'A historical comparison of the source-based theory and five schools of thought within industrial organization economics: do we have a new theory of the firm?' *Journal of Management* 17(1): 121–54.

Content Sutra (2006) 'eStatsIndia.com's estimate (Indian e-commerce market close to $1 billion)', available at: *http://www.contentsutra.com/indian-e-commerce-market-close-to-1-billion* (accessed 1 May 2008).

Convergence Plus (2004) 'Kejian launches GSM phones in India', available at: *http://www.convergenceplus.com/may04%20india% 20telecom%2001.html* (accessed 30 December 2004).

Cooke, P., Uranga, M.G. and Etxebarria, G. (1997) 'Regional innovation systems: institutional and organizational dimensions', *Research Policy* 26(4–5): 475–91.

Coonan, C. (2007) 'Brewing bootlegs', *Variety*, 13 August, pp. 3–4.

Cooper, L. and Contant, C. (2001) 'The People's Republic of China – consolidating its space power, enhancing its military might', available at: *http://www.gwu.edu/~spi/spacemilch1.html* (accessed 3 May 2008).

Cotriss, D. (2007) 'Will IPTV outstrip satellite TV?', available at: *http://www.dailyiptv.com/features/iptv-outstrip-satellite-080807* (accessed 1 January 2008).

Cottrell, T. (1994) 'Fragmented standards and the development of Japan's microcomputer software industry'. *Research Policy* 23(2): 143–74.

Covaleski, M. A. and Dirsmith, M. W. (1988) 'An institutional perspective on the rise, social transformation, and fall of a university budget category', *Administrative Science Quarterly* 33: 562–87.

Crampton, T. (2006) 'Innovation may lower net users' privacy; embraced by China, new standard helps to trace people online', *International Herald Tribune*, 20 March, p. 1.

Crane, S. (2004) 'Commentary; ghost from the past hinders China; How can we miss Jiang Zemin if he won't go away?', *Los Angeles Times*, 8 August, p. M.5.

Crowell, T. (2005) 'Do you know these brands? You will', *Christian Science Monitor*, 30 June, pp. 13–15.

Culpan, T. (2006) 'Industry awaits IP court in China', *Billboard*, 22 April, p. 14.

D'Aunno, T., Sutton, R. and Price, R. (1991) 'Isomorphism and external support in conflicting institutional environments: A study of drug abuse treatment units', *Academy of Management Journal* 34: 636–61.

Dahlman, C. J. (2007) 'China and India: emerging technological powers', *Issues in Science and Technology* 23(3): 44–53.

Dallago, B. (2002) 'The organizational effect of the economic system', *Journal of Economic Issues* 36(4): 953–79.

Dana, L. P. (1999) 'Small business as a supplement in the People's Republic of China (PRC)', *Journal of Small Business Management* 37: 76–80.

Dano, M. (2004) 'Alcatel/TCL venture underscores changes in handset market', *RCR Wireless News* 23(18): 3.

Dano, M. (2005) 'China could issue 3G permits in 2006', available at: *http://findarticles.com/p/articles/mi_hb4962/is_200510/ai_n18147238* (accessed 30 April 2008).

David, P. (1991) 'The hero and the herd in technological history: Reflections on Thomas Edison and the battle of the systems', in P. Higonnet, D. S. Landes and H. Rosovsky (eds) *Favorites of Fortune: Technology, Growth, and Economic Development since the Industrial Revolution*, Cambridge, MA: Harvard University Press, pp. 72–119.

David, P. A. (2000) 'Path dependence and varieties of learning in the evolution of technological practice', in J. Ziman (ed.) *Technological Innovation as an Evolutionary Process*, Cambridge: Cambridge University Press, pp. 118–33.

David, P. A. and Foray, D. (1994) 'Dynamics of competitive technology diffusion. Through local network structures: the case of EDI document standards', in L. Leydesdorff and P. van den Besselaar (eds) *Evolutionary Economics and Chaos Theory: New Directions in Technology Studies*, London: Pinter pp. 63–78.

David, P. and Greenstein, S. (1990) 'The economics of compatibility standards: An introduction to recent research', *Economics of Innovation- and New Technology* 1: 3–41.

David, P. and Steinmueller, W. E. (1994) 'Economics of compatibility standards and competition in telecommunication networks', *Information Economics and Policy* 6: 217–41.

David, R. (2008) 'India computer sales: it's getting personal', available at: *http://www.forbes.com/2008/02/27/india-computer-sales-markets-equity-cx_rd_0227markets04.html* (accessed 30 April 2008).

De Blij, H. (2005). *Why Geography Matters: Three Challenges Facing America: Climate Change, the Rise of China, and Global Terrorism*, New York: Oxford University Press.

Dean, J. (1999) 'Foreign companies aim to develop creativity potential of Chinese workers', *Wall Street Journal*, 1 December, p. B.7.D.

Dean, J. (2005) 'China may allow 3G mobile service in coming year', *Wall Street Journal*, 28 December, p. B.3.

Dean, J. (2006a) 'China approves a 3G standard, setting stage to issue licenses', *Wall Street Journal*, 23 January, p. B.4.

Dean, J. (2006b) 'China is set to spend billions on wireless upgrade; equipment firms are vying for deals in giant market for new-generation phones', *Wall Street Journal*, 27 February, p. A.1.

Dean, T. (2003) 'The fight for China's handset market,' *China Business Review* 30(6): 28.

Deephouse, D. L. (1996) 'Does isomorphism legitimate?' *Academy of Management Journal* 39: 1024–39.

Demsetz, H. (1967) 'Towards a theory of property rights', *American Economic Review* 57: 347–59.

Desker Shaw, S. (2003) 'Branding: Legend creates Lenovo for export', *Media Asia*, 16 May, p. 10.

Di Capua, M. (1998) 'Technology innovation in China', available at: *http://www.nautilus.org/archives/library/security/papers/Di_CapuaISO DARCO.PDF* (accessed 3 May 2008).

Dickie, M. (2005) 'Harbin harbinger delivers internet television message', *Financial Times*, 2 December, p. 29.

digitaltvnews.net (2007) 'IPTV subscriber numbers in seventh heaven', available at: *http://www.digitaltvnews.net/content/?p=2044* (accessed 1 January 2008).

Dillon, S. (2004) 'US slips in attracting the world's best students', *New York Times*, 21 December, p. A.1.

DiMaggio, P. J. (1988) 'Interest and agency in institutional theory', in L. Zucker (ed.) *Institutional Patterns and Organizations*, Cambridge, MA: Ballinger, pp. 3–22.

Dittmer, L. and Gore, L. (2001) 'China builds a market culture', *East Asia: An International Quarterly* 19(3): 9–51.

Dolven, B. (2004) 'Making the whole world listen', *Far Eastern Economic Review*, 26 February, pp. 26–29.

Dorn, J. A. (1998) 'China's future: market socialism or market Taoism?', *Cato Journal* 18(1): 131–46.

Dreyer, J. (2002) 'China rejects US push for more review of market access commitments', *Inside US Trade*, 7 September, available at: *http://www.insidetrade.com/* (accessed 4 May 2003).

dri.co.jp (2004) 'Bird exports 600,000 handsets in 1Q 2004', available at: *http://www.dri.co.jp/portelligent/techalert/2004_2Q/comp_alert_ 040511.asp* (accessed 30 April 2008).

Driedonks, C., Gregor, S., Wassenaar, A. and Van Heck, E. (2005) 'Economic and social analysis of the adoption of B2B electronic marketplaces: a case study in the Australian beef industry', *International Journal of Electronic Commerce* 9(3): 49–72.

Dunfee, T. W. and Warren, D. E. (2001) 'Is *guanxi* ethical? A normative analysis of doing business in China', *Journal of Business Ethics* 32(3): 191–204.

Dunning, J. H. (1981) 'Explaining the international direct investment position of countries: towards a dynamic or developmental approach', *Weltwirtschaftliches Archiv* 117(1): 30–64.

Ebusinessforum (2004) 'Global: Linux seen as saviour from Microsoft's grip', available at: *http://www.ebusinessforum.com/index.asp?doc_ id= 7126&layout=rich_story* (accessed 14 May 2005).

e-commercetimes (2008) 'China dismantles 44,000 sites in anti-porn offensive', available at: *http://www.ecommercetimes.com/story/China-Dismantles-44000-Sites-in-Anti-Porn-Offensive-61336.html?welcome=1201374030* (accessed 28 January 2008).

Economist (1992) 'China: new rich', *Economist*, 10 October, p. 36.

Economist (1999) 'The politics of piracy', *Economist*, 20 February, p. 64.

Economist (2002a) 'Asia: stop your searching; the internet in China', *Economist*, 7 September, p. 68.

Economist (2002b) 'Survey: The invisible green hand', *Economist*, 6 July, p. 13.

Economist (2003a) 'Business: the allure of low technology in China', *Economist*, 20 December, p. 105.

Economist (2003b) 'The big picture; Chinese brands: A Chinese television-maker goes global-TCL', *Economist*, 8 November, p. 6.

Economist (2004a) 'Business: Protecting the family jewels; Japan's technology secrets', *Economist*, 26 June, p. 81.

Economist (2004b) 'Business: A new scramble; China's business links with Africa', *Economist*, 27 November, pp. 87–8.

Economist (2005a) 'Survey: thinking for themselves', *Economist*, 22 October, p. 21.

Economist (2005b) 'Finance and economics: globalisation with a third-world face; Economics focus', *Economist*, 9 April, p. 70.

Economist (2005c) 'Survey: the next big thing', *Economist*, 18 June, p. 13.

Economist (2005d) 'A world of opportunity', *Economist*, 10 September, pp. 14–16.

Economist (2005e) 'Business: China's people problem; human resources', *Economist*, 16 April, p. 60.

Economist (2005f) 'Special report: the struggle of the champions – China's champions; China's big companies', *Economist*, 8 January, p. 58.

Economist (2005g) 'The reds in the West', 15 January, p. 50.

Economist (2006a) 'Survey: Now for the hard part', *Economist*, 3 June, p. 3.

Economist (2006b) 'Special report: the party, the people and the power of cyber-talk – China and the internet', *Economist*, 29 April, p. 28.

Economist (2006) 'Business: A grim picture; TCL-Thomson Electronics', 4 November, p. 89.

Economist Intelligence Unit (2005) 'Telecoms and technology forecast', available at: *http://store.eiu.com/country/CN.html?ref=lef_nav* (accessed 9 May 2008).

Economist Intelligence Unit (2006) 'Country commerce, China', available at: *http://www.eiu.com/index.asp?layout=AllTitles&rf=0* (accessed 19 May 2007).

Economist Intelligence Unit (2007) 'The comfort zone: Why many Chinese dealmakers will choose to stay close to home', available at: *http://eiu.ecnext.com/coms2/summary_0294-370642_ITM* (accessed 9 May 2008).

Edelman, L. B. (1990) 'Legal environments and organizational governance: the expansion of due process in the American workplace', *American Journal of Sociology* 95: 1401–40.

Efendioglu, A. M. and Yip, V. F. (2004) 'Chinese culture and e-commerce: an exploratory study', *Interacting with Computers* 16(1): 45–62.

Einhorn, B. (2003) 'Master of innovation? China aims to close its technology gap with Korea and Japan', *Business Week*, 14 April, p. 54.

Einhorn, B. (2003b) 'How Ningbo Bird became a high-flier', available at: *http://www.businessweek.com/technology/content/jan2003/tc20030121_7804.htm*.

Einhorn, B. (2004a) 'Why Beijing has 3G on hold', available at: *http://yahoo.businessweek.com/bwdaily/dnflash/nov2004/nf20041123_6420_db065.htm* (accessed 29 April 2005).

Einhorn, B. (2004b) 'For China, Linux has lots to like', available at: *http://www.businessweek.com/technology/content/mar2004/tc20040330_6891_tc167.htm* (accessed 29 April 2005).

Einhorn, B. (2006a) 'A dragon in R&D', *Business Week*, 6 November, pp. 44–50.

Einhorn, B. (2006b) 'Can China close the software gap with India?', available at: *http://www.businessweek.com/globalbiz/blog/asiatech/archives/2006/11/can_china_close.html* (accessed 30 April 2008).

Einhorn, B. (2006c) 'China's TCL: still trying to get through', available at: *http://www.businessweek.com/globalbiz/content/oct2006/gb20061025_644957.htm?site=cbs* (accessed 28 April 2008).

Einhorn, B. (2007) 'Grudge match in China', *Business Week*, 2 April, p. 42.

Einhorn, B. (2008) 'New regulations for China's YouTube wannabes', available at: *http://www.macnewsworld.com/story/web20/61268.html* (accessed 26 January 2008).

Einhorn, B. and Burrows, P. (2004) 'Huawei: Cisco's rival hangs tough; the Chinese gearmaker has not just survived but thrived since Cisco's lawsuit', *Business Week*, 19 January, p. 73.

Einhorn, B. and Kripalani, M. (2003a) 'Outsourcing: make way for China', available at: *http://businessweek.com/magazine/content/03_31/b3844132_mz033.htm* (accessed 3 May 2008).

Einhorn, B. and Kripalani, M. (2003b): 'Move over India, China is rising fast as a services outsourcing hub', *Business Week*, 11 August, p. 42.

Einhorn, B. and Reinhardt, A. (2005) 'A global telecom titan called … ZTE?', *Business Week*, 7 March, p. 58.

Einhorn, B. and Yang, C. (2000) 'Portal combat', *Business Week*, 17 January, pp. 96–7.

Einhorn. B., Carey J. and Gross, N. (2005) 'A new lab partner for the US?' *Business Week*, 22 August, p. 116.

Einhorn, B., Roberts, D. and Matlack, C. (2003) 'Bursting out of China TCL's deal with Thomson puts it in reach of a global electronics market', *Business Week*, 17 November, p. 20.

Einhorn, B., Wassener, B. and Reinhardt, A. (2005) 'No one said building a brand was easy', *Business Week*, 5 December, p. 18.

Einhorn, B., Webb, A. and Engardio, P. (2000) 'China's tangled web', *Business Week*, 17 July, pp. 56–8.

Einhorn, B., Elgin, B., Edwards, C. and Himelstein, L. (2002): 'High-tech in China: is it a threat to Silicon Valley?', available at: *http://www.businessweek.com/magazine/content/02_43/b3805001.htm* (accessed 30 April 2008).

Einhorn, B., Arndt, M., Roberts, D. and Balfour, F. (2004) 'China's power brands', *Business Week*, 8 November, p. 50.

EIU ViewsWire (2003a) 'China industry: Overseas Chinese return as entrepreneurs', available at: *http://www.viewswire.com/* (accessed 16 November 2003).

EIU ViewsWire (2003b) 'China: Software sector could learn from India: China industry: software exports to expand rapidly over next several years', available at: *http://unpan1.un.org/intradoc/groups/public/documents/APCITY/UNPAN021289.pdf* (accessed 30 April 2008).

EIU ViewsWire (2005a) 'EU economy: The brain drain', available at: *http://www.viewswire.com/index.asp?rf=0* (accessed 30 April 2006).

EIU ViewsWire (2005b) 'China economy: options for foreign studies', available at: *http://www.viewswire.com/index.asp?rf=0* (accessed 13 April 2005).

EIU ViewsWire (2005c) 'China economy: A change of focus', 12 October, available at: *http://www.viewswire.com/article1289483914.html?pubtypeId=930000293&text=china%20economy:%20a%20change%20of%20focus* (accessed 3 May 2006).

Electronic News (2002) 'Counterfeit wave rises in the east', pp. 1–2, available at: *http://www.edn.com/article/CA241771.html* (accessed 30 April 2008).

Electronics Weekly (2006) 'China seeks 3G glory', *Electronics Weekly*, 22 February, p. 1.

Elliott, M. (2006) 'India awakens', *Time Canada*, 3 July, pp. 22–5.

Elsbach, K. D. and Sutton, R. I. (1992) 'Acquiring organizational legitimacy through illegitimate actions: A marriage of institutional and impression management theories', *Academy of Management Journal* 35: 699–738.

Engardio, P., Einhorn, B., Kripalani, M. and Reinhardt, A. (2005) 'Outsourcing innovation', *Business Week*, 21 March, pp. 51–7.

Erickson, J. (2001) 'Enterprise: cover story: the next tech super power', *Asiaweek*, 27 July, p. 1.

Ericsson (2003a) 'China Mobile gets behind TD-SCDMA', available at: *http://www.ericsson.com/about/publications/contact/pdf/c06_03/09.pdf* (accessed 30 April 2004).

Ericsson (2003b) 'Ericsson in China awarded its first Telson Engine commercial contract', available at: *http://www.ericsson.com/press/20031009-104631.html* (accessed 30 April 2008).

Ernst, D. (2005) 'Complexity and internationalization of innovation – why is chip design moving to Asia?', available at: *http://www.eastwestcenter.org/stored/misc/Complexity_and_Internationalisation_Chip_design_1.3.05.pdf* (accessed 30 April 2006).

Escobar, P. (2005) 'Selling China to the world', available at: *http://www.atimes.com/atimes/China/GA15Ad01.html* (accessed 28 April 2006).

Euromonitor (2004) 'Consumer electronics in China' available at: *http://www.euromonitor.com/Consumer_Electronics_in_China* (accessed 28 April 2005).

Euromonitor (2007) 'GMID – Global Market Information Database', available at: *http://www.portal.euromonitor.com/portal/server.pt?space=Login&control=RedirectHome* (accessed 30 April 2007).

Euromonitor (2008) 'Global Market Information Database', available at: *http://www.euromonitor.com/GMID.aspx* (accessed 1 May 2008).

Evans, D. (2004) 'The great (trade) wall of China', *Wall Street Journal*, 28 June, p. A.10.

Fackler, M. (2000) 'The great fire wall of China?', available at: *http://www.abcnews.go.com/secions/tech/DailyNews/chinanet001108.html*.

Far Eastern Economic Review (2004) 'China briefing', *Far Eastern Economic Review* 167(28): 28.

Farazmand, A. (1999) 'The elite question: Toward a normative elite theory of organization', Administration and Society 31(3): 321–59.

Farrell, D. and Grant, A. J. (2005) 'China's looming talent shortage', available at: *http://www.mckinseyquarterly.com/links/19531* (accessed 3 May 2008).

Farrell, J. and Saloner, G. (1986) 'Installed base and compatibility: innovation, product preannouncements, and predation', *American Economic Review* 76(5): 940–55.

Farrell, D., Gao, P. and Orr, G. R. (2004) 'Making foreign investment work for China', available at: *http://www.mckinseyquarterly.com/ Economic_Studies/Country_Reports/Making_foreign_investment_work_ for_China_1470_abstract?gp=1* (accessed 30 April 2008).

Feigenbaum, E. A. (1999) 'Who's behind China's high technology "revolution"?: How bomb makers remade Beijing's priorities, policies, and institutions', *International Security* 24(1): 95–128.

Fenton, S. (2008) 'Cutting PC software piracy creates jobs', available at: *http://www.guardian.co.uk/feedarticle?id=7244934* (accessed 30 January 2008).

Filippo, G. D., Hou, J. and Ip, C. (2005) 'Can China compete in IT services?', available at: *http://www.mckinseyquarterly.com/High_ Tech/Software/Can_China_compete_in_IT_services_1556_abstract* (accessed 3 May 2008).

Financial Express (2004) 'New law in China to end Microsoft's dominance', available at: *http://www.financialexpress.com/fe_full_ story.php?content_id=53682* (accessed 1 April 2005).

Financial Times Information (2006) 'TD Tech reveals ambitious target', available at: *http://www.china.org.cn/english/BAT/191087.htm* (accessed 29 April 2008).

Flanigan, J. (2003) 'China's technological ambitions take flight', *Los Angeles Times*, 19 October, p. C1.

Flemings, M. C. (2007) 'China at the forefront', *Calliope* 17(6): 2–3.

Florida, R. and Kenney, M. (1994) 'The globalization of Japanese R&D: The economic geography of Japanese R&D investment in the United States', *Economic Geography* 70(4): 344–69.

Flynn, L. J. (2004) 'Intel to miss China deadline on standard for wireless', *New York Times*, 11 March, available at: *http://www.nytimes.com/2004/ 03/11/technology/11chip.html?ex=1394341200&en=c77af9aa1946d94 0&ei=5007&partner=USERLAND* (accessed 30 April 2008).

Forbes, L. C. and Jermier, J. M. (2001) 'The institutionalization of voluntary organizational greening and the ideals of environmentalism: lessons about official culture from symbolic organization theory', in Andrew J. Hoffman and Marc J. Ventresca (eds) *Organizations, Policy, and the Natural Environment: Institutional and Strategic Perspectives*, Stanford, CA: Stanford University Press, pp. 194–213.

Forrest, B. (2007) 'Silicon Siberia', *Fortune*, 2 April, p. 33.

Fors, G. and Zejan, M. (1996) 'Overseas R&D by multinationals in foreign centers of excellence', available at: *http://swopec.hhs.se/hastef/abs/hastef0111.htm* (accessed 30 April 2008).

Foushee, H. (2006) 'Gray area: the future of Chinese internet', *Harvard International Review* 8(2): 8–9.

Fowler, G. A. (2006) 'Great firewall: Chinese censors of internet face "hacktivists" in US; Programs like Freegate, built by expatriate Bill Xia, keep the web world-wide; Teenager gets his Wikipedia', *Wall Street Journal*, 13 February, p. A.1.

Frankel, M. S. (1989) 'Professional codes: why, how and with what impact?, *Journal of Business Ethics* 8: 109–15.

Frater, P. (2006) 'Asian markets speed up IPTV race', *Variety*, 9 October, p. B5–6.

Freedom House (2006) 'Freedom in the world 2006, selected data from Freedom House's annual global survey of political rights and civil liberties', available at: *http://www.freedomhouse.org/uploads/pdf/Charts2006.pdf* (accessed 9 April 2007).

French, H. W. (2005) 'A party girl leads China's online revolution', *New York Times*, 24 November, p. A.1.

French, H. W. (2006) 'Chinese discuss plan to tighten restrictions on cyberspace', *New York Times*, 4 July, p. A.3.

Friedman, T. L. (2005) 'From gunpowder to the next big bang', *New York Times*, 4 November, p. A.27.

Frieman, W. (1999) 'The understated revolution in Chinese science and technology: implications for the PLA in the 21st century', in J. R. Lilley and D. Shambaugh (eds), *China's Military Faces the Future*, Armonk, NY: M.E. Sharpe, pp. 247–67.

Fukuyama, F. (1995) *Trust: Social Virtue and the Creation of Prosperity*, New York: The Free Press.

Galaskiewicz, J. (1991) 'Making corporate actors accountable: institution building in Minneapolis–St. Paul', in W. W. Powell and P. J. DiMaggio (eds) *The New Institutionalism in Organizational Analysis*, Chicago, IL: University of Chicago Press, pp. 293–310.

Gallaway, T. and Kinnear, D. (2004) 'Open source software, the wrongs of copyright, and the rise of technology', *Journal of Economic Issues* 38(2): 467–74.

Gao, P., Woetzel, J. R. and Wu, Y. (2003) 'Can Chinese brands make it abroad?', *McKinsey Quarterly*, No. 4 (Special Edition): 54–65.

Garden, T. (2001) 'Why states pursue nuclear weapons', available at: *http://www.tgarden.demon.co.uk/writings/articles/2001/010301nuc.html* (accessed 30 April 2008).

Gardner, D. M., Johnson, F., Lee, M. and Wilkinson, I. (2000) 'A contingency approach to marketing high-technology products', *European Journal of Marketing* 34(9/10): 1053–77.

Garten, J. E. (2003) 'Chinese lessons; the best way to deliver on Bush's call for freedom in the Middle East may be found in the China model', *Newsweek*, 24 November, p. 54.

Garud, R., Jain, S. and Kumaraswamy, A. (2002) 'Institutional entrepreneurship in the sponsorship of common technological standards: the case of Sun Microsystems and Java', *Academy of Management Journal* 45: 196–214.

Gay, L. R. and Diehl, P. L. (1992) *Research Methods for Business and Management*, New York: Macmillan.

George, E., Chattopadhyay, P., Sitkin, S. B. and Barden, J. (2006) 'Cognitive underpinnings of institutional persistence and change: a framing perspective', *Academy of Management Review* 31(2): 347–85.

Gibbs, J. and Kraemer, K. L. and Dedrick, J. (2003) 'Environment and policy factors shaping global e-commerce diffusion: a cross-country comparison', *Information Society* 19(1): 5–18.

Gilboy, G. J. (2004) 'The myth behind China's miracle', *Foreign Affairs* 83(4): 33–48.

Gillmor, D. (2003) 'China tries to establish homegrown tech rules', *Mercury News*, 14 December, p. 32.

Giry, S. (2004) 'China's Africa strategy: out of Beijing, the new republic', available at: *http://www.tnr.com/doc.mhtml?i=20041115&s=giry111504* (accessed 29 April 2005).

GIS News (2001) 'China outstrips Korea in IT industry growth: Samsung', available at: *http://www.gisdevelopment.net/news/2001/oct/news111001.htm* (accessed 30 April 2008).

Global News Wire – Asia Africa Intelligence Wire (2006) 'Linux expected to soar in China', available at: *http://news.xinhuanet.com/english/2006-04/20/content_4452148.htm* (accessed 30 April 2008).

Goad, G. P. and Holland, L. (2000) 'China joins Linux bandwagon', *Far Eastern Economic Review*, 24 February, pp. 8–12.

Goldner, H. J. (1992) 'Locating your research facility', *Research and Development* 34(6): 65–8.

Goodman, P. S. (2002) How China is making the pen as mighty as the PC, *Washington Post*, 4 December, p. E.01.

Goodstein, J. (1994) 'Institutional pressures and strategic responsiveness: Employer involvement in work-family issues', *Academy of Management Journal* 37: 350–82.

Government of Quebec (2004) 'Quebec's microelectronic industry directory', available at: *http://www.mic.gouv.qc.ca/publications/secteurs-industriels/technologies-information/microelectronique_en.pdf* (accessed 30 January 2005).

Greene, J., Einhorn, B. and Hamm, S. (2007) 'A big Windows cleanup in Asia; China is discovering that it pays off to sell PCs that contain legitimate Microsoft software', *Business Week*, 4 June, p. 80.

Greenfeld, L. (2003) *The Spirit of Capitalism: Nationalism and Economic Growth*, Cambridge, MA: Harvard University Press.

Greenwood, R., and Hinings, C. R. (1996) 'Understanding radical organizational change: bringing together the old and the new institutionalism', *Academy of Management Review* 21: 1022–54.

Greenwood, R., Suddaby, R. and Hinings, C. R. (2002) 'Theorizing change: the role of professional associations in the transformation of institutionalized fields', *Academy of Management Journal* 45(1): 58–80.

Grewal, R. and Dharwadkar, R. (2002) 'The role of the institutional environment in marketing channels', *Journal of Marketing* 66(3): 82–97.

Gries, P. H. (2005) 'Chinese nationalism: challenging the state?', *Current History* 251(6).

Gu, E. X. (1997) 'Foreign direct investment and the restructuring of Chinese state-owned enterprises (1992–1995)', *China Information* 12(3): 46–71.

Guangzhou Daily (2006) 'Domestic vendors engage in war with foreign counterparts', *Guangzhou Daily*, 18 April, p. 1.

Guillén, M. F. and Suárez, S. L. (2005) 'Explaining the global digital divide: economic, political and sociological drivers of cross-national internet use', *Social Forces* 84(2): 681–708.

Gumbel, P. (2007) 'What scares executives?', *Time International*, 29 January, p. 36.

Guoguang, W. (2006) 'The peaceful emergence of a great power?' *Social Research* 73(1): 317–45.

Guthrie, D. (1998) 'The declining significance of *guaturi* in China's economic transition', *The China Quarterly* 154: 254–82.

Guthrie, D. (1999) *Dragon in a Three-Piece Suit: The Emergence of Capitalism in China*, Princeton, NJ and Oxford: Princeton University Press.

Gutmann, E. (2002) 'Who lost China's internet?' *The Weekly Standard*, 15 February, pp. 24–9.

Hachigian, N. (2001) 'China's cyber-strategy', *Foreign Affairs* 80(2): 118–33.

Haier (2007) 'Corporate profile' available at: *http://www.haier.com/Abouthaier/CorporateProfile/index.asp* (accessed 30 April 2008).

Hale, D. and Hale, L. H. (2003): 'China takes off', *Foreign Affairs* 82(6): 36–53.

Hamilton, W. (1919) 'The institutional approach to economic theory', *American Economic Review*, No. 1 (Supplement): 309–18.

Hamm, S., Roberts, D. and Lee, L. (2005) 'Lenovo and IBM: east meets west, big-time', available at: *http://www.businessweek.com/magazine/content/05_19/b3932113_mz063.htm* (accessed 9 April 2007).

Hansen, M. (1999) *Lessons in Being Chinese: Minority Education and Ethnic Identity in Southwest China*, Seattle, WA: University of Washington Press.

Harbert, T. (2003) 'China becomes top importer to United States', *Electronic Business*, 1 August, p. 18.

Hart, D. M. (1998) *Forged Consensus: Science, Technology, and Economic Policy in the United States, 1921–1953*, Princeton, NJ: Princeton University Press.

Hart-Landsberg, M. and Burkett, P. (2004) 'Contradictions of China's transformation: international', *Monthly Review: An Independent Socialist Magazine* 56(3): 81–108.

Haunschild, R. and Miner, A. S. (1997) 'Modes of interorganizational imitation: the effects of outcome salience and uncertainty', *Administrative Science Quarterly* 42: 472–500.

Hayek, F. A. (1979) *Law, Legislation and Liberty* (3 vols), Chicago, IL: University of Chicago Press.

Health Insurance Law Weekly (2006) 'The Boston Consulting Group; Study finds biopharma in China is booming and will likely leapfrog many Western European markets', *Health Insurance Law Weekly*, 19 March, p. 110.

heavyreading.com (2008) 'Ethernet Services in China', available at: *http://www.heavyreading.com/details.asp?sku_id=708&skuitem_itemid=710&promo_code=&aff_code=&next_url=%2Fdefault.asp%3F* (accessed 30 April 2008).

Heer, P. (2000) 'A house divided', *Foreign Affairs* 79: 18–24.

Heiser, S. (2007) 'IPTV in China will exceed 23 million subscribers by 2012: ABI research report finds mainland China operators see IPTV as competitive edge', available at: *http://www.telecommagazine.com/newsglobe/article.asp?HH_ID=AR_2816*.

Helm, B. and Kripalani, M. (2006) 'Life on the web's factory floor', *Business Week*, 22 May, pp. 70–1.

Hemerling, J. (2006) 'For China, it's mergers and ambition', available at: *http://my.opera.com/PRC/blog/show.dml/482436* (accessed 28 April 2008).

Hennock, M. (2003) 'Chinese phone maker's fancy footwork', available at: *http://newsvote.bbc.co.uk/mpapps/pagetools/print/news.bbc.co.uk/1/hi/business/3207829.stm.*

Herschede, F. (1991) 'Competition among ASEAN, China, and the East Asian NICs: A shift-share analysis', *ASEAN Economic Bulletin* 7(3): 290–306.

Hermida, A. (2002) 'Behind China's internet red firewall', available at: *http://news.bbc.co.uk/2/hi/technology/2234154.stm* (accessed 11 December 2006).

Herrmann-Pillath, C. (2006) 'Cultural species and institutional change in China', *Journal of Economic Issues* 40(3): 539–74.

Hertling, J. (1997) 'More Chinese students abroad are deciding not to return home', *Chronicle of Higher Education*, 28 March, pp. A51–52.

Hill, J. (2004) 'China covets African oil and trade', *Jane's Intelligence Review*, 16(11): 44–5.

Hindu Business Line (2006) 'Electronics, IT exports to grow 27 pc in 2005–06', available at: *http://www.thehindubusinessline.com/2006/03/28/stories/2006032802920400.htm* (accessed 30 April 2007).

Hirsch, P. (1997) 'Sociology without social structure: neo-institutional theory meets brave new world', *American Journal of Sociology* 102(6): 1702–23.

Hirt, M. and Orr, G. (2006) 'Helping China's companies master global M&A,' *McKinsey Quarterly*, No. 4: 38–49.

Hodgson, G. M. (2003) 'The hidden persuaders: institutions and individuals in economic theory', *Cambridge Journal of Economics* 27: 159–75.

Hoffman, A. J. (1999) 'Institutional evolution and change: environmentalism and the US chemical industry', *Academy of Management Journal* 42(4): 351–71.

Hofstede, G. and Bond, M. H. (1988) 'The Confucius connection: from cultural roots to economic growth', *Organizational Dynamics* 16(4): 4–21.

Hollings, E. F. and McMillion, C. W. (2000) 'Will the US ignore China's surging technology sector?', available at: *http://hollings.senate.gov/~hollings/opinion/2002620A58.html.*

Holm, P. (1995) 'The dynamics of institutionalization: transformation processes in Norwegian fisheries', *Administrative Science Quarterly* 40(3): 398–422.

Hoover's Company Records (2008) 'TCL Corporation', available at: *http://www.alacrastore.com/storecontent/hooversindepth/132340* (accessed 29 April 2008).

Hormats, R. D. (2001) 'Asian connection', *Across the Board* 38(1): 47–50.

Howard, M. M. (2003) *The Weakness of Civil Society in Post-Communist Europe*, Cambridge: Cambridge University Press.

Howells J. (1984) 'The location of research and development: some observations and evidence from Britain', *Regional Studies* 18: 13–29.

Huang, Z. (1993) 'Current development of the private firms in the Mainland China', *Economic Outlook* 8(32): 87–91.

Huff, A. S. (1990) 'Mapping strategic thought', in A. S. Huff (ed.) *Mapping Strategic Thought*, Chichester: Wiley, pp. 11–49.

Hughes, N. C. (2005) 'A trade war with China?', *Foreign Affairs* 84(4): 94–107.

Hui, L. (2008) '95% of villages to have access to broadband service in '08', available at: *http://news.xinhuanet.com/english/2008-02/10/content_7585629.htm* (accessed 1 May 2008).

Hull, M. (1997) 'Translating immigrant dreams into jobs', available at: *http://www.electronixstaffing.com/inews2.html* (accessed 3 May 2008).

Huus, K. (1994) 'Intellectuals: troubled waters', *Far Eastern Economic Review*, 157(19): 44–5.

Hwang, J. S. (1994) 'Economic development and internationalization as viewed through the investment development path', unpublished PhD thesis, University of Reading.

Hymer, S. (1976) *The International Operations of National Firms*, Cambridge, MA: MIT Press.

Hymowitz, C. (2005) 'Recruiting top talent in China takes a boss who likes to coach', *Wall Street Journal*, 26 April, p. B.1.

IAfrica.com (2005) 'China vows to tie up with Asia, Africa', 21 April (2005), available at: *http://business.iafrica.com/news/434784.htm* (accessed 30 April 2008).

Iammarino, S. and Michie J. (1998) 'The scope of technological globalization', *International Journal of the Economics of Business* 5(3): 335–53.

Ianchovichina, E., Suthiwart-Narueput, S. and Zhao, M. (2003) 'Regional impact of China's WTO accession', in K. Krumm and H. Kharas (eds) *East Asia Integrates: A Trade Policy Agenda for Shared Growth*, Washington DC: World Bank, pp. 57–78.

Ibrahim, G. and Galt, V. (2002) 'Bye-bye central planning, hello market hiccups: Institutional transition in Romania', *Cambridge Journal of Economics* 26(1): 105–18.

IHS (2007) 'ABI: IPTV subscribers in mainland China to exceed 23 m by 2012', available at: *http://electronics.ihs.com/news/abi-iptv-china.htm*.

Ilett, D. (2007) 'China's very different online world around the world: eBay's auction model is turned upside down; $35bn is spent on online services; and users live with censorship', *Financial Times*, 12 February, p. 10.

Indianapolis Business Journal (2004) 'Thomson venture launches ultra-thin projection sets', available at: *http://goliath.ecnext.com/coms2/gi_0199-3479814/Thomson-venture-launches-ultra-thin.html* (accessed 28 April 2008).

In-Stat (2006a) 'Chinese electronics output to exceed US$300 billion in 2006', available at: *http://www.instat.com/newmk.asp?ID=1787* (accessed 30 April 2007).

In-Stat (2006b) 'China's IPTV market developing slowly but surely', available at: *http://www.instat.com/press.asp?ID=1842&sku=IN0602776CCM* (accessed 1 January 2008).

Intermedia (2006) 'China steps onto the world telecom stage', *Intermedia* 34(4/5): 48–50.

International Financial Law Review (2005) 'TCL-Thomson Asian Awards', *International Financial Law Review* 24 (April): p. 40.

InternetWorldStats.com (2008) 'Top 20 countries with the highest number of internet users', available at: *http://www.internetworldstats.com/top20.htm* (accessed 30 April 2008).

Ionaşcu, D., Meyer, K. and Estrin, S. (2004) 'Institutional distance and international business strategies in emerging economies', available at: *http://www.wdi.umich.edu/files/Publications/WorkingPapers/wp728.pdf* (accessed 30 April 2008).

ipr.gov.cn (2006) 'Expanding business: Chinese handset vendors go abroad', available at: *http://www.ipr.gov.cn/ipr/en/info/Article.jsp?a_no=33956&col_no=927&dir=200612* (accessed 30 April 2008).

Iritani, E. (2001) 'China's WTO challenge; telecoms key test of China's accessibility; trade: US firms burned in the past worry about Beijing's willingness to pay the price of admission to the organization', *Los Angeles Times*, 5 August, p. C.1.

itfacts.biz (2006) 'IT facts', available at: *http://www.itfacts.biz/index.php?id=P448* (accessed 2 May 2006).

ITU (2005a) 'From the digital divide to digital opportunities measuring infostates for development', available at: *http://www.itu.int/ITU-D/ict/publications/dd/summary.html* (accessed 30 April 2008).

itworld.com (2007) 'India's software, services exports forecast at US$40 billion', available at: *http://www.itworld.com/Man/2701/070702india/* (accessed 30 April 2008).

Iype, G. and Prema, R. (2002) 'RAW wants 95 Chinese software engineers expelled', *India Abroad*, 23 August, p. A20.

James, D. (2001) 'China's sizzling circuits', *Upside* 13(1): 60–7.

Jasper, W. F. (2006) 'Terror in America, Made in China', *New American* 22(6): 19–21.

Jayasuriya, K. (1999) *Law, Capitalism and Power in Asia*, London: Routledge.

Jensen, R. and Szulanski, G. (2004) 'Stickiness and the adaptation of organizational practices in cross-border knowledge transfers', *Journal of International Business Studies* 35(6): 508–23.

Jepperson, R. (1991) 'Institutions, institutional effects, and institutionalism', in W. W. Powell and P. J. DiMaggio (eds) *New Institutionalism in Organizational Analysis*, Chicago, IL: University of Chicago Press, pp. 143–63.

Jesdanun, A. (2008) 'China catching up to US in number of web surfers', available at: *http://www.technewsworld.com/story/China-Catching-Up-to-US-in-Number-of-Web-Surfers-61292.html?welcome=1201370753* (accessed 26 January 2008).

Jie, L. (2006) 'TCL: Testing the thorny road for Chinese companies to go global', *Xinhua News Agency*, 28 November.

Jin, J. (2006): 'The current status of China's electronic information industry', available at: *http://jp.fujitsu.com/group/fri/en/column/economic-topics/200608/2006-08-18-1.html* (accessed 30 April 2007).

Johanson, J. and Vahlne, J.-E. (1977) 'The internationalization process of a firm – a model of knowledge development and increasing foreign market commitments', *Journal of International Business Studies* 8(1): 23–32.

John, T. W. (2002) 'The next generation: will China connect?', available at: *http://www.atimes.com/atimes/China/DK07Ad05.html* (accessed 9 May 2008).

Johnson, J. P. (2002) 'Open source software: private provision of a public good', *Journal of Economics & Management Strategy* 11(4): 637–62.

Johnson, K. (2006) 'Voices of dissent', *Time International*, 25 September, p. 50.

Johnsson, J. (2004) 'Uh-oh, Ningbo', *Crain's Chicago Business*, 26 April, p. 1.

Jones, D. K. (2004) 'The neglected role of international altruistic investment in the Chinese transition economy', *The George Washington International Law Review* 36(1): 71–146.

Jun, C. (2005) 'Overseas merger and acquisition: an inevitable trend', available at: *http://en.ce.cn/Insight/200505/30/t20050530_3933383.shtml* (accessed 20 January 2007).

Kahn, J. (2006) 'China says web controls follow the west's lead', *New York Times*, 15 February, p. A.6.

Kalafsky, R. (2006) 'Human capital in Japanese manufacturing: evidence and practices from a key capital goods sector', *The Industrial Geographer* 3(2): 13–26.

Kalathil, S. (2003) 'China's new media sector: keeping the state in', *Pacific Review* 16(4): 489–501.

Kalathil, S. and Boas, T. C. (2003) *Open Networks, Closed Regimes: The Impact of the internet on Authoritarian Rule*, Washington, DC: Carnegie Endowment for International Peace.

Kanellos, M. (2002a) 'Software comes of age: piracy crackdown pays off', available at: *http://news.com.com/2009-1001-940335.html* (accessed 11 July 2002).

Kanellos, M. (2002a) 'Tech gets set for China breakthrough', available at: *http://news.zdnet.com/2100-9595_22-942374.html* (accessed 30 April 2006).

Katz, M., and Shapiro, C. (1994) 'System competition and network effects', *Journal of Economic Perspectives* 8(2): 93–115.

Kelman, S. (1987) *Making Public Policy: A Hopeful View of American Government*, New York: Basic Books.

Kempe, F. (2006) 'Politics and economics – thinking global: Africa emerges as strategic battleground; challenges for US include terrorist ties, energy issues, countering China's inroads', *Wall Street Journal*, 25 April, p. A.8.

Kennedy, S. (2005) *The Business of Lobbying in China*, Cambridge, MA: Harvard University Press.

Kessler, M. (2004) 'US firms: doing business in China tough, but critical', *USA Today*, 17 August, p. B.01.

Kharbanda, V. P. and Suman, Y. (2002) 'Chinese initiative in the software industry – quest to leap', *Current Science* 83(12): 1450–5.

Khermouch, G., Einhorn, B. and Roberts, D. (2003) 'Chinese companies play the name game the mainland's manufacturers are building their brands to go global', available at: *http://www.businessweek.com/magazine/content/03_13/b3826120_mz033.htm* (accessed 9 May 2008).

Kima, H. J., Gursoya, D. and Leeb, S. (2006) 'The impact of the 2002 World Cup on South Korea: comparisons of pre- and post-games', *Tourism Management,* 27(1): 86–96.

Kindleberger, C. (1969) *American Business Abroad*, New Haven, CT: Yale University Press.

King, J. L., Gurbaxani, V., Kraemer, K. L., McFarlan, F. W., Raman, K. S. and Yap, C. S. (1994) 'Institutional factors in information technology innovation', *Information System Research* 5(2): 139–69.

Kirkpatrick, D. (2007) 'How Microsoft conquered China: or is it the other way', *Fortune*, 17 July, p. 78.

Kirpalani, J. (2008) 'UPDATE 1-Disney to raise stake in India's UTV', available at: *http://uk.reuters.com/article/mediaNews/idUKBOM4617 220080218* (accessed 30 April 2008).

Klare, M. T. (2005) 'Revving up the China threat', *The Nation*, 24 October, pp. 28–32.

Knight Ridder Tribune Business News (2005) 'India tops salary hike in Asia: Hewitt', *Knight Ridder Tribune Business News*, 24 November, p. 1.

Koch, B. J. and Koch, P. T. (2007) 'Collectivism, individualism, and outgroup cooperation in a segmented China', *Asia Pacific Journal of Management* 24(2): 207–25.

Kogut, B. and Zander, U. (2000) 'Did socialism fail to innovate? A natural experiment of the two Zeiss companies', *American Sociological Review* 65(2): 169–80.

Kong, E. (2006) 'China Netcom's profit falls 7.7% as growth slows', *Wall Street Journal*, 24 August, p. B.4.

Kong, X. (2001) 'Jianii yu WTO yaoqiu xiang shiying de sifa shencha zhidu' [Establish a system of judicial review that suits the needs of WTO], *Zhongguo faxue* [Chinese Jurisprudence] 6:8.

Koo, G. (1998) 'The real China', *Harvard International Review* 20(3): 68–71.

Koprowski, G. J. (2004) 'Wireless world: China's WiFi revolution', available at: *http://www.upi.com/view.cfm%3FStoryID=20040506%2D101652%2D8297r* (accessed 29 April 2005).

Koprowski, G. J. (2006) 'Report: China internet use catching up with US', available at: *http://www.technewsworld.com/story/48219.html* (accessed 1 January 2008).

Kostova, T. (1999) 'Transnational transfer of strategic organizational practices: a contextual perspective', *Academy of Management Review* 24: 308–24.

Kostova, T. and Zaheer, S. (1999) 'Organizational legitimacy under conditions of complexity: The case of the multinational enterprise', *Academy of Management Review* 24: 64–81.

Kotler, P., Jatusripitak, S. and Maesincee, S. (1997) *The Marketing of Nations*, New York: Free Press.

Kranhold, B. K. (2004) 'Tough bargain – China's price for market entry: give us your technology, too; GE shares generator plans to win $900 million deal; gray area in WTO rules; science of a turbine', *Wall Street Journal*, 26 February, p. A.1.

Krikke, J. (2002) 'Fortunes changing in Asia: China's handset makers take on the global giants', available at: *http://www.wirelessweek.com/index.asp?layout=article&articleid=CA239110* (accessed 30 April 2003).

Krikke, J. (2003) 'Linux revolution: Asian countries push open source', available at: *http://www.linuxinsider.com/story/32421.html* (accessed 30 January 2005).

Krueger, A. O. (1991) 'Benefits and costs of late development', in P. Higonnet, D. S. Landes and H. Rosovsky (eds) *Favorites of Fortune: Technology, Growth, and Economic Development Since the Industrial Revolution*, Cambridge, MA: Harvard University Press, pp. 459–481.

Kshetri, N. (2002) 'What determines internet diffusion loci in developing countries: evidence from China and India', *Pacific Telecommunications Review* 23(3): 25–34.

Kshetri, N. (2004) 'Economics of Linux adoption in developing countries', *IEEE Software* 23(1): 74–81.

Kshetri, N. (2007) 'The adoption of e-business by organizations in China: an institutional perspective', *Electronic Markets* 17(2): 113–25.

Kshetri, N. and Cheung, M. K. (2002) 'What factors are driving China's mobile diffusion?' *Electronic Markets* 12(1): 22–6.

Kshetri, N. and Dholakia, N. (2001) 'Impact of cultural and political factors on the adoption of digital signatures in Asia', paper presented at the Americas' Conference on Information Systems, Boston, MA, 3–5 August.

Kshetri, N., Dholakia, N. and Awasthi, A. (2003) 'Determinants of e-government readiness: evidence from China and India', paper presented at the First International Conference on E-governance, New Delhi, 18–20 December.

Kuemmerle, W. (1997) 'Building effective R&D capabilities abroad', *Harvard Business Review* 75(2): 61–70.

Kumaresan, N. and Miyazaki, K. (2001): 'Management and policy concerns over shifts in innovation trajectories: The case of the Japanese robotics industry', *Technology Analysis & Strategic Management* 13(3): 433–62.

Kurlantzick, J. (2005) 'Cultural revolution', *New Republic*, 27 June, pp. 16–21.

Kwong, K. K., Yau, O. H. M., Lee, J. S. Y., Sin, L.Y. M. and Tse, A. C. B. (2003) 'The effects of attitudinal and demographic factors on intention to buy pirated CDs: the case of Chinese consumers', *Journal of Business Ethics* 47(3): 223–35.

Lague, D. (2006a) 'Small steps in a long fight against piracy; Chinese measures seen as good start', *International Herald Tribune*, 18 May, p. 19.

Lague, D. (2006b) 'China begins effort to curb piracy of computer software', New York Times, 30 May, p. C.3.

Lall, S. (1983) 'The theoretical background', in S. Lall (ed.) *The New Multinationals*, Chichester: John Wiley, pp. 1–20.

Lall, S. and Albaladejo, M. (2004) 'China's competitive performance: a threat to East Asian manufactured exports?' *World Development* 32(9): 1441–66.

Lampton, D. M. (1986) *A Relationship Restored: Trends in US–China Educational Exchanges*, 1978–1984, Washington, DC: National Academy Press.

Lampton, D. M. (2005) 'Paradigm lost: the demise of "Weak China"', *The National Interest*, No. 81: 73–80.

Lan, P. and Young, S. (1996) 'Foreign direct investment and technology transfer: a case study of FDI in Northeast China', *Transnational Corporation* 5(2): 57–83.

Landay, J. S. (1997) 'Is China diverting high-technology to US foes?' *Christian Science Monitor*, 11 July, p. 1.

Lange, J. E. (2005) 'PRC investors set sights on global companies', *International Financial Law Review* 24: 45–9.

Lardy, N. R. (2003) 'The economic rise of China: threat or opportunity?', *Economic Commentary*, Federal Reserve Bank of Cleveland, 1 August: 3.

Lardy, N. R. (2002) *Integrating China into the Global Economy*, Washington, DC: Brookings Institution Press.

Laschinger, K. (2005) 'Enter the dragon', *Finance Week*, 17 January, pp. 8–10.

Lau, J. (2006a) 'China Netcom looks to IPTV Telecoms', *Financial Times*, 22 March, p. 21.

Lau, J. (2006b) 'Poor reception for China's global push TCL's problems exemplify the difficulties for mainland companies desperate to expand abroad', *Financial Times*, 3 November, p. 26.

Lau, J. and Mitchell, T. (2006) 'TCL to restructure European unit', *Financial Times*, 31 August, p. 22.

Lau, L. J., Qian, Y. and Roland, G. (2001) 'Reform without losers: An interpretation of China's dual-track approach to transition', *Journal of Political Economy* 108(1): 120–43.

Laudicina, P. and White, J. M. (2005) 'India and China: Asia's FDI magnets', *Far Eastern Economic Review* 168(9): 25–8.

Leahy, S. (2005) 'Science: a "nano-divide" widens gap between rich and poor', *Global Information Network*, 14 April, p. 1.

Lee, G. K. and Cole, R. E. (2003) 'From a firm-based to a community-based model of knowledge creation: the case of the Linux kernel development', *Organization Science* 14(6): 633–51.

Lee, Y., and O'Connor, G.C. (2003) 'New product launch strategy for network effects products', *Academy of Marketing Science Journal* 31(3): 241–55.

Lee, Y.-S., Bratton, W. and Shi, W. W. (2005) 'The role of regulators in promoting broadband in developing countries', paper presented at the

Global Symposium for Regulators, Yasmine Hammamet, Tunisia, 14–15 November.

Leggett, K. (2005) 'China forges deep alliances with war-torn nations in Africa', available at: *http://www.sudantribune.com/article.php3?id_article= 8757* (accessed 9 April 2007).

Lehrer, M., Dholakia, N. and Kshetri, N. (2002) 'National sources of leadership in 3G m-business applications: a framework and evidence from three global regions', paper presented at the Americas' Conference on Information Systems, Dallas, TX, 8–11 August.

Lemon, S. (2003) 'AMD gets supercomputing boost: China's Dawning takes wraps off Opteron-based supercomputer', available at: *http://www.infoworld.com/article/03/07/25/HNamdsupercompute_1 .html* (accessed 30 January 2005).

Lenovo (2005) 'Lenovo announces 2004/05 annual results', available at: *http://www.lenovo.com/news/ca/en/2005/06/060805annreport.html* (accessed 30 April 2008).

Lerner, J. and Tirole, J. (2002) 'Some simple economics of open source', *Journal of Industrial Economics* 50(2): 197–234.

Leung, T. (2003): 'Gartner: China needs to address software quality issue', *Asia Computer Weekly*, 14 July, p. 1.

Lewis, J. W. and Litai X. (1987) 'Strategic weapons and Chinese power: the formative years', *The China Quarterly*, No. 112: 541–54.

Li, M., Lin, Z. and Xia, M. (2004) 'Leveraging the open source software movement for development of China's software industry', *Information Technologies and International Development* 2(2): 45–63.

Li, Q. (2007) 'IPTV in Asia: lessons learned from real deployments in China and Japan', paper presented at the 2007 ChinaTel Summit and VC Forum, Honolulu, HI, 13–14 January.

Liang, Y. (2008) 'China PC sales value growth tempered by price war', available at: *http://news.xinhuanet.com/english/2008-02/26/content_ 7674477.htm* (accessed 30 April 2008).

Library of Congress (1987a) 'China: "reds" versus "experts" in the 1950s and 1960s', available at: *http://www.photius.com/countries/china/ economy/china_economy_reds_versus_exper~7317.html* (accessed 2 May 2008).

Library of Congress (1987b) 'The Cultural Revolution', 1966–76', available at: *http://countrystudies.us/china/28.htm* (accessed 2 May 2008).

Lim, K. H., Leung, K., Sia, C. L. and Lee, M. K.O. (2004) 'Is e-commerce boundary-less? Effects of individualism-collectivism and uncertainty avoidance on internet shopping', *Journal of International Business Studies* 35(6): 545–59.

Lim, L. (2006) 'In China, blogs are revolutionary tool of opinion. All things considered', available at: *http://www.npr.org/templates/story/story.php?storyId=5250144* (accessed 30 April 2008).

Lin, J.Y., Cai, F. and Li, Z. (1996) 'The lessons of China's transition to a market economy', *Cato Journal* 16(2): 201–31.

Lindblad, C. (2005) 'TCL's French meal leads to indigestion (China's TCL Multimedia reports financial losses following acquisition of Thomson's television business, France)', *Business Week*, 12 September, p. 67.

Linder, S. B. (1961) *An Essay on Trade and Transformation*, Uppsala: Almqvist and Wiksells.

Lindstrom, A. (2006) 'Citizens of China and India are hoping to reap the economic rewards of increased broadband wireless access in cities and remote areas', available at *http://www.connectwithcanopy.com/index.cfm?canopy=spotlight.story&aid=393* (accessed 1 June 2006).

Lipper, H. (1997) 'In China, firms find some employees just keep going and going and going', *Wall Street Journal*, 5 December, p. 1.

Liu, M. (2003) 'By dint of its sheer size, China may transform fields well outside of geopolitics; Technology: from chaser to maker', *Newsweek* (international edn), 15 December, p. 31.

Liu, M. (2006) 'High-tech hunger; the goal: make China a technology powerhouse – critics say by any means necessary. Inside Beijing's '863 program', available at: *http://findarticles.com/p/articles/mi_hb3335/is_200601/ai_n18038463* (accessed 30 April 2007).

Liu, M. (2007) 'Unsafe at any speed; The downside of China's manufacturing boom: deadly goods wreaking havoc at home and abroad', available at: *http://www.mywire.com/pubs/Newsweek/2007/07/16/3955630?extID=10037&oliID=229* (accessed 30 April 2008).

Liu, S. (1999) 'China promotes homegrown "smart" 3G spec', *Electronic Engineering Times*, 8 November, p. 36.

Liu, Z. (2002) 'Foreign direct investment and technology spillover: evidence from China', *Journal of Comparative Economics* 30(3): 579–602.

Lo, W. C. W. and Everett, A. M. (2001) 'Thriving in the regulatory environment of e-commerce in China: a *guanxi* strategy', *SAM Advanced Management Journal* 66(3): 17–24.

Loasby, B. (1996) 'The organisation of industry' in N. J Foss and C. Knudsen (eds) *Towards a Competence Theory of the Firm*, London: Routledge, pp. 38–53.

Loasby, B. (1998) 'The organization of capabilities', *Journal of Economic Behaviour and Organization* 35: 139–60.

Lodorfos, G. and Boateng, A. (2006) 'The role of culture in the merger and acquisition process: evidence from the European chemical industry', *Management Decision* 44(10): 1405–21.

Longo, D. (2004) 'In China, local and multinational retailers share similar problems', *Progressive Grocer* 83(18): 8–9.

Los Angeles Times (1997) 'The cutting edge; testing the boundaries; countries face cyber control in their own ways', *Los Angeles Times*, 30 June, p. 1.

Los Angeles Times (2005) 'China tops US in tech exports; A broad category of Chinese electronic goods sent abroad rises 46% in 2004, putting the Asian country in the lead', *Los Angeles Times*, 13 December, p. C.5.

Lovelock, P. (2001) 'China IP telephony country case study', available at: *http://www.itu.int/osg/spu/wtpf/wtpf2001/casestudies/chinafinal.pdf* (accessed 30 April 2002).

Lubman, S. B. (1999) *Bird in a Cage: Legal Reform in China after Mao*, Stanford, CA: Stanford University Press.

Luostarinen, R. and Hellmann, H. (1994) 'The internationalization processes and strategies of Finnish family firms', CIBR Research Papers, Helsinki: School of Economics and Business Administration.

Lyman, P. N. (2005) 'China's rising role in Africa, Council on Foreign Relations', available at: *http://www.cfr.org/publication/8436/chinas_rising_role_in_africa.html* (accessed 9 April 2007).

Lynch, D. J. (2003) 'More of China's best, brightest return home; nation's brain drain days over as it becomes land of opportunity', *USA Today*, 7 March, p. B.01.

Lynch, G. (2002) 'Are the TDMA operators really washed up?', *America's Network*, 15 November, pp. 32–5.

Lynch, G. and Chau, F. (2000) 'Surprise twists in CDMA's future', *America's Network* 104(17): 30–3.

M2 Presswire (2007) 'Research and markets: in terms of Chinese consumers expectations about how they would pay for IPTV applications, consumers interested in the applications tend to prefer paying monthly', available at: *http://goliath.ecnext.com/coms2/summary_0199-6339052_ITM* (accessed 1 January 2008).

Mackinnon, R. and Palfrey, J. (2006) 'Opinion: Censorship Inc.; If they're not careful, western tech companies could break up the web', available at: *http://findarticles.com/p/articles/mi_hb3335/is_200602* (accessed 30 April 2008).

MacLeod, C. (2006) 'China grows more aggressive in thwarting counterfeiters', *USA Today*, 21 April, p. 4B.

Madden, N. (2004) 'TCL-Thomson Electronics' advertising compaigns', *Advertising Age*, 9 August, p. 12.

Madden, N. (2006) 'Marketers in China "grab the sofa"', *Advertising Age*, 2 October, p. 16.

Madden, N. (2007) 'Faulty products could thwart the expansion plans of Chinese brands', *Advertising Age*, 20 August, p. 21.

Madrigal, A. (2008) 'The Chinese government's plans for nanotechnology', available at: *http://blog.wired.com/wiredscience/ 2008/02/the-chinese-gov.html* (accessed 30 April 2008).

Magnier, M. and Menn, J. (2005) 'As China censors the internet, money talks', *Los Angeles Times*, 17 June, p. A.1.

Mahlow, M. (2003) 'Free for all; Open-source software transforms technology in the developing world', *Newsweek*, 30 June, p. 74.

Maidment, P. (2008) 'China.com', available at: *http://www.forbes.com/ business/2008/01/09/china-internet-media-biz-media-cx_pm_ 0109notes.html* (accessed 26 January 2008).

Malerba, F. (2006) 'Innovation and the evolution of industries', *Journal of Evolutionary Economics* 16(1–2): 3–23.

Malerba, F. and Orsenigo, L. (1996): 'The dynamics and evolution of industries', *Industrial and Corporate Change* 5(1): 51–87.

Malester, J., and Tarr, G. (2003) 'Thomson merges television/DVD biz with China's TCL', *TWICE*, 10 November, p. 1.

Mallaby, S. (2005) 'In Beijing, a growing problem', *Washington Post*, 11 July, p. A.15.

Mallaby. S. (2006) 'Google and my red flag', *The Washington Post*, 30 January, p. A.17.

Malmgren, P. (2005) 'The China temptation', *International Economy* 19(2): 44–7.

Mann, C. C. (1999) 'Living with Linux', *Atlantic Monthly* 284(2): 80–6.

Mann, J. (1986) 'Life in US portrayed as difficult; China tries to cut loss of students staying abroad', *Los Angeles Times*, 1 April, p. 1.

March, J. G. and Olsen, J. P. (1989) *Rediscovering Institutions. The Organizational Basis of Politics*, New York: Free Press.

March, J. G. and Simon, H. A. (1958) *Organizations*, New York: John Wiley & Sons.

Marcial, G. G. (2006) 'A China Telecom play with a quiet ring,' 19 June, available at: *http://www.businessweek.com/magazine/content/06_25/ b3989116.htm* (accessed 30 April 2008).

Markus, M. L., Manville, B. and Agres, C. E. (2000) 'What makes a virtual organization work?' *MIT Sloan Management Review* 42(1): 13–26.

Marquand, R. (2005) 'Chinese build a high-tech army within an army', *Christian Science Monitor*, 17 November, p. 1.

Marsh, C. and Dreyer, J. T. (2003) *US-China Relations in the Twenty-First Century: Policies, Prospects, and Possibilities*, Lanham, MD: Lexington Books.

Martinsons, M. G. (2005) 'Transforming China', *Communications of the ACM* 48(4): 44–8.

Massey, J. A. (2006) 'The emperor is far away: China's enforcement of intellectual property rights protection, 1986–2006', *Chicago Journal of International Law* 7(1): 231–7.

Mathews, J. A. (2006a) 'Dragon multinationals: new players in 21st century globalization', *Asia Pacific Journal of Management* 23(1): 5–27.

Mathews, J. A. (2006b) 'Response to Professors Dunning and Narula', *Asia Pacific Journal of Management* 23: 153–5.

Matutinović, I. (2005) 'The microeconomic foundations of business cycles: from institutions to autocatalytic networks', *Journal of Economic Issues* 39(4): 867–98.

Maurer, K. (2003) 'Thomson, Chinese partner to launch new company', *Indianapolis Business Journal*, 10 November, p. 37.

Maurer-Fazio, M. (1995) 'Labor reform in China: crossing the river by feeling the stones', *Comparative Economic Studies* 37(4): 111–23.

May, C. (2006) 'Escaping the TRIPs' trap: the political economy of free and open source software in Africa', *Political Studies* 54(1): 123–46.

McAuley, A. (1999) 'Entrepreneurial instant exporters in the Scottish arts and crafts sector', *Journal of International Marketing* 7(4): 67–82.

McDaniels, I. K. and Waterman, J. (2000) 'WTO: a done deal?' *China Business Review* 27(6): 22–7.

McGill, D. C. (2001) 'Shanghai webTV is banking on broadband', available at: *http://www.virtualchina.com/archive/finance/stirfry/032300-stirfry-dcm.html* (accessed 30 April 2008).

McIllwain, J. S. (1999) 'Organized crime: a social network approach', *Crime, Law and Social Change* 32(4): 301–22.

McKinsey, K. (2001) 'Asians miss the e-biz mark', *Far Eastern Economic Review* 164(10): 42–3.

McKinsey Quarterly (2006) 'Building brands in China', available at: *http://www.mckinseyquarterly.com/article_abstract.aspx?ar=1795&L2=16&L3=14* (accessed 11 December 2006).

McLaughlin, K. E. (2005) 'China's model for a censored internet', *Christian Science Monitor* 97(210): 1–10.

Mearsheimer, J. J. (2005) 'Better to be Godzilla than Bambi foreign policy', available at: *http://www.foreignpolicy.com/story/cms.php?story_id=2740* (accessed 30 April 2007).

Media Asia (2005) 'Asia's top brands: brand legacies linger as goal posts shift portfolio', *Media Asia*, 26 August, pp. 6–75.

Medialine (2003) 'Thomson and TCL join television, DVD units', *Medialine*, 1 November, p. 56.

MediaNet Press Release Wire (2007) 'Huawei names Kasenna as partner in bringing video on-demand to China', available at: *http://www.prnewswire.com/cgi-bin/stories.pl?ACCT=109&STORY=/www/story/06-18-2007/0004609755&EDATE=* (accessed 1 January 2008).

Mehta, S. N. (2005) 'Beijing calling', *Fortune* 152(2): 131.

Mehta, S. N. (2006) 'Does it run in the family?', *Fortune*, 30 October, p. 142.

Meredith, R. (2003) '(Microsoft's) long march', *Forbes*, 17 February, pp. 78–86.

Merritt, R. (2003) 'China cellphone makers seen challenging Nokia, Motorola', available at: *http://www.electronicstimes.com/bus/news/OEG20030708S0057* (accessed 30 April 2005).

Metcalfe, J. S., Foster, J. and Ramlogan, R. (2006) 'Adaptive economic growth', *Cambridge Journal of Economics* 30(1): 7–32.

Meyer, A. and Rowan, B. (1977) 'Institutionalized organizations: formal structure as myth and ceremony', *American Journal of Sociology* 83: 340–63.

Meyer, J. W. and Scott, W. R. (1983) *Organizational Environments: Ritual and Rationality*, Beverly Hills, CA: Sage.

Meyer, J. W., Scott, W. R. and Deal, T. E. (1983) 'Institutional and technical sources of organizational structure: explaining the structure of educational organizations', in J. W. Meyer and W. R. Scott (eds) *Organizational Environments*, Beverly Hills, CA: Sage, pp. 45–67.

Meyer, P. (1987) *Ethical Journalism*, New York: Longman.

Milberg, W. (2004) 'The changing structure of trade linked to global production systems: What are the policy implications?', *International Labour Review* 143(1/2): 45–90.

Miller, D. (1996) 'Commentary: The embeddedness of corporate strategy: isomorphism vs. differentiation', *Advances in Strategic Management* 13: 283–91.

Ministry of Foreign Affairs (2006) 'China's African Policy', available at: *http://www.fmprc.gov.cn/eng/zxxx/t230615.htm* (accessed 30 April 2008).

Miroux, A. (2005) 'The globalisation of R&D by TNCs and implications for developing countries', paper presented at UNCTAD Expert Meeting on Impact of FDI on Development, Geneva, 24–26 January, available at: *http://www.unctad.org/sections/meetings/docs/miroux_en.pdf* (accessed 30 April 2006).

Miscevic, D. and Kwong, P. (2002) 'The China syndrome', available at: *http://www.thenation.com/doc/20020415/kwong* (accessed 2 May 2003).

Mishra, P. (2004) 'India to drive broadband usage *a la* Korea', *Asia Computer Weekly*, 28 June, p. 1.

Mittelstaedt, J. D. and Mittelstaedt, R.A. (1997) 'The protection of intellectual property: issues of origination and ownership', *Journal of Public Policy and Marketing* 16(1): 14–25.

Montealegre, R. (1999) 'A temporal model of institutional intervention for information technology adoption in less-developed countries', *Journal of Management Information Systems* 16(1): 207–32.

Moodley, S. and Morris, M. (2004) 'Does e-commerce fulfil its promise for developing country (South African) garment export producers?', *Oxford Development Studies* 32(2): 155–78.

Mooney, P. (2005) 'Undue fear of China Inc?', *Businessline*, 5 October, p. 1.

Mooney, P., Neelakantan, S., Birchard, K., Cohen, D. and Labi, A. (2004) 'No longer dreaming of America, in India and China, far fewer students consider the US the best place to go', *Chronicle of Higher Education*, 8 October, p. A40.

Moore, S. (1979) 'The structure of a national elite network', *American Sociological Review* 44: 673–91.

Moris, F. (2004) 'US-China R&D linkages: Direct investment and industrial alliances in the 1990s', available at: *http://www.nsf.gov/statistics/infbrief/nsf04306/* (accessed 30 April 2006).

Morris, I. (2006) 'China's broadband boom', *Telecommunications – International Edition* 40(11): 16–18.

Morse, J. (2006a) 'China continues GSM commitment with infrastructure awards', available at: *http://findarticles.com/p/articles/mi_hb4962/is_200605/ai_n18145927* (accessed 29 April 2008).

Morse, J. (2006b) 'China inching closer to 3G licensing', available at: *http://findarticles.com/p/articles/mi_hb4962/is_200612/ai_n18144663* (accessed 29 April 2008).

Mourdoukoutas, P. (2004) 'China's challenge', *Barron's*, 16 February, p. 37.

msnbc (2006) 'China says number of web users hits 123M: jump of nearly 20 per cent in past year; broadband use soars' available at: *http://www.msnbc.msn.com/id/13931834/* (accessed 1 January 2008).

Mulvenon, J. (1997) 'Chinese military commerce and US national security', available at: *http://www.softwar.net/rand.html* (accessed 30 April 2008).

Myers, W. H. (1996) 'The emerging threat of transnational organized crime from the east', *Crime, Law and Social Change* 24: 181–222.

Nanfang Daily (2006) 'Red Flag to form new Linux joint venture', *Nanfang Daily*, 19 April, p. 1.

Nanowerk (2008) 'China's 10 most significant S&T developments in 2007', available at: *http://www.nanowerk.com/news/newsid=4176.php* (accessed 23 April 2008).

National Academy of Engineering (1996) *Small Companies in Six Industries: Background Papers for the NAE Risk and Innovation Study* Washington, DC: National Academy Press.

Naughton, B. (1993) *Growing out of the Plant: Chinese Economic Reform. 1978–93*, Cambridge: Cambridge University Press.

Neale, W.C. (1994) 'Institutions' in G. M. Hodgson, W. J. Samuels and M. R. Tool (eds) *The Elgar Companion to Institutional and Evolutionary Economics*, Aldershot: Edward Elgar Publishing Limited, pp. 402–6.

Nee, V. (1992) 'Organizational dynamics of market transition: hybrid forms, property rights, and mixed economy in China', *Administrative Science Quarterly* 37(10): 1–27.

Nelson, R. R. and Winter, S. (1982) *An Evolutionary Theory of Economic Change*, Cambridge, MA: Harvard University Press.

Newman, K. L. (2000) 'Organizational transformation during institutional upheaval', *Academy of Management Review* 25(3): 602–19.

Newman, R. J., Hook, C. S. and Moothart, A. (2006) 'Can America Keep Up?; Why so many smart folks fear that the United States is falling behind in the race for global economic leadership', *US News & World Report*, 27 March, pp. 48–56.

Newsweek (2001) 'The real question isn't moral', *Newsweek*, 20 October, p. 68.

Newsweek (2006) 'High-tech hunger; the goal: make China a technology powerhouse – critics say by any means necessary. Inside Beijing's "863 program"', available at: *http://findarticles.com/p/articles/mi_hb3335/is_200601/ai_n18038463* (accessed 30 April 2008).

Nidumolu, S. R., Goodman, S. E., Vogel, D. R. and Danowitz, A. K. (1996) 'Information technology for local administration support: The Governorates Project in Egypt', *MIS Quarterly* 20(2): 196–224.

Niitamo, V. (2000) 'Making information accessible and affordable for all', available at: *http://www.wider.unu.edu/newsletter/angle2000-1.pdf* (accessed 20 July 2001).

nikkeibp.co.jp (2004) 'Chinese mobile phone makers eye global market', available at: *http://techon.nikkeibp.co.jp/NEA/archive/200403/299047* (accessed 30 April 2005).

Niosia, J. and Reidb, S. E. (2007) 'Biotechnology and nanotechnology: science-based enabling technologies as windows of opportunity for LDCs?', *World Development* 35(3): 426–38.

Nokia (2005) 'Annual report', available at: *http://www.nokia.com/ NOKIA_COM_1/About_Nokia/Financials/nokia_form_20f_2005.pdf* (accessed 20 January 2007).

Nolan, P. (2002) 'China and the global business revolution,' *Cambridge Journal of Economics* 26(1): 119–37.

Nolan, P. and Yeung, G. (2001) 'Big business with Chinese characteristics: Two paths to growth of the firm in China under reform', *Cambridge Journal of Economics* 25(4): 443–65.

Nooteboom, B. (1999) 'Innovation and inter-firm linkages: new implications for policy', *Research Policy* 28(8): 793–805.

Normile, D. (2005) 'Is China the next R&D superpower?', *Electronic Business* 31(7): 36–41.

North, D. C. (1990) *Institutions, Institutional Change and Economic Performance*, Cambridge: Cambridge University Press.

North, D. C. (1994) 'Economic performance through time', *American Economic Review* 84(3): 359–68.

North, D. C. (1996) 'Epilogue: economic performance through time', in L. J. Alston, T. Eggertsson and D. C. North (eds) *Empirical Studies in Institutional Change*, Cambridge: Cambridge University Press, pp. 342–55.

Nye, J. (2004) *Soft Power: The Means to Success in World Politics*, New York, New York: Public Affairs.

O'Driscoll, C. (2006) 'China: the new frontier', available at: *http://www.accessmylibrary.com/coms2/summary_0286 15804890_ITM* (accessed 30 April 2006).

O'Neill, M. (2005) 'Handset makers losing grip', available at: *http://www.europeanchamber.com.cn/show/details.php?id=197* (accessed 30 April 2008).

Oakes, L. S., Townley, B. and Cooper, D. J. (1998) 'Business planning as pedagogy: language and control in a changing institutional field', *Administrative Science Quarterly* 43(2): 267–302.

Ocasio, W. (1995) 'The enactment of economic diversity: A reconciliation of theories of failure-induced change and threat-rigidity', *Research in Organizational Behavior* 17: 287–331.

Ofcom (2006) 'Ofcom international communications market report', available at: *http://www.ofcom.org.uk/media/news/2006/11/nr_20061129* (accessed 30 April 2008).

Ofir, C. and Lehman, D.R. (1986) 'Measuring images of foreign products', *Columbia Journal of World Business*, 21(2): 105–109.

Ogden, J. (2006a) 'Tough call: Chinese telecoms; some analysts brush aside concern over Beijing's restructuring', *Wall Street Journal*, 15 March, p. C.14.

Ogden, J. (2006b) 'Beijing expected to tap big 3G winners', *Barron's* 86(42): M23.

Okamoto, S. (2005) 'The trade structure in East Asia: spiral pattern of development and triangular trade structure (TTS) as a regional manufacturing platform', available at: *http://www.rieti.go.jp/en/events/bbl/05071901.html* (accessed 30 April 2007).

Oksenberg, M. (1987) 'China's confident nationalism', *Foreign Affairs* 65(3): 501–23.

Olavsrud, T. (2003) 'China's Red Flag joins unbreakable Linux', available at: *http://www.internetnews.com/dev-news/article.php/2202461* (accessed 30 January 2005).

Oliver, C. (1991) 'Strategic responses to institutional processes', *Academy of Management Review* 16: 145–79.

Ollier, P. (2007) 'Why Chinese companies must build global brands', *Managing Intellectual Property*, No. 173 (October): 36–43.

Olson, M. (1965) *The Logic of Collective Action: Public Goods and the Theory of Groups*, Cambridge, MA: Harvard University Press.

Ong, A. (1997) 'Chinese modernities: narratives of nation and of capitalism', in A. Ong and D. Nonini (eds) *Underground Empires: The Cultural Politics of Modern Chinese Transformation*, New York: Routledge, pp. 171–202.

Ong, J. (2006) 'Rural laptop to cost $187, with Linux', available at: *http://www.iht.com/articles/2006/03/15/bloomberg/sxlaptop.php* (accessed 21 July 2007).

Online Reporter (2003) 'First tremors from China', available at: *http://www.onlinereporter.com/torbackissues/TOR352.htm* (accessed 30 April 2005).

Organization for Economic Cooperation and Development (2004) *Science and Technology Indicator Scoreboard*, Paris: OECD.

Ortolani, A. (2005) 'China moves from piracy to patents; more companies are trying to be product innovators rather than just imitators', *Wall Street Journal*, 7 April, p. B.4.

Osland, G. E. and Cavusgil, S. T. (1996) 'Performance issues in US-China joint ventures', *California Management Review* 38(2): 106–30.

Overholt, W. H. (1994) 'The rise of China's economy', *Business Economics* 29(2): 29–34.

Palenchar, J. (2007) 'Android to set wireless markets free: supporters', *TWICE*, 19 November, p. 6.

Palmer, K. (2005) 'Contrabandwidth', *Foreign Policy* 147 (March/April), p. 93.

Pan, H. (2006) 'Broadband in China', available at: *http://www.ptc.org/ events/ptc06/program/public/proceedings/Hui%20Pan_slides_w11.pdf* (accessed 30 April 2007).

Pang, W. (2006) 'TCL pulls plug on Euro unit to focus on China', available at: *http://www.thestandard.com.hk/news_detail.asp?pp_ cat=1&art_id=30759&sid=10646499&con_type=1&d_str=20061101* (accessed 28 April 2008).

Panitchpakdi, S. (2006) 'What's good for China is good for the world', *Global Agenda* 4(1): 102–3.

Park, Y. (2001) 'Korean multinationals in Europe', *Journal of Asian Studies* 60(3): 882–5.

Parker, B. D. (2004) 'International exodus', *Vital Speeches of the Day* 70(19): 583.

Parloff, R., Chandler, C. and Fung, A. (2006) 'Not exactly counterfeit', available at: *http://money.cnn.com/magazines/fortune/fortune_archive/ 2006/05/01/8375455/index.htm* (accessed 23 April 2008).

Parto, S. (2005) 'Economic activity and institutions: taking stock', *Journal of Economic Issues* 34(1): 21–52.

Patterson, D. A. (2001) 'The middle kingdom: technology transfer and the future of the People's Republic Of China', *Vital Speeches of the Day* 67(6): 174–9.

Pavitt, K. (1987) 'International patterns of technological accumulation', in N. Hood and J.-E. Nalhne (eds) *Strategies in Global Competition*, London: Croom Helm, pp. 126–57.

Pearce, R. D. and Singh, S. (1992) *Globalizing Research and Development*, London: Macmillan Press.

Peerenboom, R. (2002) *China's Long March toward Rule of Law*, Cambridge: Cambridge University Press.

Pei, M. (1998) 'Is China democratizing?', *Foreign Affairs* 77(1): 68–82.

Pei, M. (2003) 'The paradoxes of American nationalism', *Foreign Policy* 136 (May/June): 30–7.

Pei, M. (2006) 'The dark side of China's rise', *Foreign Policy* 153(2): 32–40.

Peng, M. W. (2006) 'Making M&A fly in China', *Harvard Business Review* 84(3): 26.

Peng, M. W. and Luo, Y. (2000) 'Managerial ties and firm performance in a transition economy: the nature of the micro-macro link', *Academy of Management Journal* 43(2): 486–501.

People's Daily (2004) 'TCL, Alcatel joint venture starts operation', available at: *http://english.peopledaily.com.cn/200410/11/eng2004 1011_159707.html* (accessed 28 April 2008).

People's Daily (2001) 'Haier rises through reform and opening up', available at: *http://english.people.com.cn/english/200108/08/eng 20010808_76828.html* (accessed 30 April 2008).

People's Daily (2005) 'Motorola's R&D investment in China tops $500 mln', available at: *http://english.people.com.cn/200511/29/eng 20051129_224450.html* (accessed 30 April 2008).

People's Daily (2006) 'China targets 10 billion dollar software exports by 2010', available at: *http://english.people.com.cn/200610/10/ eng20061010_310458.html* (accessed 30 April 2008).

People's Daily (2007) 'Secrets of "Nigcomsat-1" revealed', 18 May, available at: *http://english.people.com.cn/200705/18/eng20070518_ 375881.html* (accessed 30 April 2008).

Perotti, E. C., Sun, L. and Zou, L. (1999) 'State-owned versus township and village enterprises in China', *Comparative Economic Studies* 41(2/3): 151–79.

Peteraf, M. A. (1993) 'The cornerstone of competitive advantage: a resource-based view', *Strategic Management Journal* 14(3): 179–92.

Pfeffer, J. (1981a) 'Management as symbolic action', *Research in Organizational Behavior* 3: 1–52.

Pfeffer, J. (1981b) *Power in Organizations*, Marshfield, MA: Pitman.

Pfeffer, J. and Salancik, G. R. (1978) *The External Control of Organizations*, New York: Harper and Row.

Piller, C. (2006) 'How piracy opens doors for Windows; Bill Gates may not be entirely dismayed by software thieves. They seed the world market and make Microsoft a standard', *Los Angeles Times*, 9 April, p. C.1.

pincn.com (2005) 'Nanjing Ericsson Panda Communications Company Ltd', available at: *http://www.pincn.com/clientpages/2005/ericsson/ Business_Unit.htm* (accessed 30 April 2008).

Pistor, K. and Wellons, P. (1999) *The Role of Law and Legal Institutions in Asian Economic Development*, Hong Kong: Oxford University Press.

Pitroda, S. (1993) 'Development, democracy, and the village telephone', *Harvard Business Review* 71(6): 66–79.

Plant Engineering (2006) 'Skills gap survey: begin now to attract, train and develop manufacturing workers', *Plant Engineering* 60(1): 23–4.

Pocha, J. S. and Brown, H. (2006) 'Comparative advantage', *Forbes* 178(10): 76–8.

Political Transcript Wire (2005) 'US Representative John N. Hostettler (R-In) holds a hearing on foreign nationals and espionage', available at: *http://www.accessmylibrary.com/coms2/summary_0286-9627852_ ITM* (accessed 2 May 2008).

Popov, V. (2006) 'Foreign direct investment in Russia: why doesn't it come? Should there be more of it?' *Canadian Foreign Policy* 13(2): 51–64.

Portelligent (2002) 'Legend aims to become major cell phone brand in 5 years', available at: *http://www.portelligent.com/tas/subscribers/ archive/ wireless_archive/2002_Q2/wireless_tas_020524.asp* (accessed 30 April 2008).

Porter, M. (1990) *The Competitive Advantage of Nations*, London: Macmillan.

Potter, B. (2001) *The Chinese Legal System: Globalization and Local Legal Culture*, London: Routledge.

Potter, B. (2004) 'Legal reform in China: institutions, culture, and selective adaptation', *Law and Social Inquiry* 29(2): 465–95.

Powell, A. and DiMaggio, P. (eds) (1991) *The New Instutitionalism in Organizational Analysis*, Chicago, IL: University of Chicago Press.

Powell, B. and Roston, E. (2005) 'Why China is buying', *Time*, 27 June, p. 33.

Powell, B. and Steptoe, S. (2005) 'Can China innovate? It's already the workshop of the world, but what china really wants to be is the world's laboratory', *Time* 165(19): 27.

Powell, W. W. (1988) *Institutional Effects on Organizational Structure and Performance*. Cambridge, MA: Ballinger.

PR Newswire (2007) 'CCID report: cooperation and innovation push up China's IPTV industry', available at: *http://www.sys-con.com/read/ 349602.htm* (accessed 12 January 2008).

PR Newswire (2008a) 'CCID Consulting: TD-SCDMA the focus of china's 3G market in 2007', available at: *http://www.prnewswire.com/ cgi-bin/stories.pl?ACCT=104&STORY=/www/story/01-22- 2008/0004740186&EDATE* (accessed 9 April 2008).

PR Newswire (2008b) 'CCID consulting analyzes China's software industry', available at: *http://www.prnewswire.com/cgi-bin/stories.pl?ACCT=104& STORY=/www/story/01-15-2008/0004736467&EDATE=* (accessed 30 April 2008).

PR Newswire Europe (2006) 'Research and development (R&D) budgets set to grow with china the biggest recipient, new research for Thomson reveals', available at: *http://scientific.thomson.com/press/ 2006/8321100/* (accessed 30 April 2008).

Press, L., Foster, W. A. and Goodman, S. E. (1999) 'The internet in India and China', available at: *http://internetlab.cindoc.csic.es/autores.asp? aut=375* (accessed 30 April 2008).

PricewaterhouseCoopers (2005) 'M&A in China', available at: *http:// www.altassets.net/casefor/countries/2005/nz6504.php* (accessed 21 January 2008).

PRWeb (2007) 'China's IPTV subscribers to hit 10 million in 2008: The 2007 China Media Yearbook and Directory', available at: *http://www.prweb.com/releases/2007/2/prweb505133.htm* (accessed 12 January 2008).

Putnam, R. (1993) *Making Democracy Work: Civic Traditions in Modem Italy*, Princeton, NJ: Princeton University Press.

Putnam, R. D. (2000) *Bowling Alone: The Collapse and Revival of American Community*, New York: Touchstone Books.

Pyramid Research (2001) 'Broadband in Asia: the next big thing', available at: *http://www.pyramidresearch.com/research/APBrdbnd01.asp* (accessed 30 April 2008).

Qian, Y., Roland, G. and Xu, C. (1999) 'Why is China different from Eastern Europe? Perspectives from organization theory', *European Economic Review* 43(4–6): 1085–94.

Quan, X. (2005) 'Multinational R&D labs in China: local and global innovation', unpublished PhD dissertation, University of California, Berkeley.

R&D (2004) 'China is #1 for R&D investments', available at: *http://www.rdmag.com/ShowPR~PUBCODE~014~ACCT~1400000100~ISSUE~0410~RELTYPE~PR~Cat~14~SubCat~20~PRODCODE~00000000~PRODLETT~AN~CALLFROM~PR.html* (accessed 30 April 2008).

R&D (2005) 'China forecasts growing R&D base', *R&D* 47(4): 11.

Rabano, B. (2000) 'The penguin takes flight: But can the Linux OS take on Microsoft in Asia?', *Asiaweek*, 6 September, p. 1.

Rai, S. (2004) 'India taps China's reserve of technological talent', *New York Times*, 2 November, p. W.1.

Rajagopalan, M. (2006) 'India, China attract more R&D work', *Wall Street Journal*, 19 July, p. B.4.

Ramos, A. D. (2005a) 'China's growing appetite', available at: *http://www.cfo.com/article.cfm/5077885/c_5101083?f=insidecfo*.

Ramos, A. D. (2005b) 'From walls to bridges: how Chinese companies are redefining international M&A', available at: *http://www.cfoasia.com/archives/200511-02.htm*.

Ramstad, E. (2003) 'China's makers of cellphones thrive at home', *Wall Street Journal*, 21 August, p. B.1.

Ramstad, E. (2004) 'East meets west in television sets; huge Sino-French venture is still tuning the relationship', *Wall Street Journal*, 26 November, p. A.7.

Ramstad, E. and Li, J. (2006) 'TCL to overhaul European television manufacturing', *Wall Street Journal*, 1 November, p. A.14.

Ramstad, E. and Pringle, D. (2004) 'Alcatel shifts production to China; venture with TCL, maker of television sets, will take over manufacture of cellphones', *Wall Street Journal*, 27 April, p. B.5.

Rasiah, R. (2005) 'The competitive impact of China on Southeast Asia's labour markets', *Journal of Contemporary Asia* 35(4): 447–70.

Raymond, E. S. (2004) 'Open minds, open source', *Analog Science Fiction & Fact* 124(7/8): 100–9.

RCR Wireless News (2003) 'Firm formed to focus on TD-SCDMA phones', available at: *http://findarticles.com/p/articles/mi_hb4962/is_200301/ai_n18159725* (accessed 29 April 2008).

Redding, G. (2002) 'The capitalist business system of China and its rationale', *Asia Pacific Journal of Management* 19(2–3): 221–49.

rediff.com (2004) 'China: rising software power?', available at: *http://www.rediff.com/money/2004/jun/14spec.htm* (accessed 30 April 2008).

Reich, R. B. (2005) 'Plenty of knowledge work to go around', *Harvard Business Review* 83(4): 17.

Rein, S. (2007) 'Chinese cozy up to e-commerce', *Business Week* Online, 9 February, available at: *http://www.businessweek.com/globalbiz/content/feb2007/gb20070208_810426.htm* (accessed 26 May 2007).

Reuters (2003) 'Ningbo Bird takes wing in India', available at: *http://www.cnii.com.cn/20030526/ca182497.htm* (accessed 30 April 2008).

Reynolds, M. (2005) 'China seeks 3G glory', *Electronics Weekly*, 22 February, p. 1.

Rhodes, R. (1986) *The Making of the Atomic Bomb*, New York: Simon and Schuster.

Ridgway, N. (2005) 'Brain drain', *Forbes*, 5 September, p. 154.

Roberts, D., Balfour, F. and Engardio, P. (2005) 'China goes shopping', *Business Week*, 20 December, p. 32.

Roberts, B. (2006) 'Protect your IP: In China and elsewhere, intangible assets are more at risk than most CEOs think', available at: *http://www.edn.com/article/CA6328379.html* (accessed 4 May 2008).

Roberts, D. (2007) 'China Mobile's hot signal: it's already the world's biggest cellular carrier. Now it's planning to get even bigger', *Business Week*, 7 February, pp. 42–4.

Roberts, D., Arndt, M. and Engardio, P. (2005) 'It's getting hotter in the east', *Business Week*, 22 August, p. 78.

Roberts, D., Clifford, M. L., Crock, S. (1998) 'China's army under fire', Business Week, 10 August: 36–38.

Rodan, G. (1998) 'The internet and political control in Singapore', *Political Science Quarterly* 113(1): 63–99.

Roell, S. (2006) 'China's M&A misfortunes – in their first major mergers and acquisitions forays overseas, Chinese Companies are being buffeted by sinophobia and their own lack of nous', available at: *http://www.thebanker.com/news/fullstory.php/aid/4230/China%92s_M_A_misfortunes.html* (accessed 16 June 2008).

Romar, E. J. (2004) 'Globalization, ethics, and opportunism: a confucian view of business relationships', *Business Ethics Quarterly* 14(4): 663–78.

Rose, N. (2000) 'Governing liberty', in R. V. Ericson and N. Stehr (eds) *Governing Modern Societies*, Toronto: University of Toronto Press, pp. 141–76.

Rosett, C. (1989) 'The powder keg that is China', *Wall Street Journal*, 28 December, p. 1.

Rubin, K. (2004) 'China's students turning away or staying home', *International Educator* 13(3): 6–13.

Ruef, M. and Scott, W. R. (1998) 'A multidimensional model of organizational legitimacy: hospital survival in changing institutional environments', *Administrative Science Quarterly* 43: 877–904.

Rugman, A. M. and Li, J. (2007) 'Will China's multinationals succeed globally or regionally?' *European Management Journal* 25(5): 333–43.

Sabel, C. and Zeitlin, J. (eds) (1997) *World of Possibilities: Flexibility and Mass Production in Western Industrialization*, New York: Cambridge University Press.

Safire, W. (2002) 'Selling our secrets', *New York Times*, 30 September, p. A.25.

Salmon, P. (1995) 'Nations competing against themselves: an interpretation of European integration', in A. Breton, G. Galeotti, P. Salmon and R. Wintrobe (eds) *Nationalism and Rationality*, Cambridge: Cambridge University Press, pp. 98–115.

Sandefur, R. L. and Laumann, E. O. (1998) 'A paradigm for social capital', *Rationality and Society* 10(4): 481–510.

Sanford, J. (2004) 'Knock-off nation', *Canadian Business*, 8 November, pp. 67–71.

Sautman, B. (2001) 'Peking man and the politics of paleoanthropological nationalism in China', *Journal of Asian Studies* 60(1): 95–124.

Savan, B. (1989) 'Beyond professional ethics: issues and agendas', *Journal of Business Ethics* 8: 179–85.

Scent, B. (2008) 'Homegrown 3G trial run in eight cities', available at: *http://www.thestandard.com.hk/news_detail.asp?pp_cat=1&art_id=65191&sid=18713203&con_type=1* (accessed 30 April 2008).

Schafer, S. (2004) 'Microsoft's Cultural Revolution; battered by pirates and struggling to turn a profit, the brash American software giant is no longer trying to change China. Instead China's changing the company', *Newsweek*, 21 June, p. 33.

Scharpf, F. (2000) 'Interdependence and democratic legitimation', in Susan J. Pharr and Robert D. Putnam (eds). *Disaffected Democracies: What's Troubling the Trilateral Countries*, Princeton, NJ: Princeton University Press.

Schillings, B. (2006) 'Tapping Linux as an application framework for consumer electronics', *EDN*, 26 October, p. 93.

Schmid, A. A. (1987) *Property, Power, and Public Choice: An inquiry into Law and Economics* (2nd edition), New York: Praeger Publisher.

Schneider, A. (1999) 'US neo-conservatism: cohort and cross-cultural perspective', *International Journal of Sociology and Social Policy* 19(12): 56–86.

Schoenberger, K. (1996) 'Motorola bets big on China', *Fortune*, 27 May, pp. 40–5.

Schuiling, I. and Kapferer, J.-N. (2004) 'Real differences between local and international brands: strategic implications for international marketers', *Journal of International Marketing* 12(4): 97–112.

Schwartz, E. (2005) 'Linux über alles', *InfoWorld* 27(34): 6.

Schwartz, M. and Paul, S. (1992) 'Resource mobilization versus the mobilization of people', in A. D. Morris and C. McClurg (eds) *Frontiers in Social Movement Theory*, New Haven, CT: Yale University Press, pp. 205–23.

Scott, W. R. (1995) *Institutions and Organizations*, Thousand Oaks, CA: Sage.

Scott, W. R. (2001) *Institutions and Organizations* (2nd edn), Thousand Oaks, CA: Sage.

Scott, W. R., Ruef, M., Mendel, P. J. and Caronna, C. A. (2000) *Institutional Change and Healthcare Organizations: From Professional Dominance to Managed Care*, Chicago, IL: University of Chicago Press.

Sender, H. (2003) 'A rude awakening', *Far Eastern Economic Review*, 13 November, pp. 52–5.

Secured Lender (2004) 'E-commerce news', *Secured Lender* 60(5): 62–4.

seekingalpha.com (2007) 'Why is Baidu expanding to Japan?', available at: *http://seekingalpha.com/article/31045-why-is-baidu-expanding-to-japan* (accessed 21 January 2008).

Segal, A. (2004) 'Is America losing its edge?', available at: *http://www.foreignaffairs.org/20041101facomment83601-p20/adam-segal/is-america-losing-its-edge.html* (accessed 2 May 2005).

Selznick, P. (1984) *Leadership in Administration*, Evanston, IL: Peterson.

Sen, S. (2005) 'Internet advertising in India boots for boom time', available at: *http://www.monstersandcritics.com/tech/news/printer_1031212.php* (accessed 1 May 2008).

Sender, H. (2003) 'A rude awakening', *Far Eastern Economic Review*, 13 November, pp. 52–5.

Sengupta, S. (2006) 'Ban in India incurs wrath of bloggers', available at: *http://www.chicagotribune.com/news/nationworld/chi-0607190202jul19,1,2375353.story?coll=chi-newsnationworld-hed* (accessed 30 April 2008).

Seno, A. A. (2004) 'The buy side; homebody: the race to establish China's first global brand is now led by a little-known contender that wants to purchase a name, not be one', *Newsweek*, 1 December, p. 66.

Shannon, V. (2006) 'China largely silent on telecom strategy', *New York Times*, 5 December, p. C.13.

Shen, X. (2005) 'A dilemma for developing countries in intellectual property strategy? Lessons from a case study of software piracy and Microsoft in China', *Science and Public Policy* 32(3): 187–98.

Shenshen, Z. (2007) 'Tech tiff won't slow expansion of IPTV', available at: *http://www.shanghaidaily.com/sp/article/2007/200708/20070829/article_329082.htm.*

Shie, T. R. (2004) 'The tangled web: does the internet offer promise or peril for the Chinese Communist Party?' *Journal of Contemporary China* 13(40): 523–40.

Shih, T. H. (2006a) 'TCL placing seeks 1.1b yuan', *South China Morning Post*, 28 October, p. 1.

Shih, T. H. (2006b) 'Thomson to cut stake in TCL by year's end; move follows warning multimedia firm overpaid for French brand in decline', *South China Morning Post*, 2 November, p. 4.

Siddiqi, M. (2006) 'Is Africa a poor trader? Just why is Africa's international – and internal – trade at such a dismally low level? Is market access the problem or is it what we take (or don't take) to market that is the real culprit?', *African Business*, No. 318 (March): 24–5.

Siemens (2003) 'China makes space for TD-SCDMA', available at: *http://www.siemens.ie/i&cworld/dec02/tdscdma.htm* (accessed 30 April 2004).

Sigurdson, J. (2004) 'Technological superpower China?', *R&D Management* 34(4): 345–7.

Sikorski, D. and Menkhoff, T. (2000) 'Internationalization of Asian business', *Singapore Management Review* 22(1): 1–17.

Silva, J. (2006) 'Beyond voice, substitution on steroids', *RCR Wireless News*, 25 December, pp. 8–9.

Simon, D. (2001) 'The microelectronics industry crosses a critical threshold', *China Business Review* 28(6): 8–20.

Simons, S. (2006) 'The Huawei way; The telecom giant is either a security menace or a real comer – or it could be a house of cards. Or all of the above', available at: *http://www.newsweek.com/id/47446?tid=relatedcl* (accessed 30 April 2008).

Sin, K. (2007) 'Bandwidth cost issues slow uptake of HD in Asia', *Telecom Asia* 18 (March Supplement): 8–9.

Singer, A., Salamanca-Buentello, F. and Daar, A. S. (2005) 'Harnessing nanotechnology to improve global equity', *Issues in Science and Technology* 21(4): 57–64.

Singer, M. (2002) 'Google blocked in China', available at: *http://siliconvalley.internet.com/news/article.php/1455921* (accessed 11 December 2003).

Singer, P. A., Salamanca-Buentello, F. and Daar, A. S. (2005) 'Harnessing nanotechnology to improve global equity', *Issues in Science and Technology* 21(4): 57–64.

Singer, P., Daar, A., Salamanca-Buentello, F. and Court, E. (2006) 'Nano-diplomacy', *Georgetown Journal of International Affairs* 7(1): 129–38.

Singh, J., House, R. and Tucker, D. (1986) 'Organizational change and organizational mortality', *Administrative Science Quarterly* 31: 587–611.

SinoCast China (2003a) 'China to boost information technology education', *Business Daily News*, 23 September, p. 1.

SinoCast China (2003b) 'Salary of returned overseas students declines', *Business Daily News*, 7 August, p. 1.

SinoCast China (2003c) 'WiFi makes internet plus coffee possible in Beijing and Tianjin', *Business Daily News*, 25 July, p. 1.

SinoCast China (2004a) 'MII unveiled 19 government-sponsored IT projects', *Business Daily News*, 19 July, p. 1.

SinoCast China (2004b) 'China to promote Linux operating system', *Business Daily News*, 4 August, p. 1.

SinoCast China (2004c) 'Datang selects Linux as 3G Handset operating system', *Business Daily News*, 23 July, p. 1.

SinoCast China (2004d) 'China becomes top DSL country', *Business Daily News*, 9 March, p. 1.

SinoCast China (2004e) 'China's mobile phone export to grow to 100 mln in 2004', *Business Daily News*, 30 April, p. 1.

SinoCast China (2004f) 'Chinese cellphone makers expand by allying with foreign partners', *Business Daily News*, 12 May, p. 1.

SinoCast China (2006a) 'TD-SCDMA sees new helper', *IT Watch*, 31 August, p. 1.

SinoCast China (2006b) 'Nortel debuts a new hi-tech park in China', *IT Watch*, 10 November, p. 1.

SinoCast China (2006c) 'Latest report on China's internet', *IT Watch*, 19 January, p. 1.
SinoCast China (2006d) 'Netcom signs new Olympic deal', *IT Watch*, 11 August, p. 1.
SinoCast China (2006e) 'ZTE wins IPTV contracts in Shanghai', *IT Watch*, 7 September, p. 1.
SinoCast China (2006f) 'Two telecom carriers expanding IPTV biz', *IT Watch*, 23 November, available at: *http://www.cn-c114.net/newschina_html/20061123104134-1.Html* (accessed 9 May 2008).
SinoCast China (2006g) 'IPTV subscribers to reach new high', *IT Watch*, 12 December, p. 1.
SinoCast China (2006h) 'China's export to Africa grew by over 35% for 3 years', *Business Daily News*, 3 March, p. 1.
SinoCast China (2006i) 'TV giant undergoing transformation', *IT Watch*, 1 November, p. 1.
SinoCast China (2006j) 'TCL to reshuffle European operations', *IT Watch*, 9 November, p. 1.
SinoCast China (2007a) 'TD stirs up listing spree', *IT Watch*, 8 January, p. 1.
SinoCast China (2007b) 'China Mobile: the most possible TD net builder', *IT Watch*, 10 January, p. 1.
SinoCast China (2007c) 'China to release draft IPTV standard in August', available at: *http://findarticles.com/p/articles/mi_hb5562/is_200706/ai_n22745121* (accessed 12 January 2008).
SinoCast China (2007d) 'China plans deployment of AVS-supported IPTV network', *IT Watch*, 19 March, p. 1.
SinoCast China (2007e) 'China Telecom to test AVS', available at: *http://findarticles.com/p/articles/mi_hb5562/is_200704/ai_n22743500* (accessed 12 January 2008).
SinoCast China (2007f) 'UTStarcom, ZTE argue for no. 1 title in Chinese IPTV market', *IT Watch*, 6 August, p. 1.
Sitkin, S. B. and Sutcliffe, K. M. (1991) 'Dispensing legitimacy: professional, organizational and legal influences on pharmacist behavior', in P. Tolbert and S. Barley (eds) *Research in the Sociology of Organizations, Vol. 8*, Greenwich, CT: JAI Press, pp. 269–95.
Siu, R. C. S. (2006) 'Evolution of Macao's casino industry from monopoly to oligopoly: social and economic reconsideration', *Journal of Economic Issues* 40(4): 967–90.
Skeldon, R. (1996) 'Migration from China', *Journal of International Affairs* 49(2): 434–55.
Sklair, L. (1994) 'The culture-ideology of consumerism in urban China: some findings from a survey in Shanghai', *Research in Consumer Behavior* 7: 259–92.

Smart, D. and Ang, G. (1993) 'Exporting education: From aid to trade to internationalization?' *IPA Review* 46(1): 31–3.

Smith, C. S. (2000) 'China moves to cut power of Microsoft', *New York Times*, 8 July, p. A1.

Solinger, D. J. (1995) 'China's urban transients in the transition from socialism and the collapse of the communist urban public goods regime', *Comparative Politics* 27(2): 127–46.

Sornn-Friese, H. (2005) 'Interfirm linkages and the structure and evolution of the Danish trucking industry', *Transportation Journal* 44(4): 10–26.

South China Morning Post (1997) 'Warning issued on brain drain', *South China Morning Post* (Hong Kong), 11 March, p. 1.

Southerland, D. (1990) 'China to add restrictions on overseas study; students must work 5 years before going abroad to attend school', *Washington Post*, 7 February, p. A.1.

Sovich, N. (2006) 'Politics & economics: western firms find hiring, retention in China surprisingly tough', *Wall Street Journal*, 11 August, p. A.4.

Sparshott, J. (2005) 'China moves past the US in tech exports; think-tank report looks at Beijing's rising power', *Washington Times*, 13 December, p. A01.

SPC Asia (2004) 'Amway makes moves in China', *SPC Asia* 37 (August): 3–4.

Spierer, J. C. (1999) 'Intellectual property in China: prospects for new market entrants', available at: *www.harvard.edu/~asiactr/haq/199903/9903a010.htm* (accessed 21 July 2003).

Srivastava, A. and Thomson, S. B. (2007) 'E-business law in China: strengths and weaknesses', *Electronic Markets* 17(2): 126–31.

State Council Information Office (2000) 'Fifty years of progress in China's human rights', available at: *http://www.china.org.cn/e-white/3/3.1.htm* (accessed 12 May 2005).

Steinbock, D. (2007) 'New innovation challengers: the rise of China and India', *National Interest*, No. 87 (Jan/Feb): 67–73.

Steinfield, C. (2002) 'Conceptualizing the role of collaborative e-commerce in geographically defined business clusters', available at: *www.msu.edu/~steinfie/B2Bsocialcapital.pdf* (accessed 11 December 2006).

Stephen, C. (2007) 'IPTV: focus on Asia: in the slow lane', available at: *http://www.bnsltd.com/image/0605.pdf* (accessed 1 May 2008).

Stevenson-Yang, A. (2006) 'China's online mobs: the new red guard?', *Far Eastern Economic Review* 169(8): 53–7.

Stevenson-Yang, A. and DeWoskin, K. (2005) 'China destroys the IP paradigm', *Far Eastern Economic Review* 168(3): 9–18.

Stokes, B. (2005) 'China's high-tech challenge', *National Journal*, 30 July, pp. 2448–57.

Stoner-Weiss, K. (2006) 'Russia: authoritarianism without authority', *Journal of Democracy* 17(1): 104–18.

Storper, M. and Walker, R. (1989) *The Capitalist Imperative: Territory, Technology and Industrial Growth*, London: Basil Blackwell.

Stout, K. L. (2002) 'China sites count cost of cyber-control', available at: http://www.cnn.com/2002/TECH/11/03/china.content/ (accessed 11 December 2006).

Strang, D. and Meyer, J. W. (1993) 'Institutional conditions for diffusion', *Theory and Society* 22: 487–511.

Studt, T. (2006) 'R&D writing is on the wall; more research operations plan to open R&D facilities in Asia – mostly China – at the expense of their US facilities', available at: http://www.rdmag.com/pdf/RD66 EXE_Global.pdf (accessed 30 April 2008).

Su, F. and Yang, D. L. (2000) 'Political institutions, provincial interests, and resource allocation in reformist China', *Journal of Contemporary China* 9 (24): 215–30.

Suchman, M. C. (1995) 'Managing legitimacy: strategic and institutional approaches', *Academy of Management Review* 20: 571–610.

Sun, C. (2006) 'The right way to handle IP in M&A deals in China', *Managing Intellectual Property*, No. 155 (Dec/Jan), pp. 40–2.

Sun, L. (1997) 'Emergence of unorthodox ownership and governance structures in East Asia: an alternative transition path,' paper presented at the WIDER Workshop on Transition Strategies, Alternatives and Outcomes, Helsinki, 30 June to 2 July.

Sun, L. (2007) 'IPTV spells uncertainty' *Telecommunications* 41(9): 10.

Sun, L. H. (1992) 'China tries to lure overseas students to stop `brain drain'; member of democracy movement arrested', *Washington Post*, 1 September, p. A.3.

Suttmeier, R. P. (2004) 'Assessing China's technology potential', *Georgetown Journal of International Affairs* 5(2): 97–105.

Szulanski, G. (1996) 'Exploring internal stickiness: Impediments to the transfer of best practice within the firm', *Strategic Management Journal* 17: 27–43.

Taft, D. K. (2006) 'China business; technology vendors are not just selling products into the Chinese market these days', *eWeek* 23(2): 20.

Tagliabue, J. (2003) 'Thomson and TCL to join television units', *New York Times*, 4 November, p. W.1.

Takahashi, D. (2003) 'China gives boost to Sun Microsystems', *Knight Ridder Tribune Business News*, 18 November, p. 1.

Tan, J. (2003) 'Tapping China's market', *Business Times*, 4 December, p. 15.

Tan, Z., Meuller, M. and Foster, W. (1997) 'China's new internet regulations', *Communications of the ACM* 40(12): 11–16.

Tanner, J. C. (2007) 'The year in 3G: steady as she goes', *Telecom Asia*, 18(1): 18–20.

Taylor, C. T. and Silberston Z. A. (1973) *The Economic Impact of the Patent System: A Study of the British Experience*, Cambridge: Cambridge University Press.

tdscdma-forum.org (2003) 'Motorola Releases TD-SCDMA Software Libraries for the MRC6011 RCF Device', available at: *http://www.tdscdma-forum.org/nenglish/readnews.asp?id=2583* (accessed 9 April 2004).

Teece, D. J. (1998) 'Capturing value from knowledge assets: the new economy, markets for know-how, and intangible assets', *California Management Review* 40(3): 55–79.

Teece, D. J. and Chesbrough, H. W. (2005) 'The globalization of R&D in the Chinese semiconductor industry', report to the Alfred P. Sloan Foundation, available at: *http://web.mit.edu/ipc/sloan05/chesbrough.pdf* (accessed 30 April 2006).

Telecomasia.net (2006) 'Ningbo Bird becomes top Chinese mobile phone exporter', available at: *http://china-netinvestor.blogspot.com/2006/02/ningbo-bird-becomes-top-chinese-mobile.html* (accessed 30 April 2008).

Tempest, R. (1995) 'China tries to lure its best home; many students who flock to overseas colleges don't return. Now Beijing is willing to pay up to regain some of its lost brainpower', *Los Angeles Times*, 3 January, p. 1.

Terrill, R. (1977) 'China and the world: self-reliance or interdependence?' *Foreign Affairs* 55(2): 295–305.

Terrill, R. (2005) 'What does China want?' *Wilson Quarterly* 29(4): 50–61.

Theil, S. (2006) 'Companies: cultural confusion; what goes around, comes around. It's China's turn', available at: *http://findarticles.com/p/articles/mi_hb3335/is_200610/ai_n18037723* (accessed 9 May 2008).

Thompson, D. (2005) 'China's soft power in Africa: from the 'Beijing Consensus' to Health Diplomacy', *China Brief*, 13 October, p. 1–4.

Tilton, J. E. (1971) *International Diffusion of Technology: The Case of Semiconductors*, Washington, DC: Brookings Institution.

Tindal, R. (2003) 'Building *guanxi* on the web: 10 tips for creating an online presence in China', *Public Relations Tactics* 10(7): 6.

Tolbert, P. S., and Zucker, L. G. (1983) 'Institutional sources of change in the formal structure of organizations: the diffusion of civil service reform, 1880–1935', *Administrative Science Quarterly* 28: 22–39.

Tolbert, P. S., and Zucker, L. G. (1996) 'The institutionalization of institutional theory', in S. R. Clegg, C. Hardy, and W. R. Nord (eds) *Handbook of Organization Studies*, London: Sage, pp. 175–90.

Tolentino, P. E. E. (1993) *Technological Innovation and Third World Multinationals*, London & New York: Routledge.

Tong, T. (2005) 'China: time to roll out the R&D infrastructure; Relying solely on cost competitiveness won't make China a high technology dynamo', *Business Times Singapore*, 11 October, p. 5.

Toon, J. (2008) 'Study shows China as world technology leader: technology indicators show China ahead of the US in technological standing', available at: *http://www.gatech.edu/news-room/release.php?id=1682* (accessed 26 January 2008).

trackthat.wordpress.com (2007) 'Evolving networks, evolving models — IDC predicts top 10 most impactful telecommunications trends across Asia-Pacific (ex. Japan) in 2007', available at: *http://trackthat.wordpress.com/2007/01/22/evolving-networks-evolving-models-idc-predicts-top-10-most-impactful-telecommunications-trends-across-asia-pacific-ex-japan-in-2007/* (accessed 1 May 2008).

Tremblay, J. F. (2006) 'R&D takes off in Shanghai', *Chemical and Engineering News*, 21 August, pp. 15–22.

Trombly, M. and Marcus, B. (2006) 'Bridging the Chinese skills gap', *Computerworld*, 26 June, pp. 37–8.

Tsang, E., W K. (1998) 'Can *guanxi* be a source of sustained competitive advantage for doing business in China?', *Academy of Management Executive* 12(2): 64–73.

Tschang, C.-C. (2008) 'Mozilla takes on Microsoft in China', available at: *http://www.zdnetasia.com/news/internet/0,39044908,62036532,00.htm* (accessed 30 April 2008).

Tucker, P. (2006) 'Made in China: branding a new image', *Futurist* 40(1): 12–13.

Ulfelder, S. (2003) 'China: Low-level work at lower-than-average cost: Low-level IT work is its forte now, but this Asian giant is eyeing a future as a major outsourcing player', available at: *http://www.computerworld.com/action/article.do?command=viewArticleTOC&specialReportId=360&articleId=84863* (accessed 30 April 2008).

Un, K. (2006) 'State, society and democratic consolidation: the case of Cambodia', *Pacific Affairs* 79(2): 225–48.

United Nations Conference on Trade and Development (UNCTAD) (2003) *World Investment Report 2002: Transnational Corporations and Export Competitiveness*, New York: United Nations.

United Nations Conference on Trade and Development (UNCTAD) (2005) 'World Investment Report: Transnational Corporations and the

Internationalization of R&D', available at: *http://www.unctad.org/en/docs/wir2005_en.pdf* (accessed 9 April 2007).

United Nations Development Program (UNDP) (1997) *Human Development Report*, New York: Oxford University Press.

United Nations Development Program (UNDP) (2001) 'Human development report 2001', available at: *http://www.undp.org/hdr2001/completenew.pdf* (accessed 11 July 2001).

United Nations Development Program (UNDP) (2002) 'Human development report 2002: Deepening Democracy in a Fragmented World', available at: *http://hdr.undp.org/en/reports/global/hdr2002/* (accessed 30 April 2008).

United Nations Development Program (UNDP) (2004a) 'Human development report 2004', available at: *http://hdr.undp.org/reports/global/2004/pdf/hdr04_HDI.pdf* (accessed 2 May 2005).

United Nations Development Program (UNDP) (2004b) 'Regional human development report: promoting ICT for human development in Asia 2004', available at: *http://hdrc.undp.org.in/APRI/Publication/PBriefings/summary.htm* (accessed 30 April 2005).

United Nations Development Program (UNDP) (2005) 'Human development report 2005: International cooperation at a crossroads: Aid, trade and security in an unequal world', available at: *http://hdr.undp.org/en/reports/global/hdr2005/* (accessed 30 April 2008).

United Nations Development Program (UNDP) (2006) 'Human Development Report 2006: Beyond scarcity: Power, poverty and the global water crisis', available at: *http://hdr.undp.org/en/reports/global/hdr2006//* (accessed 30 April 2008).

United Press International (2006) 'India clamps down on bloggers, cell users,' available at: *http://www.physorg.com/news72460837.html* (accessed 21 April 2008).

untius.com (2008) 'Digital Wireless Communications Technology', available at: *http://www.nuntius.com/technology2.html* (accessed 30 April 2008).

US Embassy Beijing (1996) 'Chinese challenges in absorbing and producing new technology: a report from US Embassy Beijing', available at: *http://www.usembassy-china.org.cn/sandt/stnutek7.htm* (accessed 26 January 2008).

US Fed News Service, including US State News (2006a) 'VOA News: Group accuses internet companies in China of rights violations', available at: *http://www.allbusiness.com/media-telecommunications/internet-www/7540754-1.html* (accessed 30 April 2008).

US Fed News Service, including US State News (2006b) 'Democracy and human rights: Somalia, Mauritania, Internet', available at: *http://www*

.europarl.europa.eu/news/expert/infopress_page/015-9503-187-07-28-902-20060629IPR09390-06-07-2006-2006-false/default_de.htm (accessed 30 April 2008).

Veblen, T. (1912/1959) *The Theory of the Leisure Class: An Economic Study of Institutions*, New York: Mentor Books.

Verma, R. (2005) 'China's IT development not a threat', *Beijing Review*, 7 July, p. 20.

Vernon, R. (1966) International investment and international trade in the product cycle, *Quarterly Journal of Economics* 80: 190–207.

Vogelstein, F. (2004) 'How Intel got inside', *Fortune*, 4 October, pp. 127–32.

Vogelstein, F., Boyle, M., Lewis, P. and Kirkpatrick, D. (2004) '10 tech trends to bet on', *Fortune*, 23 February, pp. 75–80.

Von Hippel, E. and von Krogh, G. (2003) 'Open source software and the 'private-collective' innovation model: issues for organization science', *Organization Science* 14(2): 209–23.

Von Krogh, G. and Haefliger, S. (2007), 'Nurturing respect for IP in China', *Harvard Business Review* 85(4): 23–4.

Voon, J. P. and Yue, R. (2003) 'China-ASEAN export rivalry in the US market: the importance of the HK-China production synergy and the Asian financial crisis', *Journal of the Asia Pacific Economy* 8(2): 157–79.

Wall Street Journal (2006) 'China develops chip for 3G phones', *Wall Street Journal*, 4 January, p. 1.

Wallace, R. (2005) 'Intel spies opportunities aplenty for China growth', *Electronic Engineering Times*, 8 August, pp. 1–3.

Walsh, K. A. (2005) 'China's high-technology development', available at: http://www.uscc.gov/hearings/2005hearings/written_testimonies/05_21_22wrts/walsh_kathleen_wrts.pdf (accessed 30 April 2007).

Wang, B. (2006) 'Hype or hope: IPTV reality in Asia', *Telecom Asia* 17(7): 28–9.

Wang, L. (1993) 'The Chinese traditions inimical to the patent law', *Northwestern Journal of International Law & Business* 14(1): 15–65.

Wang, Y. F. (1993) *China Science and Technology Policy: 1949–1989*, Aldershot: Aylesbury.

Ward, S. P., Ward, R., Deck, D. R. and Alan, B. (1993) 'Certified public accountants: ethical perceptions, skills and attitudes on ethics education', *Journal of Business Ethics* 12: 601–10.

Watanabe, T. (1994) 'The Asia boom training go-go growth points up shortage of skilled labor many experts say Asia needs education more than it does infrastructure', *Los Angeles Times*, 29 November, p. 4.

Watson, J. (2005) 'Rising sun', *Harvard International Review* 26(4): 46–9.

Weaver, L. R. (2002) 'Report: China blocks another search engine', available at: *http://edition.cnn.com/2002/TECH/internet/09/06/china.internet.block/index.html* (accessed 28 January 2008).

Weber, M. (1978) *Economy and Society*, Translated by G. Roth and C. Wittich, Berkeley and Los Angeles, CA: University of California Press.

Wehrfritz, G., Hewitt, D. and Ansfield, J. (2005) 'One billion couch potatoes; Chinese by the millions are channel surfing their way into the future. How reforms and the onslaught of new technologies are enhancing their ride', *Newsweek*, 6 June, p. 72.

Wei, P. (2002) 'China Unicom, a rising telecom carrier', available at: *http://www.cnii.com.cn/20020808/ca90669.htm*.

Weidenbaum, M. (2006a) 'China's progress in developing modern business practices', *Vital Speeches of the Day* 72(20/21): 576–8.

Weidenbaum, M. (2006b) 'Doing business with China', *USA Today Magazine*, 1 November, pp. 18–20.

Weitao, L. (2003) 'Taiwan bars mainland handsets', available at: *http://www1.chinadaily.com.cn/en/doc/2003-06/24/content_242331.htm* (accessed 30 April 2004).

Weitao, L. (2005) 'China tuning into IPTV biz', available at: *http://www.chinadaily.com.cn/english/doc/2005-03/31/content_429829.htm* (accessed 1 May 2008).

Weitao, L. (2007) 'Regulator urged to decontrol Net TV', available at: *http://www.chinadaily.com.cn/cndy/2007-06/06/content_887941.htm* (accessed 1 May 2008).

Welch, L. S. and Luostarinen, R. K. (1993) 'Inward–outward connections in internationalization', *Journal of International Marketing* 1(1): 44–56.

Wells, L. T., Jr. (1983) *Third World Multinationals*, Cambridge, MA: MIT Press.

Wernerfelt, B. (1984) 'A resource-based view of the firm', *Strategic Management Journal* 5(2): 171–80.

West, D. M. (2002) 'Global e-government, 2002', available at: *http://www.insidepolitics.org/egovt02int.PDF* (accessed 22 June 2003).

Westphal, J. D. and Zajac, E. J. (1994) 'Substance and symbolism in CEOs' long-term incentive plans', *Administrative Science Quarterly* 39: 367–90.

Westphal, J. D. and Zajac, E. J. (1998) 'The symbolic management of stockholders: corporate governance reforms and shareholder reactions', *Administrative Science Quarterly* 43: 127–53.

Westphal, J. D. and Zajac, E. J. (2001) 'Explaining institutional decoupling: the case of stock repurchase programs', *Administrative Science Quarterly* 46: 202–28.

Whitcomb, L. L., Erdener, C. B. and Li, C. (1998) 'Business ethical values in China and the US', *Journal of Business Ethics* 17(8): 839–52.

White, H. (1992) *Identity and Control: A Structural Theory of Social Interaction*, Princeton, NJ: Princeton University Press.

Wieland, K. (2004) 'China: a question of standards', available at: *http://www.itudaily.com/new/printarticle.asp?articleid=4090906* (accessed 9 April 2005).

Wikipedia (2008) 'Communications in the People's Republic of China', available at: *http://en.wikipedia.org/wiki/Communications_in_the_ People%27s_Republic_of_China* (accessed 30 April 2008).

Williamson, P. and Zeng, M. (2004) 'Strategies for competing in a changed China', *MIT Sloan Management Review* 45(4): 85–92.

Wilson Center (2007) 'Nanotechnology in China: ambitions and realities', available at: *http://www.wilsoncenter.org/index.cfm? topic_id=166192&fuseaction=topics.event_summary&event_id=2188 54* (accessed 23 April 2008).

Wilson, C. (2005) 'Municipal networks gain ground', *Telephony* 246(8): 6–7.

Wilson, C. (2007a) 'China grabs global IPTV leadership', *Telephony*, 7 May, pp. 8–9.

Wilson, C. (2007b) 'IPTV lessons from China', available at: *http://telephonyonline.com/iptv/technology/iptv_lessons_china_051007/* (accessed 1 May 2008).

Wilson, C. (2007c) 'IPTV in China: Shanghai surprises', available at: *http:// telephonyonline.com/iptv/news/china_iptv_050707/* (accessed 1 May 2008).

Wilson, C. (2007d) 'Chinese put their own spin on IPTV', available at: *http://telephonyonline.com/iptv/news/china_iptv_050707/* (accessed 1 May 2008).

Wilson, C. (2007e) 'China IPTV market ready to bust out?', available at: *http://telephonyonline.com/iptv/technology/china_iptv_utstarcom_04 2307/* (accessed 1 May 2008). .

Wilson, E. J. and Segal, A. (2005) 'Trends in China's transition toward a knowledge economy', *Asian Survey* 45(6): 886.

Wilson, R. W. (1992) *Compliance Ideologies: Rethinking Political Culture*, New York: Cambridge University Press.

Winter, S. (1990) 'Survival, selection, and inheritance in evolutionary theories of organization', in J. V. Singh (ed.) *Organizational Evolution: New Directions*, Newbury Park, CA: Sage, pp. 269–97.

Wireless News (2004) 'Motorola, Piscel collaborate on smartphone', available at: *http://findarticles.com/p/articles/mi_hb5558/is_200404/ai_ n22065348* (accessed 30 April 2008).

Wolfe, J. (2007) 'Nanotech gets big in China', available at: http://www.forbes.com/2007/02/01/nanotech-china-veeco-pf-guru-in_jw_0201soapbox_inl.html (accessed 23 April 2008).

World Bank (1999) *China: Weathering the Storm and Learning the Lessons,* Washington, DC: World Bank.

World IT Report (2003) 'Are Chinese Linux developers holding back on code?', available at: http://findarticles.com/p/articles/mi_qn4175/is_20030408/ai_n12926150 (accessed 30 April 2008).

World Trade Organization (2001) 'Report of the Working Party on Accession of China', available at: http://unpan1.un.org/intradoc/groups/public/documents/APCITY/UNPAN002144.pdf (accessed 30 April 2008).

Wu, H. and Chen, C. (2004) 'Changes in the foreign market competitiveness of East Asian exports', *Journal of Contemporary Asia* 34(4): 503–22.

Wuzhou, L. (2003) 'China's burgeoning mobile phone industry', available at: http://www.chinatoday.com.cn/English/9p12.htm (accessed 30 April 2004).

Xinhua News Agency (2004) 'China develops first nano-satellite', available at: http://www.china.org.cn/english/scitech/93341.htm (accessed 23 April 2008).

Xinhua News Agency (2006a) 'China's State Post Bureau to cooperate with Alibaba in e-commerce', available at: http://fec2.mofcom.gov.cn/aarticle/news/200611/20061103838882.html (accessed 30 April 2008).

Xinhua News Agency (2006b) 'Exports and imports of China's major high-tech companies increase in 2005', available at: http://english.peopledaily.com.cn/200602/18/eng20060218_243788.html (accessed 30 April 2008).

Xinhuanet (2007) 'China becomes Japan's biggest software outsourcing base', available at: http://en-1.ce.cn/National/Politics/200704/13/t20070413_11025663.shtml (accessed 30 April 2008).

Yam, S. (2006) 'Cash-strapped TCL Multimedia surveys the difficult road home', *South China Morning Post*, 4 November, p. 12.

Yan, C. (2007) 'China Telecom to launch AVS-based IPTV trial', available at: http://www.eetimes.com/news/latest/showArticle.jhtml?articleID=199702112 (accessed 1 May 2008).

Yan, X. (2003) 'The economic context of 3G development in China', available at: http://www.tdscdma-forum.org/EN/pdfword/200461412400489465.pdf (accessed 9 April 2005).

Yan, X. (2004) '3G mobile policy: the case of China and Hong Kong', available at: http://www.itu.int/osg/spu/ni/3G/casestudies/china/China_3g_Final.doc (accessed 9 April 2005).

Yang, B. (2002) 'Cultural conflicts and convergences: impacts of Confucianism, socialism and capitalism on managerial philosophy and practice in the P. R. China', paper presented at the Fourth International Symposium on Multinational Business Management, 19–21 May, Nanjing.

Yang, B., Zhang, D. and Zhang, M. (2004) 'National human resource development in the People's Republic of China', *Advances in Developing Human Resources* 6(3): 297–306.

Yang, D. L. (2001) 'The great net of China', *Harvard International Review* 22(4): 64–9.

Yang, D. M. (2002) 'Can the Chinese state meet its WTO obligations? Government reforms, regulatory capacity, and WTO membership', *American Asian Review* 20(2): 191–221.

Yee, C.M. (2001) 'E-business: in Asia, it's not a wide-open web: the big internet companies often censor their sites to please local officials', *Wall Street Journal*, 9 July, p. B1.

Yegin, C. E. (1995) 'Technological development and cooperation in greater China', *Managerial and Decision Economics* 16(5): 565–79.

Ying, H. (2002) 'Software 'made in China' becomes procurement target of Chinese government', available at: *http://english.people.com .cn/200207/22/eng20020722_100153.shtml* (accessed 30 April 2008).

Yong, W. (2006) 'China in the WTO: a Chinese view', *China Business Review* 33(5): 42–52.

Yongxiang, L. (2001) 'The role and contributions of the Chinese Academy of Sciences to China's science and technology development', available at: *http://www.asiasociety.org/speeches/lu.html* (accessed 2 May 2003).

Young, S., Huang, C.-H. and McDermott, M. (1996) 'Internationalization and competitive catch-up processes: case study evidence on Chinese multinational enterprises', *Management International Review* 36(4): 295–314.

Yu, V. (2006) 'Chinese FM's African visit to boost energy ties', available at: *http://www.mg.co.za/articlePage.aspx?articleid=261006&area=/ breaking_news/breaking_news__business/* (accessed 9 April 2007).

Yu, Z. and Xin, T. (2003) 'An innovative region in China: Interaction between multinational corporations and local firms in a high-tech cluster in Beijing', *Economic Geography* 79(2): 129–52.

Yuan, L. (2006) 'How China's 3G telecom initiative could work against western firms', *Wall Street Journal*, 28 November, p. C.1.

Yuan, Z. (2005) 'Features and impacts of internationalization of transnational corporations' R&D: China's case', paper presented at the Expert Meeting on FDI Development, Geneva, 24–26 January, available at: *http://www.unctad.org/sections/meetings/docs/zhou_en.pdf* (accessed 30 April 2008).

Yue, C. S. (2003) 'China Seen as 'second engine' of growth for ASEAN nations', available at: *http://search.japantimes.co.jp/cgi-bin/nb20030425c2.html* (accessed 3 May 2008).

Zaheer, S. (1995) 'Overcoming the liability of foreignness', *Academy of Management Journal* 18: 439–64.

Zainulbhai, A. S. (2005) 'What executive are asking about India', *McKinsey Quarterly*, Special Edition, pp. 26–33.

Zajac, E. J. and Westphal, J. D. (1995) 'Accounting for the explanations of CEO compensation: substance and symbolism', *Administrative Science Quarterly* 40: 283–308.

Zamiska, N. (2006) 'Novartis to establish drug R&D center in China', *Wall Street Journal* 6 November, p. A.3.

ZDNet Research (2006) 'In Q1 2006 China added 3.7 mln new broadband subscribers, US added 3.3 mln, 7/12', available at: *http://blogs.zdnet.com/ITFacts/index.php?cat=15* (accessed 11 December 2006).

ZDNet Asia (2003) 'Can China's homegrown 3G measure up?', available at: *http://www.zdnetasia.com/news/business/0,39044229,39150203,00.htm* (accessed 11 December 2006).

Zeng, M. and Williamson, P. J. (2003) 'The hidden dragons', *Harvard Business Review* 81(10): 92–9.

Zhang, J. (2001) 'China's "Government Online" and attempts to gain technical legitimacy', available at: *http://web.syr.edu/~ztan/Gov2.pdf* (accessed 11 December 2006).

Zhang, X. (1992) 'Residential preferences: A brain drain study on Chinese students in the United States', unpublished PhD dissertation, Harvard University, Graduate School of Education.

Zhao, M. (2004) 'China equation: No money = no innovation', *Electronic Engineering Times*, 22 November, p. 6.

Zhao, S. (2000) 'Chinese nationalism and its international orientations', *Political Science Quarterly* 115(1): 1–33.

Zhou, L and Hui, M. K. (2003) 'Symbolic value of foreign products in the People's Republic of China', *Journal of International Marketing* 11(2): 36–58.

Zhou, N. and Belk, R.W. (1993) 'China's advertising and the export marketing learning curve: the first decade', *Journal of Advertising Research* 33(6): 50–66.

Zim Online (2005) 'Mugabe to crack down on internet use', available at: *http://www.zimbabwesituation.com/jun10_2005.html#link10* (accessed 30 April 2008).

Zita, K. (1991) 'China's telecommunications and American strategic interests', in US Congress, Joint Economic Committee, *China's Economic Dilemmas in the 1990s: The Problems of Reforms, Modernization, and Interdependence*, Washington, DC: US Congress Joint Economic Committee, pp. 482–94.

zte.com.cn (2006) 'ZTE Profile', available at: *http://www.zte.com.cn/English/01about/index.jsp* (accessed 10 April 2006).

Zucker, L. G. (1988) 'Where do institutional patterns come from? Organizations as actors in social systems', in L. G. Zucker (ed.) *Institutional Patterns and Organizations: Culture and Environment*, Cambridge, MA: Ballinger, pp. 23–49.

Zweig, D. (1997) 'To return or not to return? Politics vs. economics in China's brain drain', *Studies in Comparative International Development* 32(1): 92–125.

Zweynert, J. and Goldschmidt, N. (2006) 'The two transitions in Central and Eastern Europe as processes of institutional transplantation', *Journal of Economic Issues* 40(4): 895–918.

Index

Africa, 68, 121, 183, 187–9, 193–4, 198, 200, 204, 221, 230, 240
  Sub-Saharan Africa, 198
Alcatel, 12, 38–9, 66, 203–4, 217, 219–20, 229–31
Alcatel Shanghai Bell, 166
Altavista, 24, 28, 33, 40, 42, 50
American Civil Library Association, 29
Amnesty International, 35, 38–9, 47
ASEAN, 237–8

Barrett, Craig, 81
Beijing Software Industry Production, 14, 107
Bloomberg, Michael, 241
broadband, 143–54, 157, 159–60

cathode ray tube television, 227, 233
CDMA 2000, 53, 56, 62, 64, 67, 69, 195
CDMA450, 68, 186, 188, 193, 215
China, Japan and South Korea (CJK) open-source alliance, 58
China Development Bank, 68
Chinese 2000 Mobile Linux Operating System for handheld devices, 140
Chinese Academy of Engineering, 93

Chinese Academy of Sciences, 93, 96, 106, 115–16, 133–4, 138, 140, 174–5
Chinese Communist Party, 5, 32, 48, 94, 101–2, 128, 160, 211
Chinese government, 5–8, 11, 13, 15–17, 19, 21–2, 26–7, 29, 35–7, 42, 45, 47–9, 53–7, 60–2, 64, 66–8, 83, 95, 100–2, 106, 108, 113, 116, 118–19, 124, 133–4, 138, 142, 148, 160, 170, 180, 194, 214–15, 218, 224, 236
China Netcom, 54, 69, 87, 122, 144, 149, 155–6, 161, 163, 166, 168
complementarity, 7–8, 10, 12, 19, 169
Confucianism, 17–18, 30, 96, 123, 130, 133, 135, 141
Consumer Federation of America, 29
Cultural Revolution, 96–8, 102, 105, 138
cyber-control, 37–42, 44–7, 50
cyber police, 41, 43, 50

Datang, 55–6, 60, 65–6, 69, 118, 137, 206
Dawning 4000A, 123–4
Dawning 4000L, 138
democracy, 26, 38–9, 41, 43, 51, 198
demonstration effects, 84, 86

Deng, Xiaoping, 98, 128, 142, 211
digital and electronic signatures, 41

e-business, 21–4, 27–36, 38, 44–6, 49
Electronic Privacy Information Center, 129
Ericsson, 65–6, 81, 104, 193, 203–4, 206, 210, 218, 220
EU, 119, 121, 205, 233, 238
Europe, 12, 53–4, 72, 75, 79–80, 83, 94, 100, 109, 116, 129, 135, 177, 179, 184, 188, 193, 204, 216–17, 219, 222, 225–6, 228–9, 231
  Eastern Europe, 123, 129
  Eastern and Central Europe, 193
Europe International Business School, 94
externality, 136

France, 185, 187, 204, 206, 227, 229, 230

Gates, Bill, 2
Germany, 43, 82, 113, 124, 139, 160, 173, 206, 217, 220, 225–8, 236
Global Online Freedom Act 2006, 46
Google, 24, 28–9, 31, 33, 35, 38, 40, 42, 47, 50, 124
Government Procurement Law, 6, 116, 131
GSM, 12, 67, 195, 203–4, 210, 216
guanxi, 24–5, 30–1, 33–5

Haier, 68, 87, 173–4, 184–5, 187
Henan province, 103
Hongqi Linux, 116, 140
Hu, Qiheng, 28

Huawei, 59, 65–6, 68, 87, 152, 166, 184, 187–8, 193, 197, 202–3, 205, 209, 213, 215, 241
Huizhou city government, 220
Hussein, Saddam, 197, 241

IBM, 74, 77–9, 104, 115–16, 123, 140, 176, 179
India, 29, 73, 75, 80–3, 90–3, 95, 107–9, 115, 117–21, 143, 147–51, 154, 172, 187–9, 216, 233
Indian Telecom Ministry, 241
institutions, 1–3, 5–6, 18–19, 22–4, 33–7, 47, 49, 51, 54, 57, 75, 111, 121, 123, 125, 128, 130–1, 141–2, 157, 165, 170, 180–1, 190–2, 196, 198, 235, 238, 240–1
  cognitive institutions, 23–4, 37, 40, 46, 51, 75, 130–1, 134, 141, 191, 198
  de-institutionalisation, 165
  institutional actors, 1–2, 4–6, 9, 13, 39, 46, 49, 110
  institutional change, 34, 112
  institutional distance, 191, 196–8
  institutional evolution, 40, 47
  institutional field, 37, 39, 41, 47
  institutional pillars, 22–3, 40, 131
  international pressure, 19, 50
  institutional theory, 2–3, 7, 21, 125
  institutional war, 3, 34, 40
  normative institutions, 17, 23–4, 33–4, 37, 40, 44, 51, 75, 130–2, 141, 191, 198
  regulative institutions, 23, 37, 40–1, 51, 75, 130–1, 141, 170, 180, 191, 198
  re-institutionalisation, 165

Index

intellectual property rights, 1–4, 8, 10, 13, 16, 18, 158, 164–5, 169
internationalisation, 72, 76, 182–6, 199, 201–15
internet protocol television, 151–9, 165–70
Internet Society of China, 28–30, 33, 44–6, 48
Iran, 39
ISC – *see* Internet Society of China
Italy, 185, 187, 204, 228

Japan, 43, 54, 56, 58, 65, 82, 84–5, 96, 98, 103, 109, 117, 119–21, 123, 136, 141, 158, 172, 177, 179, 186, 195, 223, 236–9
Jiang, Zemin, 134, 218

kaleidoscopic comparative advantage, 176
Kingsoft, 116, 141
Korea, 56, 58, 117, 141, 175, 237, 239
  North Korea, 195
  South Korea, 58, 66, 103, 120, 136, 165, 169, 172, 204, 238–9

legitimacy, 8–9, 13–14, 16–18, 23–4, 31–5, 48, 50, 57, 66, 160, 181–2, 191–2, 196–7, 199
Lenovo, 2, 13, 87, 118, 182, 184, 189, 217
Linux, 14–16, 60, 74, 79, 104, 115–16, 118, 123–4, 133–4, 136–42
LinuxLab of the USA, 140
Linux operating system, 74, 79, 115, 137–8, 142
Long March 3B, 189

Mao, Tse Tung, 17, 106, 113, 133, 224
Ministry of Commerce, 73, 214
Ministry of Education, 95
Ministry of Information Industry, 3, 14, 23, 42, 53–4, 64, 162–3
Ministry of Public Security, 152
Ministry of Railways, 152
Microsoft, 2, 7, 13–15, 18, 29, 33, 35, 38, 47, 59–60, 74, 78, 87, 104, 106, 115, 131, 133, 137–8, 142, 163, 169, 211, 220, 222
Miracle Linux, 116
MNC, 71–85, 88, 115–16, 118–19, 126, 130, 134, 139, 141, 181–2, 191, 199, 201–3, 208–10, 212
Mobile Linux Operating System, 116
Motorola, 12, 57, 65, 69, 74, 77–9, 81, 87, 104, 116, 137, 139–40, 152, 192, 212, 214, 222, 230
Mugabe, Robert, 194

Nanometer Technology Center, 12, 175, 180
nanotech bone scaffold, 177
nanotechnology, 82, 171–80
nationalism, 21–2, 25, 32–3, 45, 105, 160
Nigeria, 68, 188, 198, 215
Nokia, 12, 65, 74, 77–89, 104, 137, 212, 214, 216, 230–1

OECD, 144, 153, 182
Olympics, 53–4, 61–2, 152
online game, 149, 159
open source software, 14, 16, 18–19, 114–21, 123–44
outsourcing, 73, 80, 108, 120–1, 151, 154, 207, 224, 233

313

Pakistani government, 241
People's Liberation Army, 132, 138, 142, 211, 239, 241
Pfizer, 11, 88
PLA – see People's Liberation Army

R&D, 55, 71–88, 104, 118, 162, 175–6, 178–9, 182, 184–5, 187, 208, 212–13, 221–2, 227, 231, 235–6, 238, 241
  adaptive R&D, 72, 74–9, 86
  innovative R&D, 72–6, 79–82, 84, 86
Red Flag Linux, 14, 16, 107, 115–16, 136, 140
Reporters without Borders, 48
research and development – see R&D
resource-based view, 192
reverse-engineering, 84, 86

Schneider, 204, 217, 219, 225–6, 230
Siemens, 55–6, 65–6, 77–9, 104, 205–6
Sohu, 87, 144, 165
South Africa, 189, 221, 240
Soviet Union, 96, 98, 100, 123, 129, 158
spillover effects, 85–6, 235
Super 301, 6

Taliban, 241
TCL, 13, 68, 185, 187, 202, 204–7, 210, 213, 215–16, 219–33
TD-SCDMA, 53–69, 117–19, 186, 189, 195, 202, 206, 214
technology transfer, 83, 86, 147, 175, 207–11
Telecom Kenya, 193

theory of comparative advantage, 176
Third Generation Partnership Project, 55
Thomson, 217, 219–20, 226–9
TRIPS, 10, 15, 112, 132
TurboLinux, 116, 138

UK, 43–4, 76, 82, 92, 98, 113, 120, 136, 151, 160, 175, 204, 206, 226
UK Mobile Marketing Association, 29
USA, 6, 29, 36, 38, 44, 47, 56, 59, 72–3, 75–6, 80, 82–3, 87, 91–2, 96–8, 100–1, 116, 118–21, 135–6, 138, 140, 143–5, 171–3, 175, 177–9, 184, 198, 213, 216, 219, 222–3, 228, 236, 238–9, 241
USSR – see Soviet Union
UTStarcom, 166, 168

W-CDMA, 53–4, 57, 62, 64, 67, 69, 188, 195
Windows, 2, 9, 13–16, 18, 60, 74, 78, 132, 134, 136–7, 211
World Cup, 61
World Trade Organization, 2–3, 5–6, 9, 112, 131, 211

Xu Guanhua, 59, 133

Yang, Jerry, 28, 46, 50
Yangfan Linux, 14, 116

Zimbabwe, 186, 188, 194, 196
ZTE, 13, 59, 66, 68, 124, 152, 166, 184, 187–8, 192, 203, 205, 209, 213, 215